MW00578668

REVIEW COPY
from Science

PROGRESS IN ENVIRONMENTAL RESEARCH

PROGRESS IN ENVIRONMENTAL RESEARCH

IRMA C. WILLIS
EDITOR

Nova Science Publishers, Inc.
New York

Copyright © 2007 by Nova Science Publishers, Inc.

All rights reserved. No part of this book may be reproduced, stored in a retrieval system or transmitted in any form or by any means: electronic, electrostatic, magnetic, tape, mechanical photocopying, recording or otherwise without the written permission of the Publisher.

For permission to use material from this book please contact us:
Telephone 631-231-7269; Fax 631-231-8175
Web Site: http://www.novapublishers.com

NOTICE TO THE READER

The Publisher has taken reasonable care in the preparation of this book, but makes no expressed or implied warranty of any kind and assumes no responsibility for any errors or omissions. No liability is assumed for incidental or consequential damages in connection with or arising out of information contained in this book. The Publisher shall not be liable for any special, consequential, or exemplary damages resulting, in whole or in part, from the readers' use of, or reliance upon, this material.

Independent verification should be sought for any data, advice or recommendations contained in this book. In addition, no responsibility is assumed by the publisher for any injury and/or damage to persons or property arising from any methods, products, instructions, ideas or otherwise contained in this publication.

This publication is designed to provide accurate and authoritative information with regard to the subject matter covered herein. It is sold with the clear understanding that the Publisher is not engaged in rendering legal or any other professional services. If legal or any other expert assistance is required, the services of a competent person should be sought. FROM A DECLARATION OF PARTICIPANTS JOINTLY ADOPTED BY A COMMITTEE OF THE AMERICAN BAR ASSOCIATION AND A COMMITTEE OF PUBLISHERS.

LIBRARY OF CONGRESS CATALOGING-IN-PUBLICATION DATA

Progress in environmental research / Irma C. Willis (editor).
   p. cm.
Includes index.
ISBN-13: 978-1-60021-618-3 (hardcover)
ISBN-10: 1-60021-618-8 (hardcover)
1. Environmental sciences--Research. I. Willis, Irma C.
GE70.P785   2007
628--dc22   2007006569

Published by Nova Science Publishers, Inc. ✦ New York

CONTENTS

PREFACE

The environment consists of the surroundings in which an organism operates, including air, water, land, natural resources, flora, fauna, humans and their interrelation. It is this environment which is both so valuable, on the one hand, and so endangered on the other. And it is people which are by and large ruining the environment both for themselves and for all other organisms. This new book presents important new research on a wide variety of topics in this field.

Chapter 1 - The federal government owns about 671.8 million acres (29.6%) of the 2.27 billion acres of land in the United States. Four agencies administer 628.4 million acres (93.5%) of this land: the Forest Service in the Department of Agriculture, and the Bureau of Land Management, Fish and Wildlife Service, and National Park Service, all in the Department of the Interior. Most of these lands are in the West, including Alaska. They generate revenues for the U.S. Treasury, some of which are shared with states and localities. The agencies receive funding from annual Interior and Related Agencies appropriations laws, trust funds, and special accounts.

Chapter 2 - The Uruguayan coastal zone is bathed by the waters of the Southwest Atlantic Ocean (SWAO, 230 km) and the Río de la Plata Estuary (450 km), one of the largest estuaries in the world. The main tributaries of this estuary are the Paraná-Paraguay and Uruguay rivers, which drain the second largest basin in South America and provide the major source of freshwater runoff to the SWAO. Typical coastal ecosystems are sandy beaches with rocky points, sub-estuaries flowing along the Río de la Plata one and coastal lagoons in the Atlantic region. The estuarine portion is characterised by muddy sediments while sandy-shell debris are the dominant sediment type in the Atlantic portion. One of the most relevant features in this coastal zone is the interaction between the SWAO and the Río de la Plata waters, being that salinity is of primary importance in regulating the benthic biodiversity. In general, autochthonous fauna of the Uruguayan coast is characteristic of the temperate zone with temperate-cold and temperate-warm components, which correspond to the Patagonic biogeographic province, and show a break in the region of the Río de la Plata Estuary influence. Several studies had demonstrated that the Uruguayan coastal zone is under the effects of different kinds of human related stressors. Urbanisation, harbour, shipping and industrial activities are the main perturbation factors for the Río de la Plata portion, while, agricultural and tourism affect preferentially the Atlantic one. In addition, these studies had shown that petroleum hydrocarbons, heavy metals and the organic enrichment of bottom sediments have a direct influence on biodiversity patterns. Furthermore, some morphological anomalies have been detected in benthic foraminifera, which seem to be related to the heavy metal and organic content of the sediments. Despite the existence of a clear salinity gradient in this estuarine area; the utilization of different approaches, together with the integration of physical, chemical and biological data, demonstrated the occurrence of an environmental quality gradient with the improvement of conditions from the inner stations of Montevideo Bay to the outer coastal ones. Recent studies warned about the presence of non indigenous invasive species in both zones of the Uruguayan littoral, however, the degree of incidence seems to be greater in the estuarine portion. Their introduction would be related to the discharge of ballast water, and their distribution determined by salinity patterns. The available information about marine biodiversity and environmental perturbation in the Uruguayan coastal zone was improved during the last decades; however, it is still restricted to isolated areas along it and to some aspects of aquatic ecosystems. The implementation and development of integrative baseline studies on these topics are highly relevant, in order to contribute to the conservation of benthic biodiversity in the coastal zone of Uruguay.

Chapter 3 - Field studies monitoring herbicide pollution in the vineyards of Burgundy (France) have revealed that the drinking water reservoirs are contaminated with several herbicides. The purpose of this work is to assess the effectiveness of alternative soil management practices, such as grass cover, for reducing the leaching of oryzalin, diuron, glyphosate and some of their metabolites in soils, and subsequently in preserving groundwater quality. The mobility of these herbicides and their metabolites was studied in structured soil columns (15 x 20 cm) under outdoor conditions. In the first experiment the leaching of diuron, oryzalin and glyphosate was monitoring from May 2001 to May 2002. The soil was a calcaric cambisol under two vineyard soil management practices: a bare soil chemically-treated, and a vegetated soil in which grass was planted between the vine rows. A second experiment was conducted with diuron applied in early May 2002 on three calcaric cambisols and monitored until November 2002. Grass was planted on three of the six columns. In the second experiment, higher amounts of diuron residues were found in the effluents from the grass-covered columns (from 0.10% to 0.48%) than from the bare-soil columns (0.09% - 0.16%). At the end of the monitoring period, less of the diuron residues were recovered in the vegetated soil profiles, from 4.3% to 10.6%, than in the bare soils, from 10.8% to 34.9%. In the first experiment, greater amounts of herbicide residues were measured in the percolates of the bare soil, 0.96%, 0.10% and 0.21% for diuron, oryzalin and glyphosate respectively than in the percolates of the vegetated soil, 0.16%, 0.10% and 0.05%, respectively. At the end of the monitoring period, more residues from diuron, oryzalin and glyphosate were recovered in the bare soil profiles: respectively 7.6%, 2.4% and 0.007% than in the vegetated soil, 12.5%, 4.4% and 0.01%, respectively. The dissipation of the three herbicide residues seems faster in the bare soil than in the vegetated soil, contrary to the diuron dissipation in the experiment 2.

This study showed that diuron, oryzalin and to a lesser extent, glyphosate, AMPA, DCPMU and DCPU, leach through the soils; thus, these molecules may be potential contaminants of groundwater. The alternative soil management practice of planting a grass cover under the vine rows could reduce groundwater contamination by pesticides by reducing the infiltrating amounts of diuron, oryzalin, glyphosate and some of their metabolites. Nevertheless, after one year, persistence of herbicide residues was greater in the grass cover soil than in the bare soil. The difference in the results for diuron may be explained by the difference in the duration of the grass cover, 4 years in the experiment 1 versus 4 weeks in the experiment 2.

Chapter 4 - The 109th Congress is considering issues related to the public lands managed by the Bureau of Land Management (BLM) and the national forests managed by the Forest Service (FS). The Administration is addressing issues through budgetary, regulatory, and other actions. Several key issues of congressional and administrative interest are covered here.

Chapter 5 - The time course in the expression of sublethal effects that precede mortality events in mussels exposed to municipal wastewaters is not well understood. Our study thus sought to examine such a time course in biomarker responses of freshwater mussels exposed to two final aeration lagoons for the treatment of domestic wastewater. Mussels were caged in two aeration lagoons (AL1 and AL2) for different time periods (1, 15, 30, 40 and 60 days) and examined for biomarkers representative of biotransformation, metabolism of endogenous ligands, tissue damage and gametogenesis. Results showed that mortality events occurred after day 29 of exposure, gradually reaching 30% and 45% mortality for AL1 and AL2, respectively. These aeration lagoons were estrogenic, as evidenced by the rapid induction of vitellogenin-like proteins, and they were also good inducers of cytochrome P340 3A activity, considered as one of the major metabolizing enzymes of pharmaceutical products. Several biomarkers were expressed before the manifestation of mortality (vitellogenin-like proteins, aspartate transcarbamoylase, metallothioneins, monoamine oxidase, heme peroxidase, cytochrome P450A activity, glutathione S-transferase, DNA strand breaks and mitochondrial electron transport activity), while others were expressed during mortality events (xanthine oxydoreductase, cytochrome P4501A1 activity, lipid peroxidation, gonad lipids and cyclooxygenase). A factorial analysis revealed that temperature-dependent mitochondrial electron transport activity, hemoprotein oxidase, monoamine and aspartate transcarbamoylase had the highest factorial weights. Furthermore, canonical analysis of biomarkers revealed that reproduction (gametogenesis) was more significantly related with those linked to tissue damage and biotransformation, while energy status (i.e. increased energy expenditure and decreased lipid reserves) was not significantly related to gametogenesis or tissue damage in AL1, the least toxic and estrogenic lagoon. In AL2, the more toxic and estrogenic lagoon, energy status was significantly correlated with tissue damage and gametogenesis, in addition to biotransformation activity and metabolism of endogenous substrates. Domestic wastewaters treated in aeration lagoons display a complex pattern of sublethal responses in mussels prior to the manisfestation, in some instances, of mortality events.

Chapter 6 - The stress imposed on ecosystems has produced important biodiversity loss. In terrestrial ecosystems, loss of plant cover leads to soil degradation impairing the system capacity to recover after perturbations. Plant – microbe interactions are the key factor for productivity because of the beneficial role played by the microbial community living in the root zone. Soil degradation may negatively affect this microbial community and, indirectly affect the microbial predators such as the amoeboid protozoa. The authors aim to describe the changes in the amoebae community when soil is transiting to badlands formation and to correlate this variation with the soil physicochemical factors at the semiarid Valley of Tehuacán, México. Samples from the root zone of mesquite (*Prosopis laevigata*), the columnar cactus *Pachycereus hollianus*, and from the interspace soil were taken at 0 – 10 cm layer in both terraces. Samplings were done every four weeks from July to October 2002 to observe the variation change during the transition from dry to wet season. The physicochemical factors and the community of amoebae were significantly different between terraces and microenvironments. *Vahlkampfia, Platyamoeba, Mayorella, Hartmannella* and *Acanthamoeba* were the most frequent genera found in soil. However amoebae community was different between microenvironments. Species richness was smaller in soil forming badlands; even though plants played the role of fertility islands in this soil. Principal component analysis showed that pH, porosity and real density explained only 23.9 % of the amoebae community variation. The response of amoebae to soil degradation was the reduction of species richness and change of the dominant species in the community. The physicochemical factors correlated poorly with the observed community variance.

Chapter 7 - The Bureau of Land Management (BLM) is taking a two-pronged approach to grazing reform, by proposing changes to grazing regulations (43 C.F.R. Part 4100) and considering other changes to grazing policies. On June 17, 2005, BLM issued a final environmental impact statement (FEIS) that analyzes the potential impact of proposed changes in the regulations, a slightly different alternative, and the status quo. On August 9, 2005, BLM announced its intent to prepare a supplement to the FEIS. The delay is intended to allow the agency time to address public comment received after the closing date of March 2, 2004, primarily from the Fish and Wildlife Service. The agency anticipates developing the supplement in the fall of 2005, soliciting and reviewing public comment, and issuing it in final form in the spring of 2006. No deadline for the final rule has been announced.

Chapter 8 - Forest loss and fragmentation over the past decades in the west and central Africa region is having a direct effect on the habitats of valuable plants, driving species isolation, reductions in species populations and in some cases, increasing extinction rates of potentially useful plants. Furthermore, some tropical rainforest plants exhibit hampered seed germination or seedling establishment through hampered natural regeneration in disturbed ecosystems.

Nevertheless, these forests in west and central Africa remain important sites, habitats and sources of potentially useful plantdiversity. Many tropical tree species and their products have been documented regarding the roles they play as food, medicine and in terms of other services they provide to local peoples. The exploitation, uses and commercialisation of these tree products, constitute an important activity to people living around forests and beyond within the region. For some of these species, existing markets have expanded within and outside their wide ecological range as well as great potentials that exist for their further development at industrial level.

Since 1998, the World Agroforestry Centre, Africa humid Tropics, in partnership with several local and regional stakeholders in West and Central Africa have been implementing a Tree domestication programme aimed at diversifying smallholder livelihood options through the selection, multiplication, integration, management and marketing of indigenous trees/plants and their products, ensuring that they provide both livelihood and environmental services. As tree domestication itself depends on existing plant diversity, biodiversity at genetic, species and ecosystem levels have been important considerations in cultivar selection, farming systems diversification and contributing towards ecosystems resilience, respectively.

This tree domestication programme is being implemented in Cameroon, Nigeria, Gabon, Equatorial Guinea and more recently in the Democratic republic of Congo.

The programme started with the prioritisation of a range of indigenous fruit and medicinal tree species at local community levels. Emphasis then moved to capacity-building:

training, follow-up and information dissemination focussing on a range of low-tech and adaptable propagation, marketing, selection, cultivation and management techniques for local level stakeholders, and training, backstopping and dissemination, for a range of regional government and non-governmental partners, in order to enhance ownership and adoption of the process.

So far, the programme has contributed firstly, in building of both natural assets of resource-poor farmers to increase their access to a diverse range of agroforestry trees and products, and human assets for perpetuating the knowledge and experience in the region, as well develop mechanisms for increasing and diversifying household revenue through better marketing of indigenous agroforestry tree products, protecting biodiversity on-farm and recognizing the value in maintaining both intra and inter-specific diversity on-farms. The programme has also developed multi-species, on-farm needs-based live gene banks as well as classical ones of regionally important high-value indigenous tree species in both Cameroon and Nigeria.

As the programme develops in the region, increasing emphasis is being placed on building strategic partnerships in order to achieve greater and more far-reaching impact by increasing the potential contribution that diverse agroforestry trees make to household revenue, and environmental management at farm and at landscape scale.

Chapter 9 - Serious environmental problems occurred in the process of rapid growth of economy after World War II in Japan, including three water pollution (Kumamoto Minamata Disease, Niigata Minamata Disease, Itai-Itai Disease) and one air pollution (Yokkaichi Asthma), which had been called the Four Major Pollution in Japan.

Attention to Yokkaichi Asthma came from increasing patients of chronic obstructive pulmonary disease, such as bronchial asthma and chronic bronchitis, caused by very high concentration of SO_2 in the air. This problem occurred since around 1957, and was continued for about 20 years. By regulation of total emission of sulfur oxides from 1972, SO_2 concentration decreased greatly and the pollution problem was solved. In this article, many precious experiences in Yokkaichi Asthma are described.

Chapter 10 - We aim here to review biological monitoring which has been carried out nation-wide in Japan during the last 15 years and its future directions. Biological monitoring is an assessment of overall exposure to chemicals that are present in our environment, including workplaces, through measurement of the appropriate determinants in biological specimens collected from humans. The measurement of lead concentration in blood (Pb-B) and *delta*-aminolevulinic (ALA) acid in urine of workers who handle lead, and urinary metabolites of workers who handle organic solvents (xylene, N,N-dimehyleformamide, styrene, tetrachloroethylene, 1,1,1-trichloroethane, trichloroethylene, toluene, and n-hexane) have been some of the indispensable items of occupational health examinations since 1989. During the last 15 years, the total number of Pb-B and ALA analyses decreased about 10 – 20%. On the other hand, annual analyses of urinary metabolites have been constant for the last 15 years (500,000 to 600,000 per year). Among analyses of urinary metabolites of organic solvents, the analyses of total trichlorocompounds and trichloroacetic acid have decreased, where as N-methylformamide and 2,5-hexanedione analyses have increased. The measurement results are classified into one of three categories, that is, distribution 1, 2 or 3. The percentages of distribution 2 (more than 1/3 - 1/2 of the biological exposure index (BEI) but less than BEI) and distribution 3 (over BEI) have deceased for the last 15 years. About 90% of institutes of occupational health examination had the samples analyzed in laboratories specializing in clinical chemistry outside of their institutes (outside orders) in 2004. The increase of outside orders resulted in an oligopoly by a few laboratories specializing in clinical chemistry. The development of new analytical methods has affected the measurement results. Especially, the change of ALA analysis from a colorimetric method to an HPLC method lowered measurement results. Also, Threshold limit value-time weighted average (TLVs-TWA) and occupational exposure levels (OELs) of some organic solvents have been altered during the last 15 years. In spite of such circumstances, the criteria for classifying distribution 1, 2 and 3, which are related to BEIs, have not been alterd because the criteria should be constant in order to compare the measurement results on the same scale over a long-term period. The determinants of occupational exposure should be specific to the occupational exposure itself. Therefore, chemical substances themselves and their metabolites are useful as determinants of biological monitoring in the occupational health field. It has been attempted to expand the number of chemical substances subjected to biological monitoring, and more than 15 determinants of exposure to organic solvents have recently been proposed. In addition, nonspecific early biological health effects which reflect combined exposure or long-term exposure are suitable as determinants of environmental exposure. Hemoglobin adducts, the Comet assay and urinary 8-hydroxyguanine have been reported to have some usefulness for monitoring the exposure to 7, 20 and 9 chemical substances, respectively, in humans.

In: Progress in Environmental Research
Editor: Irma C. Willis, pp. 1-2

ISBN 978-1-60021-618-3
© 2007 Nova Science Publishers, Inc.

Expert Commentary

PENGUINS – USEFUL MONITOR FOR ENVIRONMENTAL CHANGES IN ANTARCTICA

Roumiana Metcheva

Institute of Zoology, Bulgarian Academy of Sciences,
Bd.Tzar Osvoboditel 1, 1000 Sofia, Bulgaria

The Antarctic environment is a relatively new topic in the world ecology, economy, and policy. The study of Antarctica is of great importance in the context of the investigation of the Earth's problems. The scientists call Antarctica a natural laboratory. The remoteness and the extreme conditions in this area make it an object of special interest for studying the fauna, the biogeochemical cycles of trace elements, etc. For a long time the Antarctic environment was considered to be a pristine and unpolluted one. Recently, however, research has shown that pollutants have occurred there. Heavy metals were detected in sea birds and other animals. Anthropogenic sources seem a likely explanation for a significant excess of Pb. Relatively high levels of Cd have been established in tissues of Antarctic birds and seals but this may be due to a natural enrichment of this element in the Antarctic food chain because of greater concentration of Cd in krill and squid – important food components of seals' and penguins' diets. The increase of the bioavailability of metals in the Antarctic environment has been explained by upwelling of Cd-rich waters and local volcanism. Thus, the exploration of the metals dynamics in Antarctic food chain is important. In spite of the Cd-enrichment of the Antarctic food web, Cd levels measured in Antarctic birds are significantly lower compared to those in other continents. This fact confirms the idea that relatively uncontaminated Antarctic ecosystems could provide data about baseline concentrations of metals and these levels could be used as a database for pollution detection. In order to run an adequate management in Antarctica, it is necessary first of all to do a continual monitoring of the Antarctic environment.

It is of great concern to monitor the Antarctica sea zones for toxic elements and heavy metals in the fauna. For that purpose molting penguin feathers are an excellent subject for biomonitoring because molting plumage accumulates toxicants during one year of live.

The disulfide bonds in the feather's structural protein readily reduce to sulfhydryl groups, for which metals have high affinity. Element accumulation in feathers takes place during the entire process of feather growth. The levels of metals in the feather reflect the amount of metals circulating in the blood during the 3- to 4-week period when the feather is forming. Thereafter atrophing of the blood supply the feather remains as a permanent record of blood composition for many years or centuries if specimens are in permanent collections. Even feathers in museum collections are possible to be used to examine changes in pollution levels of large geographical areas.

Penguins have potential to be a standard biological indicator in monitoring programs of Antarctica nearshore ecosystems because they have long life span, permanent ecological niche, dominate the aviafauna in Antarctica and molt annually. The concentrations of toxic elements and heavy metals in plumage could be define precisely every year. A continual environmental biomonitoring using penguins feather could establish a possible trend to contamination of the Antarctica sea zones.

Penguins are members of the order *Sphenisciformes*, which contains only one family: *Spheniscidae*. Six genera, containing different number of species, belong to this family. The genus *Pygoscelis* is the most widely distributed and the most examined. It contains three species: Chinstrap (*P. Antarctica*), Gentoo (*P. papua*), and Adelie (*P. adeliae*). We have focused our studies on *P. papua* and *P. Antarctica*.

Lead measured in feathers of pygoscelid penguins showed levels 3–8 times higher compared with those of cadmium. Lead is known as a calcium-formations-seeking element, readily accumulating in bone, hair and different compartments such as nails, feathers etc. It is not metabolically regulated and shows great bioaccumulation in bird feathers. This makes Pb one of the most suitable metals for monitoring of anthropogenic pollution. Among the heavy metals Cd has the lowest concentration in feathers. Really, feathers could be considered as a significant Pb pool in the cases of lead contamination. Lead and mercury levels are much higher in feathers than they are in other tissues. This fact underlines the usefulness of feathers as a biomonitoring tool.

Antarctic animals are least investigated compared with animals from other continents, hence it is important to determine the concentrations of the major essential elements and trace elements in marine fauna in Antarctica. This could provide valuable information about the natural levels of those elements in the bird's diet. The data regarding some trace elements could be used as a baseline database for pollution detection. Physiologists are interested mainly in chemical composition of different body tissues and fluids. When ecology problems are concerned, the determination of element concentrations in epidermal compartments and plumage becomes also very significant. So the penguin are a key tool for monitoring of Antarctic environment.

In: Progress in Environmental Research
Editor: Irma C. Willis, pp. 3-73

ISBN 978-1-60021-618-3
© 2007 Nova Science Publishers, Inc.

Chapter 1

FEDERAL LAND MANAGEMENT AGENCIES: BACKGROUND ON LAND AND RESOURCES MANAGEMENT[*]

Carol Hardy Vincent

SUMMARY

The federal government owns about 671.8 million acres (29.6%) of the 2.27 billion acres of land in the United States. Four agencies administer 628.4 million acres (93.5%) of this land: the Forest Service in the Department of Agriculture, and the Bureau of Land Management, Fish and Wildlife Service, and National Park Service, all in the Department of the Interior. Most of these lands are in the West, including Alaska. They generate revenues for the U.S. Treasury, some of which are shared with states and localities. The agencies receive funding from annual Interior and Related Agencies appropriations laws, trust funds, and special accounts.

The lands administered by the four agencies are managed for a variety of purposes, primarily related to preservation, recreation, and development of natural resources. Yet, each of these agencies has distinct responsibilities for the lands and resources it administers. The Bureau of Land Management (BLM) manages 261.5 million acres, and is responsible for 700 million acres of subsurface mineral resources. BLM has a multiple-use, sustained-yield mandate that supports a variety of uses and programs, including energy development, timber harvesting, recreation, grazing, wild horses and burros, cultural resources, and conservation. The Forest Service (FS) manages 192.5 million acres also for multiple use and sustained yields of various products and services, for example, timber harvesting, recreation, grazing, watershed protection, and fish and wildlife habitats. Most of the lands are designated national forests, but there are national grasslands and other lands. National forests now are created and modified by acts of Congress. Both the BLM and FS have several authorities to acquire and dispose of lands.

The Fish and Wildlife Service (FWS) manages 95.4 million acres, primarily to conserve and protect animals and plants. The 793 units of the National Wildlife Refuge System include

---

[*] Excerpted from CRS Report 1-60021-151-8, dated August 2, 2004.

refuges, waterfowl production areas, and wildlife coordination units. Units can be created by an act of Congress or executive order, and the FWS also may acquire lands for migratory bird purposes. The National Park Service (NPS) manages 79.0 million acres of federal land (and oversees another 5.4 million acres of nonfederal land) to conserve and interpret lands and resources and make them available for public use. Activities that harvest or remove resources generally are prohibited. The National Park System has diverse units ranging from historical structures to cultural and natural areas. Units are created by an act of Congress, but the President may proclaim national monuments.

There also are three special management systems that include lands from more than one agency. The National Wilderness Preservation System consists of 105.2 million acres of protected wilderness areas designated by Congress. The National Wild and Scenic Rivers System contains 11,303 miles of wild, scenic, and recreational rivers, primarily designated by Congress and managed to preserve their free-flowing condition. The National Trails System contains four classes of trails managed to provide recreation and access to outdoor areas and historic resources.

ABBREVIATIONS

RSI: Resources, Science, and Industry;
INF: Information Research;
ALD: American Law

INTRODUCTION [1]: SCOPE AND ORGANIZATION

This article provides an overview of how federal lands and resources are managed, the agencies that manage the lands, the authorities under which these lands are managed, and some of the issues associated with federal land management. The article is divided into nine sections. The introduction provides a brief historical review and general background on the federal lands. "Federal Lands Financing" describes revenues derived from activities on federal lands; the appropriation processes and the trust funds and special accounts that fund these agencies; federal land acquisition funding, especially from the Land and Water Conservation Fund; and programs that compensate state and local governments for the tax-exempt status of federal lands. The next sections pertain to the four major federal land management agencies: the Forest Service (FS) in the Department of Agriculture, and the Bureau of Land Management (BLM), Fish and Wildlife Service (FWS), and National Park Service (NPS), all in the Department of the Interior. The sections relate each agency's history; organizational structure; management responsibilities; procedures for land acquisition, disposal, and designation, where relevant; current issues; and statutory authorities. The final sections provide essentially the same information for the three major protection systems that are administered by more than one agency and hence cross agency jurisdictions: the National Wilderness Preservation System, the National Wild and Scenic Rivers System, and the National Trails System. Relevant CRS reports are listed following each section. The article concludes with an appendix of acronyms used in the text, and another defining selected terms used.

BACKGROUND

The federal government owns and manages approximately 671.8 million acres of land in the United States — 29.6% of the total land base of 2.27 billion acres. [2] Table 1 identifies the acreage of federal land located in each state and the District of Columbia. The figures range from 5,318 acres of federal land in Rhode Island to 243,847,037 federal acres in Alaska. Further, while a dozen states contain less than ½ million acres of federal land, another dozen have more than 10 million federal acres within their borders. Table 1 also identifies the total size of each state, and the percentage of land in each state that is federally owned. These percentages point to significant variation in the size of the federal presence within states. Specifically, the figures range from 0.5% of Connecticut land that is federally owned to 91.9% of land in Nevada that is federally owned. All 12 states where the federal government owns the most land are located in the West (including Alaska).

Four agencies administer about 628.4 million acres (93.5%) of the 671.8 million acres of federal land. [3] These four agencies are the Forest Service, Bureau of Land Management, Fish and Wildlife Service, and National Park Service. [4] The BLM has jurisdiction over approximately 261.5 million acres (38.9%) of the federal total. The FS has jurisdiction over approximately 192.5 million acres (28.7%) of the total federal acreage. The FWS administers approximately 95.4 million acres (14.2%). The National Park Service (NPS) administers about 79.0 million acres of federal land (11.8%), and oversees another 5.4 million acres of nonfederal land, for a total of about 84.4 million federal and nonfederal acres. Figure 1 shows the percent of land managed by each agency, and Table 2 displays the acreage for each of these four agencies in each state, the District of Columbia, and the territories. The lands administered by these four agencies are managed for a variety of purposes, primarily relating to the preservation, recreation, and development of natural resources. Although there are some similarities among the agencies, each agency has a distinct mission and special responsibilities for the lands under its jurisdiction.

The majority of the 671.8 million acres of federal lands are in the West, a result of early treaties and land settlement laws and patterns. Management of these lands is often controversial, especially in states where the federal government is a predominant or majority landholder and where competing and conflicting uses of the lands are at issue.

HISTORICAL REVIEW

The nation's lands and resources have been important in American history, adding to the strength and stature of the federal government, serving as an attraction and opportunity for settlement and economic development, and providing a source of revenue for schools, transportation, national defense, and other national, state, and local needs.

Table 1. Federally Owned Land by State, as of September 30, 2003

State	Total Acreage in State	Acreage of Federally Owned Land in State	% of Land Federally Owned in State
Alabama	32,678,400	1,202,614	3.7
Alaska	365,481,600	243,847,037	66.7
Arizona	72,688,000	36,494,844	50.2
Arkansas	33,599,360	3,955,959	11.8
California	100,206,720	46,979,891	46.9
Colorado	66,485,760	23,174,340	34.9
Connecticut	3,135,360	15,212	0.5
Delaware	1,265,920	29,488	2.3
District of Columbia	39,040	10,284	26.3
Florida	34,721,280	4,605,762	13.3
Georgia	37,295,360	2,314,386	6.2
Hawaii	4,105,600	671,580	16.4
Idaho	52,933,120	35,135,709	66.4
Illinois	35,795,200	651,603	1.8
Indiana	23,158,400	534,126	2.3
Iowa	35,860,480	302,601	0.8
Kansas	52,510,720	641,562	1.2
Kentucky	25,512,320	1,706,562	6.7
Louisiana	28,867,840	1,501,735	5.2
Maine	19,847,680	164,003	0.8
Maryland	6,319,360	192,692	3.0
Massachusetts	5,034,880	105,973	2.1
Michigan	36,492,160	3,638,588	10.0
Minnesota	51,205,760	3,534,989	6.9
Mississippi	30,222,720	2,101,204	7.0
Missouri	44,248,320	2,237,951	5.1
Montana	93,271,040	29,239,058	31.3
Nebraska	49,031,680	1,458,802	3.0
Nevada	70,264,320	64,589,139	91.9
New Hampshire	5,768,960	830,232	14.4
New Jersey	4,813,440	180,189	3.7
New Mexico	77,766,400	26,518,360	34.1
New York	30,680,960	242,441	0.8
North Carolina	31,402,880	3,602,080	11.5
North Dakota	44,452,480	1,333,375	3.0
Ohio	26,222,080	457,697	1.7
Oklahoma	44,087,680	1,331,457	3.0
Oregon	61,598,720	30,638,949	49.7
Pennsylvania	28,804,480	724,925	2.5
Rhode Island	677,120	5,318	0.8
South Carolina	19,374,080	1,236,214	6.4

Table 1. (Continued)

State	Total Acreage in State	Acreage of Federally Owned Land in State	% of Land Federally Owned in State
South Dakota	48,881,920	2,314,007	4.7
Tennessee	26,727,680	2,016,138	7.5
Texas	168,217,600	3,171,757	1.9
Utah	52,696,960	35,024,927	66.5
Vermont	5,936,640	450,017	7.6
Virginia	25,496,320	2,617,226	10.3
Washington	42,693,760	13,246,559	31.0
West Virginia	15,410,560	1,266,422	8.2
Wisconsin	35,011,200	1,981,781	5.7
Wyoming	62,343,040	31,531,537	50.6
Total	**2,271,343,360**	**671,759,298**	**29.6**

Source: U.S. General Services Administration, Overview of the United States Government's Owned and Leased Real Property: Federal Real Property Profile as of September 30, 2003. See Table 16, GSA website at [http://www.gsa.gov/gsa/cm_attachments/GSA_DOCUMENT/Annual%20Report %20%20FY2003-R4_R2M -n11_0Z5RDZ-i34K-pR.pdf], visited March 8, 2004. The data do not include trust properties or Department of Defense land outside the United States.

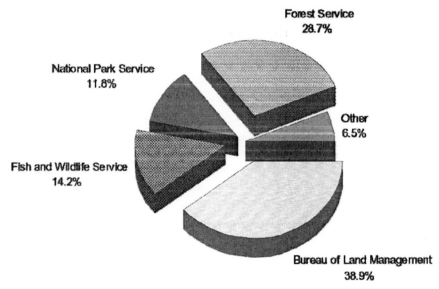

Note: Percentages do not add to 100% due to rounding.

Figure 1. Agency Jurisdiction Over Federally Owned Land in the United States.

The formation of our current federal government was particularly influenced by the struggle for control over what were known as the "western" lands — the lands between the Appalachian Mountains and the Mississippi River claimed by the original colonies. Prototypical land laws enacted by the Continental Congress, such as the Land Ordinance of 1785 [5] and the Northwest Ordinance of 1787, [6] established the federal system of rectangular land surveying for disposal and set up a system for developing territorial governments leading to statehood. During operation of the Articles of Confederation, the states that then owned the western lands were reluctant to cede them to the developing new government, but eventually acquiesced. This, together with granting constitutional powers to the new federal government, including the authority to regulate federal property and to create new states, played a crucial role in transforming the weak central government under the Articles of Confederation into a stronger, centralized federal government under our Constitution.

The new Congress, which first met in 1789, enacted land statutes similar to those enacted by the Continental Congress. Subsequent federal land laws reflected two visions: reserving some federal lands (such as for national forests and national parks) and selling or otherwise disposing of other lands to raise money or to encourage transportation, development, and settlement. From the earliest days, these policy clashes took on East/West overtones, with easterners more likely to view the lands as national public property, and westerners more likely to view the lands as necessary for local use and development. Most agreed, however, on measures that promoted settlement of the lands to pay soldiers, to reduce the national debt, and to strengthen the nation. This settlement trend accelerated after the Louisiana Purchase in 1803, the Oregon Compromise with England in 1846, and cession of lands by treaty after the Mexican war in 1848. [7]

During the mid- to late 1800s, Congress passed numerous laws that encouraged and accelerated the settlement of the West by disposing of federal lands. Examples include the Homestead Act of 1862 [8] and the Desert Lands Entry Act of 1877.

Approximately 815.9 million acres of the public domain lands were transferred to private ownership between 1781 and 2002. Another 328.5 million acres were granted to the states generally, and an additional 127.5 million were granted in Alaska under state and native selection laws. [9] Most transfers to private ownership (97%) occurred before 1940; homestead entries, for example, peaked in 1910 at 18.3 million acres but dropped below 200,000 acres annually after 1935, until being totally eliminated in 1986. [10]

Certain other federal laws were "catch up" laws designed to legitimize certain uses that already were occurring on the federal lands. These laws typically acknowledged local variations and customs. For example, the General Mining Law of 1872 recognized mineral claims on the public domain lands in accordance with local laws and customs, and provided for the conveyance of title to such lands. In addition, early land disposal laws allowed states to determine the rights of settlers to use and control water. The courts later determined, however, that the federal government could also reserve or create federal water rights for its own properties and purposes.

Although some earlier laws had protected some lands and resources, such as timber needed for military use, other laws in the late 1800s reflected the growing concern that rapid development threatened some of the scenic treasures of the nation, as well as resources that would be needed for future use. A preservation and conservation movement evolved to ensure that certain lands and resources were left untouched or reserved for future use. For example, Yellowstone National Park was established in 1872 [11] to preserve its resources in a natural condition, and to dedicate recreation opportunities for the public. It was the world's first national park, [12] and like the other early parks, Yellowstone was protected by the U.S. Army — primarily from poachers of wildlife or timber. In 1891, concern over the effects of timber harvests on water supplies and downstream flooding led to the creation of forest reserves (renamed national forests in 1907).

Table 2. Acreage Managed by Federal Agencies, by State

State	Forest Service	National Park Service	Fish and Wildlife Service	Bureau of Land Management
Alabama	665,978	16,917	59,528	111,369
Alaska	21,980,905	51,106,274	76,774,229	85,953,625
Arizona	11,262,350	2,679,731	1,726,280	11,651,958
Arkansas	2,591,897	101,549	361,331	295,185
California	20,741,229	7,559,121	472,338	15,128,485
Colorado	14,486,977	653,137	84,649	8,373,504
Connecticut	24	6,775	872	0
Delaware	0	0	26,126	0
Dist. of Col.	0	6,949	0	0
Florida	1,152,913	2,482,441	977,997	26,899
Georgia	864,623	40,771	480,634	0
Hawaii	1	353,292	299,380	0
Idaho	20,465,345	761,448	92,165	11,846,931
Illinois	293,016	12	140,236	224
Indiana	200,240	11,009	64,613	0
Iowa	0	2,708	112,794	378
Kansas	108,175	731	58,695	0
Kentucky	809,449	94,169	9,078	0
Louisiana	604,505	14,541	545,452	321,734
Maine	53,040	76,273	61,381	0
Maryland	0	44,482	45,030	548
Massachusetts	0	33,891	16,797	0
Michigan	2,865,103	632,368	115,244	74,807
Minnesota	2,839,693	142,863	547,421	146,658
Mississippi	1,171,158	108,417	226,039	56,212
Missouri	1,487,307	63,436	70,859	2,094
Montana	16,923,153	1,221,485	1,328,473	7,964,623
Nebraska	352,252	5,909	178,331	6,354
Nevada	5,835,284	777,017	2,389,616	47,874,294
New Hampshire	731,942	15,399	15,822	0
New Jersey	0	38,505	71,197	0
New Mexico	9,417,693	379,042	385,052	13,362,538
New York	16,211	37,114	29,081	0
North Carolina	1,251,674	394,833	423,948	0
North Dakota	1,105,977	71,650	1,566,026	59,642
Ohio	236,360	20,552	8,875	0
Oklahoma	399,528	10,200	170,032	2,136
Oregon	15,665,881	197,301	572,590	16,125,145
Pennsylvania	513,399	51,239	10,048	0
Rhode Island	0	5	2,179	0
South Carolina	616,970	27,488	162,958	0
South Dakota	2,013,447	263,644	1,300,465	274,960

Table 2. (Continued)

State	Forest Service	National Park Service	Fish and Wildlife Service	Bureau of Land Management
Tennessee	700,764	362,133	116,966	0
Texas	755,363	1,184,046	534,319	11,833
Utah	8,180,405	2,099,083	112,027	22,867,896
Vermont	389,200	21,513	33,230	0
Virginia	1,662,124	336,950	132,989	805
Washington	9,273,381	1,933,972	344,956	402,355
West Virginia	1,033,882	62,707	18,595	0
Wisconsin	1,525,978	74,010	236,470	159,982
Wyoming	9,238,067	2,393,281	101,857	18,354,151
Territories	28,149	33,179	1,766,965	0
Total	**192,511,012**	**79,005,557**	**95,382,237**	**261,457,325**

Sources: For FS: See the FS website at [http://www.fs.fed.us/land/staff/lar/LAR03/table4.htm], visited April 1, 2004. Data are current as of September 30, 2003. They reflect land managed by the FS that is within the National Forest System, including national forests, national grasslands, purchase units, land utilization projects, experimental areas, and other land areas, water areas, and interests in lands.

For NPS: U.S. Dept. of the Interior, National Park Service, Land Resources Division, *National Park Service, Listing of Acreage by State, as of 12/31/2003*, unpublished document. The data consist of all federal lands managed by the NPS. For information on acreage by type of unit as of September 30, 2003, see the NPS website at [http://www2.nature.nps.gov/stats/acresum03cy.pdf], visited April 1, 2004.

For FWS: U.S. Dept. of the Interior, Fish and Wildlife Service, *Annual Report of Lands Under Control of the U.S. Fish and Wildlife Service, as of September 30, 2002*. They comprise all land managed by the FWS, whether the agency has sole, primary, or secondary jurisdiction, and include acres under agreements, easements, and leases. For more information, see the FY2002 Annual Report of Lands on the FWS website at [http://realty.fws.gov/brochures.html], visited April 1, 2004.

For BLM: U.S. Dept. of the Interior, Bureau of Land Management, *Public Land Statistics, 2002*, and are current as of September 30, 2002. The data consist of lands managed exclusively by BLM, including certain types of surveyed and unsurveyed public and ceded Indian lands as well as withdrawn or reserved lands. For more information, see the BLM website at [http://www.blm.gov/natacq/pls02/], visited April 1, 2004.

The creation of national parks and forest reserves laid the foundation for the current development of federal agencies with primary purposes of managing natural resources on federal lands. For example, in 1905, responsibility for management of the forest reserves was joined with forestry research and assistance in a new Forest Service within the Department of Agriculture. The National Park Service was created in 1916 [13] to manage the growing number of parks established by Congress and monuments proclaimed by the President. The first national wildlife refuge was proclaimed in 1903, although it was not until 1966 that the refuges coalesced into the National Wildlife Refuge System. The Grazing Service (Department of the Interior, first known as the Grazing Division) was established in 1934 to administer grazing on public rangelands. It was combined with the General Land Office in 1946 to form the Bureau of Land Management (BLM). [14]

In addition to the conservation laws and activities noted above, emphasis shifted during the 20th century from the disposal and conveyance of title to private citizens to the retention and management of the remaining federal lands. Some laws provided for sharing revenues from various uses of the federal lands with the states containing the lands. Examples include the Mineral Leasing Act of 1920, [15] which provides for the leased development of certain federal minerals, and the Taylor Grazing Act of 1934, which provides for permitted private livestock grazing on public lands. [16]

During debates on the Taylor Grazing Act, some western Members of Congress acknowledged the poor prospects for relinquishing federal lands to the states, but language included in the act left this question open. It was not until the passage of the Federal Land Policy and Management Act of 1976 (FLPMA, P.L. 94-579, 43 U.S.C. §§1701, et seq.) that Congress expressly declared that the remaining public domain lands generally would remain in federal ownership. [17] This declaration of policy was a significant factor in what became known as the Sagebrush Rebellion, an effort that started in the late 1970s to take state or local control of federal land and management decisions. To date, judicial challenges and legislative and executive attempts to make significant changes to federal ownership have proven unsuccessful. Current authorities for acquiring and disposing of federal lands are unique to each agency, and are described in subsequent chapters of this article.

ISSUES

Since the cession to the federal government of the western lands of several of the original 13 colonies, many issues and conflicts have recurred. Ownership continues to be debated, with some advocating increased disposal of federal lands to state or private ownership, and others supporting retention of federal lands by the federal government. Still others promote acquisition by the federal government of additional land, including through an increased, and more stable, funding source. A related issue is determining the optimal division of resources between federal acquisition of new lands and maintenance of existing federal lands and facilities.

Another focus is whether federal lands should be managed primarily to produce national benefits or benefits primarily for the localities and states in which the lands are located. Who decides these issues, and how the decisions are made, also are at issue. Some would like to see more local control of land and a reduced federal role, while others seek to maintain or enhance the federal role in land management to represent the interests of all citizens.

The extent to which federal lands should be made available for development, preserved, and opened to recreation has been controversial. Significant differences of opinion exist on the amount of traditional commercial development that should be allowed, particularly involving energy development, grazing, and timber harvesting. How much land to accord enhanced protection, what type of protection to accord, and who should protect federal lands are continuing questions. Whether and where to restrict recreation, either generally or for such uses as motorized off-road vehicles, also is a focus of debate.

The debate over land uses perhaps has intensified with the increase over the decades in visitors to federal lands. Current agency figures on visitor use point to recreation as a fast-growing use of agency lands overall. For FY2003, recreation visits totaled 265 million for the

National Park System, 53 million for BLM lands, and 39 million for the National Wildlife Refuge System. For FY2002, recreation visits to the National Forest System totaled 211 million.

CRS REPORTS AND COMMITTEE PRINTS [18]

CRS Issue Brief IB10076, *Bureau of Land Management (BLM) Lands and National Forests*, coordinated by Ross W. Gorte and Carol Hardy Vincent.

CRS Report RS20002, *Federal Land and Resource Management: A Primer*, coordinated by Ross W. Gorte.

CRS Report RL30126, *Federal Land Ownership: Constitutional Authority; the History of Acquisition, Disposal, and Retention; and Current Acquisition and Disposal Authorities*, by Ross W. Gorte and Pamela Baldwin.

CRS Issue Brief IB10093, *National Park Management and Recreation*, coordinated by Carol Hardy Vincent.

U.S. Congress, Committee on Interior and Insular Affairs, *Multiple Use and Sustained Yield: Changing Philosophies for Federal Land Management? The Proceedings and Summary of a Workshop Convened on March 5-6, 1992*, committee print prepared by the Congressional Research Service, No. 11 (Washington, DC: GPO, Dec. 1992).

U.S. Congress, Committee on Energy and Natural Resources, *Outdoor Recreation: A Reader for Congress*, committee print prepared by the Congressional Research Service, S.Prt. 105-53 (Washington, DC: GPO, June 1998).

U.S. Congress, Committee on Environment and Public Works, *Ecosystem Management: Status and Potential. Summary of a Workshop Convened by the Congressional Research Service, March 24-25, 1994*, committee print prepared by the Congressional Research Service, S.Prt. 103-98 (Washington, DC: GPO, Dec. 1994).

FEDERAL LANDS FINANCING [19]

Financial issues are a persistent concern for federal agencies, including the land management agencies. However, the sale or lease of the lands and resources being managed provides these agencies with an opportunity to recover some of their operations and capital costs. This section summarizes the revenues of the four land management agencies and provides a brief overview of annual appropriations, the trust funds and special accounts funded from revenues, and land acquisition funding. It concludes with a discussion of the programs that compensate state and local governments for the tax-exempt status of federal lands.

Revenues from Activities on Federal Lands

The federal land management agencies are among the relatively few federal agencies that generate revenues for the U.S. Treasury. However, none of these four agencies consistently collects more money than it expends. Revenues are derived from the use or sale of lands and resources. Major revenue sources include timber sales, grazing livestock fees, energy and mineral leases, and fees for recreation uses. The FY2003 revenues collected by these four agencies, excluding deposits to trust funds and special accounts, are shown in Table 3.

**Table 3. Revenues from the Sale and Use of Agency Lands and Resources for FY2003
(thousands of dollars; excluding deposits to trust funds and special accounts)**

Resource	BLM	FWS	NPS	FS
Mineral Leases and Permits	$103,857[a]	n/a[b]	$0	$187,114[c]
Sales of Timber and Other Forest Products	$11,501	n/a[b]	$12	$58,548
Grazing Leases, Licenses, and Permits	$11,828	n/a[b]	—[d]	$4,351
Recreation, Admission, and User Fees	$0[e]	n/a[b]	$0[f]	$44,381
Other	$135,941[g]	n/a[b]	$15	$12,072
Total	**$263,127**	**$6,895**	**$27**	**$306,466**

Sources: For BLM: U.S. Dept. of the Interior, Budget Justifications and Performance Information, Fiscal Year 2005: Bureau of Land Management, p. II-1.

For FWS: U.S. Dept. of the Interior, Budget Justifications and Performance Information, Fiscal Year 2005: U.S. Fish and Wildlife Service, p. 445.

For NPS: U.S. Dept. of the Interior, Budget Justifications and Performance Information, Fiscal Year 2005: National Park Service, p. Overview-26.

For FS: U.S. Dept. of Agriculture, Forest Service, USDA Forest Service FY2005 Budget Justification, pp. A-9 - A-10.

a. Includes mineral leasing on national grasslands, the Naval Oil Shale Reserve, and the National Petroleum Reserve-Alaska, and mining claim and holding fees.
b. n/a: data are not available in published form.
c. Includes estimated $154.5 million collected by Departments of the Interior and Energy for mineral leases and power licenses.
d. Included with revenues for sales of timber and forest products.
e. All BLM recreation fees are now deposited in its Recreation Fee Demonstration Account, totaling $10 million.
f. The NPS is now authorized through several permanently appropriated accounts to retain all such fees in permanently appropriated accounts, totaling $245 million.
g. Includes Treasury deposits from land sales ($13 million), sale of helium ($87 million), other fees, charges, and collections ($33 million), and earnings on investments ($2 million).

Agency Appropriations

Annual Appropriations

Funding for all four of the federal land management agencies is contained in the annual Department of the Interior and Related Agencies appropriations bill. The FS is a USDA agency, but has been included in the Interior bill as a "related agency" since 1955. It receives the largest appropriation of any agency in the Interior bill, with funding of $4.54 billion (including emergency fire funding) in the Interior Appropriations Act for FY2004 (P.L. 108-108). The NPS receives the next largest appropriations of the federal land management agencies, with FY2004 funding of $2.26 billion. For FY2004, the BLM received $1.79 billion (including emergency fire funding). The FWS has the lowest funding of the land management agencies, with FY2004 appropriations at $1.31 billion. For more information on annual funding for these agencies, see CRS Report RL32306, *Appropriations for FY2005: Interior and Related Agencies*, available on the CRS website at [http://www.crs.gov/products/appropriations/apppage.shtml].

Trust Funds and Special Accounts

The federal land management agencies also have a variety of trust funds and special accounts. Some require annual appropriations; most of these are small, but the Land and Water Conservation Fund used for federal land acquisition is relatively large and controversial, and is discussed separately below.

A number of the trust funds and special accounts are permanently appropriated (also known as mandatory spending). This means that the agencies can spend the receipts deposited in the accounts without annual appropriations by Congress. Many of these accounts (15) were established to compensate state and local governments for the tax-exempt status of federal lands; these accounts will be discussed separately below. Others receive funds from particular sources (e.g., excise taxes, timber sales, recreation fees) for grants or for agency operations. The receipts deposited in these accounts are *in addition to* the Treasury receipts shown in Table 3.

The FWS has the largest annual funding in permanently appropriated trust funds and special accounts, with FY2003 budget authority of $661 million. The two largest accounts are the Sport Fish Restoration Trust Fund ($330 million), established by the Federal Aid in Sport Fish Restoration Act; [20] and the Wildlife Restoration Special Account ($235 million), established by the Federal Aid in Wildlife Restoration Act. [21] These accounts are largely funded by excise taxes on equipment related to fishing and hunting, respectively, and the money is distributed to the states mostly to fund fish and wildlife restoration activities by state agencies. The third largest account is the Migratory Bird Conservation Fund ($44 million), which uses the revenues from selling duck stamps to hunters, refuge visitors, stamp collectors, and others to acquire lands for the National Wildlife Refuge System (as noted below, under "Land Acquisition Funding").

The BLM and NPS have numerous permanently appropriated trust funds and special accounts, with total budget authority of $305 million for each in FY2003. Most of the BLM accounts are much smaller than for the other federal land management agencies, but the one largest account — Southern Nevada public land sales — had FY2003 budget authority of $279 million (92% of BLM permanent appropriations for operations).

The NPS permanently appropriated special accounts and trust funds allow the agency to retain 100% of its recreation and admission fees. The largest is the Recreational Fee Demonstration Program, described below. Two funds are unique to the NPS: the concessions improvement account and park concessions franchise fees (a combined total of $54 million in FY2003). Two other funds are common to all four land management agencies, but are significantly larger for the NPS. One is the fund for maintaining employee quarters ($16 million for the NPS, less than $11 million total for the other three agencies) paid by rent from employees. Another consists of contributions and donations from interested individuals and groups ($29 million for the NPS; less than $3 million total for the other three agencies).

The FS has the least annual funding in permanently appropriated trust funds and special accounts. The FS has 20 accounts with FY2003 budget authority of $285 million. Six of the eight largest are directly or substantially related to timber sales, including the Salvage Sale Fund ($58 million), the Knutson-Vandenberg Fund ($48 million), other cooperative deposits ($41 million), the Reforestation Trust Fund ($30 million), National Forest roads and trails ($12 million), [22] and brush disposal ($12 million).

Finally, two programs were established to authorize the four agencies to retain recreation fees. The first, recreation fee collection costs (P.L. 103-66, §10002(b)), allows the agencies to retain up to 15% of recreation fees to cover the costs to collect the fees. The second, much larger program is the Recreational Fee Demonstration Program, created to allow the agencies to test the feasibility and public acceptability of user fees to supplement appropriations for operations and maintenance (P.L. 104-134, §315). This "Fee Demo" program authorized new or increased entrance fees at federal recreation sites from FY1996 through FY1998; it has been extended multiple times, and now is authorized for fee collections through December 31, 2005 (with expenditures through FY2008). FY2003 collections are $124 million for the NPS, $37 million for the FS, $9 million for the BLM, and $4 million for the FWS.

Land Acquisition Funding

The largest source of funding for federal land acquisition is the Land and Water Conservation Fund. LWCF is a special account created in 1964 specifically to fund federal land acquisition and state recreation programs. It can be credited with revenues from federal recreation user fees (other than those collected under the Recreational Fee Demonstration Program and the Fee Collection Cost Program), the federal motorboat fuel tax, and surplus property sales; these are supplemented with revenues from federal offshore oil and gas leases, up to the authorized level of $900 million annually.

LWCF does not operate the way a "true" trust fund would in the private sector. The fund is *credited* with deposits from specified sources, but Congress must enact appropriations annually for the agencies to spend money from the fund. Through FY2004, $27.2 billion has been credited to the LWCF, and $13.8 billion has been appropriated. Unappropriated funds remain in the U.S. Treasury and can be spent for other purposes.

The 105th, 106th, and 107th Congresses considered legislation that would have supplemented or supplanted the LWCF and fully funded it for 15 years. The Clinton Administration successfully pursued another avenue (the Lands Legacy Initiative that led to the creation of the Conservation Spending Category) to increase funding for LWCF federal land acquisition through the annual appropriations process and to use some of the LWCF authorization for other (non-acquisition) federal programs. President Bush has expanded on this latter approach, proposing in FY2005 to fully fund LWCF — requesting $900.2 million — but use more than half of the total for non-acquisition federal programs, including several Fish and Wildlife Service and Forest Service programs. In FY2003, LWCF appropriations for federal land acquisition alone totaled $313.0 million, and in FY2004 they declined to $169.7 million, both down from the FY1998 peak of $897.1 million. For FY2005, President Bush has requested $220.2 million for LWCF federal land acquisition.

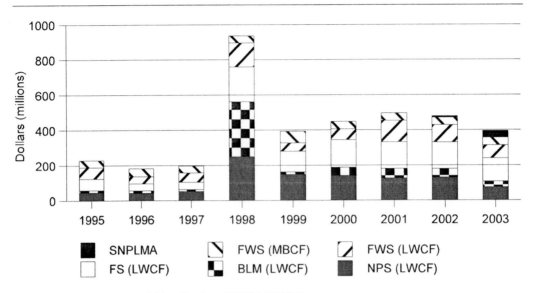

Figure 2. Federal Land Acquisition Funding, FY1995-FY2003.

Other federal programs also provide funding for federal land acquisition. The largest is the FWS's Migratory Bird Conservation Fund (MBCF). Receipts from the sale of duck stamps to hunters, refuge visitors, stamp collectors, and others are deposited in this account. The funds are permanently appropriated to the FWS to acquire lands for the National Wildlife Refuge System, and often provide more than half the total FWS land acquisition funding. In FY2003, the FWS used $43.8 million of MBCF for land acquisition.

The BLM has a mandatory spending program for land acquisition and other activities in Nevada, funded from sales of BLM land in that state (Southern Nevada Public Land Management Act, SNPLMA, P.L. 105-623). This program allows money from BLM land sales in Nevada to be used for land acquisition by the federal land management agencies, but also for capital improvements on federal lands and state and local government purposes. Since 2000, this program has generated more than $400 million, and it is projected to generate $338 million in FY2004 and $846 million in FY2005. The portion spent on federal land acquisition varies, and totaled $38.6 million in FY2003. This relatively small amount is attributable in part to the newness of the program and it is expected to increase in coming years. In addition, the FS has a very small program (about $1 million annually) for acquiring lands in certain parts of Utah and California.

Figure 2 shows federal land acquisition funding since FY1995. Total funding rose from a low of $181.5 million in FY1996 to a peak of $936.7 million in FY1998, then declined to $395.4 million in FY2003. Funding for federal land acquisition (excluding SNPLMA) is estimated at $212.0 million for FY2004, and at $263.4 million under President Bush's FY2005 budget request. [23]

Compensation to State and Local Governments

Because federal property is exempt from state and local taxation, Congress has enacted mechanisms to compensate state and local governments for tax revenues that would have been collected if the lands were privately owned. Many of the mechanisms provide for sharing revenues from federal lands with state and/or local governments; only the NPS has no agency-specific compensation system. The Payments In Lieu of Taxes (PILT) Program provides additional revenues.

Revenue-Sharing

The amount and percentage of federal revenues that are shared with state and/or local governments depends upon the history of the land and the type of activities generating the revenues. Congress created the simplest system for revenue-sharing for FS lands. Since 1908, the agency has returned 25% of its gross revenues to the states for use on roads and schools in the counties where the national forests are located. The states determine which road and school programs are to be funded, and how much goes to each program, but the amount allocated to each county is determined by the FS and the states cannot retain any of the funds. For the national grasslands, 25% of *net* revenues go directly to the counties. In addition, three counties in Minnesota receive a special payment of 0.75% of the appraised value of the Superior NF lands in the county. Payments for these FS programs are permanently appropriated from any FS revenues; in FY2003, total FS payments were $393 million.

Because of concerns over declining timber revenues in many areas, and the approaching end of the special "spotted owl payments" program, [24] the 106th Congress debated bills to modify the FS revenue-sharing program. In the Secure Rural Schools and Community Self-Determination Act of 2000 (P.L. 106-393), Congress enacted a six-year program allowing counties to supplant the historic 25% payment with the average of the three highest payments to the state between 1986 and 1999. Of these high-3 payments, 15%-20% must be spent on certain county programs or on projects on federal lands recommended by a local advisory committee or chosen by the FS. This program accounted for 72% of the $393 million in FS payments in FY2003.

For BLM lands and revenues, the revenue-sharing system is more complicated. The share going to state and local entities ranges from 0% to 90% of gross program revenues, as specified in individual statutes. For example, states and counties receive 12.5% of revenues from grazing within grazing districts (under §3 of the Taylor Grazing Act of 1934) and 50% of revenues from grazing outside grazing districts (under §15 of the Taylor Grazing Act). Another example is timber sale revenues. The states and counties receive 4% of timber revenues from most BLM lands. However, the counties receive up to 75% from the heavily timbered Oregon and California (OandC) railroad grant lands in Western Oregon. [25] Counties with the Coos Bay Wagon Road (CBWR) grant lands (adjoining and usually identified with the OandC lands) similarly receive up to 75%, but actual payments are limited by county tax assessments. Because the OandC and CBWR payments have been largely from timber sales, which have declined since the late 1980s, they were included with national forest lands (see above) in the spotted owl payments program and the six-year program of payments at the average of the three highest, under P.L. 106-393.

These examples demonstrate the complexity of the legal direction to share BLM revenues with state and local governments. The BLM revenue-sharing payments are permanently appropriated, with 10 separate payment accounts; FY2003 budget authority was $157 million, of which $111 million was for the OandC and CBWR lands and $38 was related to oil leasing in the National Petroleum Reserve-Alaska.

Finally, the FWS has a revenue-sharing program, but payments depend on the history of the land. For refuges reserved from the public domain, the payments are based on 25% of *net* revenues (in contrast to 25% of *gross* revenues from FS lands other than national grasslands). For refuges which have been created on lands acquired from other landowners, payments are based on the *greatest* of: 25% of net revenues, 0.75% of fair market value of the land, or $0.75 per acre. The National Wildlife Refuge Fund is permanently appropriated for making these payments, but net revenues have been insufficient to make the authorized payments. Although payments have been supplemented with annual appropriations, total payments — $14 million in FY2003 — consistently have been less than the authorized level.

Payments in Lieu of Taxes

The most comprehensive federal program for compensating local governments for the tax-exempt status of federal lands was created in the 1976 Payments in Lieu of Taxes (PILT) Act. PILT payments are made in addition to any revenue-sharing payments, although the payments may be reduced by such revenue-sharing payments, as discussed below. Federal lands encompassed by this county-compensation program include lands in the National Forest System, lands in the National Park System, and those administered by the BLM, plus the National Wildlife Refuge System lands reserved from the public domain, and a few other categories of federal lands.

In 1994, Congress amended the PILT Act to more than double the authorized payments over five years, to adjust for inflation between 1976 and 1994, and to build in adjustments for future inflation. The two formulae used to calculate the FY2003 authorized payment level for each county with eligible federal lands are:

1. Which is *less*: (a) the county's eligible acres times $0.27 per acre; or (b) the county's payment ceiling (determined by county population level). Pick the lesser of these two. This option is called the *minimum provision.*

2. Which is *less*: (a) the county's eligible acres times $2.02 per acre; or (b) the county's payment ceiling (determined by county population level). Pick the lesser of these two, and from it subtract the previous year's total payments under other payment or revenue-sharing programs of the agencies that control the eligible land (as reported by each state to the BLM). This option is called the *standard provision.*

The county is authorized to receive whichever of the above calculations (1 or 2) is *greater*. This calculation must be made for all counties individually to determine the national authorization level.

In contrast to most of the revenue-sharing programs, PILT requires annual appropriations from Congress. Those appropriations generally had been sufficient to compensate the counties at the authorized level prior to the 1994 amendments. Those amendments raised the authorization; however, subsequent appropriations have been substantially below the increased authorization. Figure 3 compares the level of authorization and appropriation for each year since FY1993.

Issues

Several financing themes are perennial issues for Congress, involving fees charged (or not charged) and how these revenues relate to agency activities. One issue has been the question of whether prices set administratively (rather than by markets) subsidize some resource users. This issue typically has focused on fees for private livestock grazing on federal lands and for hardrock (locatable) minerals that are currently available for private development under a claims system without royalty payments. Another issue is whether "below-cost" timber sales should continue if the government is losing money on them. In addition, whether to permanently authorize the Recreational Fee Demonstration Program, and which agencies' lands and programs to include, is a continuing congressional focus.

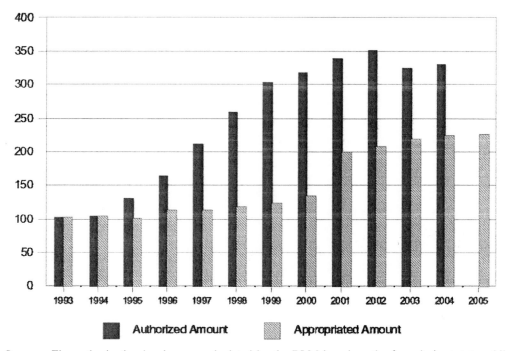

Authorized Amount   Appropriated Amount

Sources: The authorization levels were calculated by the BLM based on the formula in statute, while the appropriation levels were taken from laws appropriating funds for the Department of the Interior.

Notes: The FY2004 authorized amount is an estimate; the FY2005 authorized amount is not yet estimated. The FY2005 appropriation level reflects the Administration's request. Authorization for a given year depends on receipts from the previous year from the agencies that administer the eligible lands. Consequently, no authorization level can be determined for FY2005.

Figure 3. PILT: Authorized and Appropriated Amounts, FY1993-FY2005 (in millions of $).

Another persistent issue is determining the annual appropriations for the Department of the Interior and related agencies (including the FS). The budget levels for the agencies often are controversial, especially in today's climate of increasing budget deficits and expenditures for the war on terrorism. Legislative provisions and directions/restrictions on spending contained in appropriations bills, commonly referred to as environmental and resource "riders," often are the most controversial parts of these bills.

Funding for wildfire protection has grown significantly in recent years, following the severe fire seasons of 2000 and 2002. Annual appropriations for fire suppression operations have not been sufficient, and the agencies have used their authority to borrow from other accounts to fund fire suppression. These borrowings typically are repaid in an emergency supplemental appropriation bill or in the subsequent annual appropriations bill. However, the borrowed funds are not always repaid promptly, leading to funding shortfalls in the accounts from which the funds were borrowed (such as land acquisition).

Major Statutes

Department of the Interior and Related Agencies Appropriations Act for FY2004 (the most recent in the annual series of such acts): Act of Nov. 10, 2003; P.L. 108-108.

Forest Service Revenue-Sharing Act: Act of May 23, 1908; ch. 192, 35 Stat. 251. 16 U.S.C. §500.

Land and Water Conservation Fund Act of 1965: Act of Sept. 3, 1964; P.L. 88-578, 78 Stat. 897. 16 U.S.C. §460*l*.

Payments in Lieu of Taxes Act: Act of Oct. 20, 1976; P.L. 94-565, 90 Stat. 2662. 31 U.S.C. §§6901-6907.

Secure Rural Schools and Community Self-Determination Act of 2000: Act of Oct. 19, 2000; P.L. 106-393.

CRS Reports and Committee Prints [26]

CRS Report RL32306, *Appropriations for FY2005: Interior and Related Agencies*, coordinated by Carol Hardy Vincent and Susan Boren. (The most recent in an annual series of such reports.)

CRS Report RL30335, *Federal Land Management Agencies' Permanently Appropriated Accounts*, by Ross W. Gorte, M. Lynne Corn, and Carol Hardy Vincent.

CRS Report 98-980, *Federal Sales of Natural Resources: Pricing and Allocation Mechanisms,* by Ross W. Gorte.

CRS Report 90-192, *Fish and Wildlife Service: Compensation to Local Governments*, by M. Lynne Corn.

CRS Report RL30480, *Forest Service Revenue-Sharing Payments: Legislative Issues*, by Ross W. Gorte.

CRS Report RS21503, *Land and Water Conservation Fund: Current Status and Issues,* by Jeffrey Zinn.

CRS Issue Brief IB10093, *National Park Management and Recreation*, coordinated by Carol Hardy Vincent.

CRS Report RL31392, *PILT (Payments in Lieu of Taxes): Somewhat Simplified*, by M. Lynne Corn.

THE NATIONAL FOREST SYSTEM [27]

The National Forest System (NFS) is administered by the Forest Service (FS) in the U.S. Department of Agriculture. The NFS is comprised of national forests, national grasslands, and various other designations. Although NFS lands are concentrated (87%) in the West, the FS administers more federal land in the East than all other federal agencies combined. NFS lands are administered for sustained yields of multiple uses, including outdoor recreation (camping, hiking, hunting, sightseeing, etc.), livestock grazing, timber harvesting, watershed protection, and fish and wildlife habitats.

Background [28]

In 1891, Congress granted the President the authority (now repealed) to establish forest reserves from the public domain. Six years later, in 1897, Congress stated that the forest reserves were:

> to improve and protect the forest within the reservation, or for the purpose of securing favorable conditions of water flows, and to furnish a continuous supply of timber for the use and necessities of the citizens of the United States.

Initially, the reserves were administered by the Division of Forestry in the General Land Office of the Department of the Interior. In 1905, this division was combined with the USDA Bureau of Forestry, renamed the Forest Service, and the administration of the 56 million acres of forest reserves (renamed *national forests* in 1907) was transferred to the new agency within the Department of Agriculture. NFS management is one of the three principal FS programs. [29]

In 1906 and 1907, President Theodore Roosevelt more than doubled the acreage of the forest reserves. In 1907, Congress limited the authority of the President to add to the system in certain states. [30] Then in 1910, Congress repeated the limitation in the Pickett Act. In 1911, Congress passed the Weeks Law, authorizing additions to the NFS through the purchase of private lands. Under this and other authorities, the system has continued to grow slowly, from 154 million acres in 1919 to 192.5 million acres in 2003. This growth has resulted from purchases and donations of private land and from land transfers, primarily from the BLM.

Organization

The NFS includes 155 national forests with 188 million acres (97.6% of the system); 20 national grasslands with 4 million acres (2.0%); and 121 other areas, such as land utilization projects, purchase units, and research and experimental areas, with 0.8 million acres (0.4%). [31] The NFS units are arranged into nine administrative regions, each headed by a regional forester. The nine regional foresters report to the NFS Deputy Chief, who reports to the Chief of the Forest Service. In contrast to the heads of other federal land management agencies, the Chief traditionally has been a career employee of the agency. The Chief reports to the Secretary through the Undersecretary for Natural Resources and Environment.

The NFS regions often are referred to by number, rather than by name. Table 4 identifies the number, states encompassed, and acreage for each of the regions. Although the NFS lands are concentrated in the seven western FS regions, including Alaska (87%), the FS manages more than half of all federal land in the East. *Inholdings* shown in Table 4 is land (primarily private) within the designated boundaries of the national forests (and other NFS units) which is not administered by the FS. Inholdings sometimes pose difficulties for FS land management, because the agency generally does not regulate the development and use of the inholdings. The uses of private inholdings may be incompatible with desired uses of the federal lands, and constraints on crossing inholdings may limit access to some federal lands. Many private landowners, however, object to federal restrictions on the use of their lands and to unfettered public access across their lands. This is particularly true in the Southern and Eastern Regions, where nearly half of the land within the NFS boundaries is inholdings.

Table 4. The National Forest System

Forest Service Region		States containing NFS lands[a]		National Forest System Acreage[b]	
Region Name	No.	States		Federal	Inholdings
Northern	1	ID, MT, ND		25,441,585	2,727,271
Rocky Mountain	2	CO, NE, SD, WY		22,069,840	2,380,838
Southwestern	3	AZ, NM		20,805,767	1,668,087
Intermountain	4	ID, NV, UT, WY		32,003,788	2,250,034
Pacific Southwest	5	CA		20,137,345	3,629,680
Pacific Northwest	6	OR, WA		24,737,016	2,660,525
Southern	8	AL, AR, FL, GA, KY, LA, MS, NC, OK, PR, SC, TN, TX, VA		13,273,000	12,324,182
Eastern	9	IL, IN, ME, MI, MN, MO, NH, NY, OH, PA, VT, WI, WV		12,061,766	9,895,489
Alaska	10	AK		21,980,905	2,375,273
National Forest System Total				192,511,012	39,911,379

Source: U.S. Dept. of Agriculture, Forest Service, *Land Areas of the National Forest System, as of Sept. 30, 2004*, Tables 1 and 2, from [http://www.fs.fed.us/land/staff/lar/LAR03/], visited Feb. 20, 2004.

Notes: In 1966, Region 7, the Lake States Region, was merged with Region 9, the Northeastern Region, to form the current Eastern Region. Although this merger left 9 regions, the numbering sequence skips 7 and ends with 10, as shown in the table.

a. This column lists only states (and territories) that currently contain NFS lands.

b. *Federal* is federally owned land managed by the FS. *Inholdings* are private and other government lands within NFS boundaries that are not administered or regulated by the FS.

MANAGEMENT

The management goals for the national forests were first established in 1897, as described above. Management goals were further articulated in §1 of the Multiple-Use Sustained-Yield Act of 1960 (MUSYA), which states:

> It is the policy of the Congress that the national forests are established and shall be administered for outdoor recreation, range, timber, watershed, and wildlife and fish purposes. The purposes of this Act are declared to be supplemental to, but not in derogation of, the purposes for which the national forests were established as set forth in the Act of June 4,

1897.... The establishment and maintenance of areas as wilderness are consistent with the purposes and provisions of this Act.

MUSYA directs land and resource management of the national forests for the combination of uses that best meets the needs of the American people. Management of the resources is to be coordinated for *multiple use* — considering the relative values of the various resources, but not necessarily maximizing dollar returns, nor requiring that any one particular area be managed for all or even most uses. The act also calls for *sustained yield* — a high level of resource outputs maintained in perpetuity but without impairing the productivity of the land. Other statutes, such as the Endangered Species Act, that apply to all federal agencies also apply.

NFS planning and management is guided primarily by the Forest and Rangeland Renewable Resources Planning Act (RPA) of 1974, as amended by the National Forest Management Act (NFMA) of 1976. Together, these laws encourage foresight in the use of the nation's forest resources, and establish a long-range planning process for the management of the NFS. RPA focuses on the national, long-range direction for forest and range conservation and sustainability. [32] RPA requires the FS to prepare four documents for Congress and the public: an Assessment every 10 years to inventory and monitor the status and trends of the nation's natural resources; a Program every five years to guide FS policies; a Presidential Statement of Policy to accompany the Program and guide budget formulation; and an Annual Report to evaluate implementation of the Program. [33]

NFMA requires the FS to prepare a comprehensive land and resource management plan for each unit of the NFS, coordinated with the national RPA planning process. [34] The plans must use an interdisciplinary approach, including economic analysis and the identification of costs and benefits of all resource uses. Planning regulations (36 C.F.R. §219) were issued in 1979, then revised in 1982. Revision of the 1982 regulations was begun with an advance notice of proposed rulemaking in 1991, and proposed revised regulations were issued in 1995. In 1997, the Secretary of Agriculture chartered a Committee of Scientists to review the planning process, and its March 1999 report, *Sustaining the People's Lands*, made numerous recommendations. [35] On October 5, 1999, the Clinton Administration proposed new regulations (64 *Federal Register* 54073), with final regulations revising the planning process on November 9, 2000 (65 *Federal Register* 67514). These regulations would have increased emphasis on ecological sustainability, and would have been implemented over several years.

On December 6, 2002, in response to concerns about whether the Clinton regulations could be implemented and about the lack of emphasis on economic and social sustainability, the Bush Administration proposed new regulations (67 *Federal Register* 72700) to supplant the Clinton regulations before they were implemented. The proposed Bush regulations seek to balance ecological sustainability with economic and social considerations, and would reduce national direction in FS decision-making. Final regulations have not been issued.

Congress has provided further management direction within the NFS by creating special designations for certain areas. Some of these designations — wilderness areas, wild and scenic rivers, and national trails — are part of larger management systems affecting several federal land management agencies; these special systems are described in later chapters of this article.

In addition to these special systems, the NFS includes several other types of land designations. The NFS contains 21 national game refuges and wildlife preserves (1.2 million acres), 20 national recreation areas (2.9 million acres), 4 national monuments (3.7 million acres), 2 national volcanic monuments (167,427 acres), 6 scenic areas (130,435 acres), a scenic-research area (6,637 acres), a scenic recreation area (12,645 acres), a recreation management area (43,900 acres), 3 special management areas (91,265 acres), 2 national protection areas (27,600 acres), 2 national botanical areas (8,256 acres), a primitive area (173,762 acres) and a national historic area (6,540 acres). [36] Resource development and use is generally more restricted in these specially designated areas than on general NFS lands, and specific guidance typically is provided with each designation.

Land Ownership

Designation

As noted above, in 1891, the President was authorized to reserve lands from the public domain as forest reserves (16 U.S.C. §471, now repealed), but this authority was subsequently limited by Congress, and it appears that no new NFS lands were reserved in the West after 1907. However, many proclamations and executive orders subsequently have modified boundaries and changed names, including establishing new national forests from existing NFS lands. National forests in the East generally were established between 1910 and 1950, with the Hoosier and Wayne Forests (in Indiana and Ohio, respectively) the last proclaimed, in 1951.

Presidential authority to proclaim forest reserves from the public domain was restricted piecemeal. The 1897 Act established management direction by restricting the purposes for the reserves. The 1907 Act that renamed the forest reserves as the national forests also prohibited the establishment of new reserves in six western states, although President Theodore Roosevelt did not sign the law until he had reserved 16 million acres in those states. Presidential authority to withdraw public lands to establish new national forests was not formally repealed until 1976. [37] Today, establishing a new national forest from public domain lands or significantly modifying the boundaries of an existing national forest created from the public domain requires an act of Congress. [38]

Acquisition Authority

The Secretary of Agriculture has numerous authorities to add lands to the NFS. The first and broadest authority was in the Weeks Law of 1911 (as amended by NFMA; 16 U.S.C. §515):

> The Secretary is hereby authorized and directed to examine, locate, and purchase such forested, cut-over, or denuded lands within the watersheds of navigable streams as in his judgment may be necessary to the regulation of the flow of navigable streams or for the production of timber.

Originally, the acquisitions were to be approved by a National Forest Reservation Commission, but the Commission was terminated in 1976 by §17 of NFMA.

Other laws also authorize land acquisition for the national forests, typically in specific areas or for specific purposes. For example, §205 of FLPMA authorizes the acquisition of access corridors to national forests across nonfederal lands (43 U.S.C. §1715(a)). The

Federal Land Management Agencies 25

Southern Nevada Public Land Management Act of 1998 authorizes acquisition of environmentally sensitive lands in Nevada, some of which have been added to the National Forest System. Also, under the Federal Land Transaction Facilitation Act, the Secretary of Agriculture may acquire inholdings and other nonfederal land. (See discussion of BLM "Disposal Authority," below.)

Finally, the Bankhead-Jones Farm Tenant Act of 1937 authorizes and directs the Secretary of Agriculture to establish (7 U.S.C. §1010):

> a program of land conservation and land utilization, in order to correct maladjustments in land use, and thus assist in controlling soil erosion, reforestation, preserving natural resources, protecting fish and wildlife, developing and protecting recreational facilities, mitigating floods, preventing impairment of dams and reservoirs, developing energy resources, conserving surface and subsurface moisture, protecting the watersheds of navigable streams, and protecting public lands, health, safety, and welfare ....

Initially, the act authorized the Secretary to acquire submarginal lands and lands not primarily suitable for cultivation (§1011(a)); this provision was repealed in 1962. This authority allowed the agency to acquire and establish the 20 national grasslands and 6 land utilization projects that account for 2% of the NFS. In addition, millions of acres acquired under this authority have been transferred to the BLM.

Disposal Authority

The Secretary of Agriculture has numerous authorities to dispose of NFS lands, all constrained in various ways and seldom used. In 1897, the President was authorized (16 U.S.C. §473):

> to revoke, modify, or suspend any and all Executive orders and proclamations or any part thereof issued under section 471 of this title, from time to time as he deems best for the public interests. By such modification he may reduce the area or change the boundary lines or may vacate altogether any order creating a national forest.

The 1897 Act also provided for the return to the public domain of lands better suited for agriculture or mining. These provisions have not been repealed, but §9 of NFMA prohibits the return to the public domain of any land reserved or withdrawn from the public domain, except by an act of Congress (16 U.S.C. §1609).

The 1911 Weeks Law authorizes the Secretary to dispose of land "chiefly valuable for agriculture" which was included in lands acquired (inadvertently or otherwise), if agricultural use will not injure the forests or stream flows and the lands are not needed for public purposes (16 U.S.C. §519).

The Bankhead-Jones Farm Tenant Act authorizes the disposal of lands acquired under its authority, with or without consideration, "under such terms and conditions as he [the Secretary of Agriculture] deems will best accomplish the purposes of this" title, but "only to public authorities and only on condition that the property is used for public purposes" (7 U.S.C. §1011(c)). Yet the grasslands were included in the NFS in 1976 and current regulations (36 C.F.R. §213) refer to them as being "permanently held."

The 1958 Townsites Act authorizes the Secretary to transfer up to 640 acres adjacent to communities in Alaska or the 11 western states for townsites, if the "indigenous community objectives ... outweigh the public objectives and values which would be served by maintaining such tract in Federal ownership" (16 U.S.C. §478a). There is to be a public notice of the application for such transfer, and upon a "satisfactory showing of need," the Secretary may offer the land to a local governmental entity at "not less than the fair market value."

The 1983 Small Tracts Act authorizes the Secretary to dispose of three categories of land, by sale or exchange, if valued at no more than $150,000 (16 U.S.C. §521e):

> (1) tracts of up to 40 acres interspersed with or adjacent to lands transferred out of federal ownership under the mining laws *and* which are inefficient to administer because of their size or location; (2) tracts of up to 10 acres encroached upon by improvements based in good faith upon an erroneous survey; or (3) road rights-of-way substantially surrounded by nonfederal land and not needed by the federal government, subject to the right of first refusal for adjoining landowners.

The land can be disposed of for cash, lands, interests in land, or any combination thereof for the value of the land being disposed (16 U.S.C. §521d) plus "all reasonable costs of administration, survey, and appraisal incidental to such conveyance" (16 U.S.C. §521f).

Finally, in Title II (the Education Land Grant Act) of P.L. 106-577, Congress authorized the FS to transfer up to 80 acres of NFS land for a nominal cost upon written application of a public school district. Section 202(e) provides for reversion of title to the federal government if the lands are not used for the educational purposes for which they were acquired.

Issues

In the past few years, the focus of discussions and legislative proposals on FS management of the NFS has been forest health and wildfires, especially in the intermountain West. The 2000 and 2002 fire seasons were, by most standards, among the worst since 1960. Many believe that excessive accumulations of biomass — dead and dying trees, heavy undergrowth, and dense stands of small trees —reflect degraded forest health and make forests vulnerable to conflagrations. These observers advocate rapid action to improve forest health — including prescribed burning, thinning, and salvaging dead and dying trees — and that rapid action is needed to protect NFS forests and nearby private lands and homes. Critics counter that authorities to reduce fuel levels are adequate, treatments that remove commercial timber degrade forest health and waste taxpayer dollars, and expedited processes for treatments are a device to reduce public oversight of commercial timber harvesting.

In September 2000, President Clinton requested an additional $1.6 billion (for the FS and the BLM) for fire protection including funds to pay for the 2000 summer's fire suppression efforts and for fuel treatment to address forest health in the *wildland-urban interface* (i.e., wildlands near communities threatened by potential wildfire conflagrations). Congress included much of this funding in the FY2001 Interior Appropriations Act (P.L. 106-291), and has continued to fund FS and BLM wildfire programs at more than double the level of the 1990s. Nonetheless, fuel treatment funding is still far below the amount that would be needed to reduce fuels on the federal lands many identify as at high risk of significant ecological damage from wildfire. (For further information, see "Current Issues" section of CRS Report RL30755, *Forest Fire/Wildfire Protection*, by Ross W. Gorte.)

In August 2002, President Bush proposed a Healthy Forests Initiative to expedite fuel reduction treatments for federal forests. Because the 107th Congress did not enact legislation on this initiative, portions of it were accomplished through regulatory changes. These include categorically excluding some fuel reduction treatments from NEPA environmental reviews and public involvement (68 *Federal Register* 33814, June 5, 2003); modifying the FS administrative appeal process (68 *Federal Register* 33582, June 4, 2003); categorically excluding small timber sales from NEPA environmental reviews and public involvement (68 *Federal Register* 44598, July 29, 2003); and allowing agencies to consult their own personnel on ESA impacts, known as *counterpart* regulations (68 *Federal Register* 68254, December 8, 2003).

On December 2, 2003, Congress enacted the Healthy Forests Restoration Act of 2003 (P.L. 108-148) containing parts of the President's Healthy Forests Initiative. One title, which garnered most of the attention in debates over the legislation, established an expedited process for fuel reduction activities. Other titles provide research and financial assistance in using forest biomass; direction on surveying and controlling insects and diseases; watershed forestry assistance to states and private landowners; and payments to private landowners for protecting special forestlands.

Another major issue concerns whether, when, and where to build forest roads. Road construction is supported by those who use the roads for access to the national forests for timber harvesting, fire control, recreation (including hunting and fishing), and other purposes. New roads are opposed by others, on the grounds that they can degrade the environment both during and after construction, exacerbate fire risk and spread invasive species, alter areas that some wish to preserve as pristine wilderness, and be expensive to build and maintain. Decisions over road building and protecting roadless areas generally have been made locally, which led to much local litigation.

In October 1999, the Clinton Administration proposed a nationwide rule to provide "appropriate long-term protection for ... 'roadless' areas." Final regulations were to become effective on March 13, 2001, but the Bush Administration delayed the effective date and subsequent court actions have prevented implementation. On July 15, 2003, the Bush Administration issued an advanced notice of proposed rulemaking to gather comments on roadless area management (68 *Federal Register* 41864). On December 30, 2003, the Administration provided a temporary exemption from the roadless rule for the Tongass NF in Alaska (68 *Federal Register* 75136). Final regulations on roadless area protection are still in development. However, interim guidance has returned decisions about roadless area protection to the local or regional level, raising the possibility of litigation over local decisions.

Major Statutes

Cooperative Forestry Assistance Act of 1978: Act of July 1, 1978; P.L. 95-313, as amended, 92 Stat. 365. 16 U.S.C. §§2101, et seq.

Forest and Rangeland Renewable Resources Planning Act of 1974 (RPA): Act of August 17, 1974; P.L. 93-378, 88 Stat. 476. 16 U.S.C. §§1600, et seq.

Forest and Rangeland Renewable Resources Research Act of 1978: Act of June 30, 1978; P.L. 95-307, 92 Stat. 353. 16 U.S.C. §§1641, et seq.

Healthy Forests Restoration Act of 2003: Act of December 3, 2003; P.L. 108-148, 117 Stat. 1887. 16 U.S.C. §§6501-6591.

Multiple-Use Sustained-Yield Act of 1960 (MUSYA): Act of June 12, 1960; P.L. 86-517, 75 Stat. 215. 16 U.S.C. §§528, et seq.

National Forest Management Act of 1976 (NFMA): Act of October 22, 1976; P.L. 94-588, 90 Stat. 2949. 16 U.S.C. §§1601, et al.

Organic Administration Act of 1897: Act of June 4, 1897; ch. 2, 30 Stat. 11. 16 U.S.C. §§473, et seq.

Pickett Act: Act of June 25, 1910; ch. 421, 36 Stat. 847.

Weeks Law of 1911: Act of March 1, 1911; ch. 186, 36 Stat. 961. 16 U.S.C. §§515, et al.

CRS Reports and Committee Prints [39]

CRS Issue Brief IB10076, *Bureau of Land Management (BLM) Lands and National Forests*, coordinated by Ross W. Gorte and Carol Hardy Vincent.

CRS Report 98-917, *Clearcutting in the National Forests: Background and Overview*, by Ross W. Gorte.

CRS Report 98-233, *Federal Timber Harvests: Implications for U.S. Timber Supply*, by Ross W. Gorte.

CRS Report RS20822, *Forest Ecosystem Health: An Overview*, by Ross W. Gorte.

CRS Report RL30755, *Forest Fire/Wildfire Protection*, by Ross W. Gorte.

CRS Report RL30647, *The National Forest System Roadless Areas Initiative*, by Pamela Baldwin.

CRS Report RS21544, *Wildfire Protection Funding*, by Ross W. Gorte.

CRS Issue Brief IB10124, *Wildfire Protection in the 108th Congress*, by Ross W. Gorte.

CRS Report RS21880, *Wildfire Protection in the Wildland-Urban Interface*, by Ross W. Gorte.

BUREAU OF LAND MANAGEMENT [40]

The Bureau of Land Management (BLM) manages 261.5 million acres of land, nearly 12% of the land in the United States. Most of this land is in the West, with about one-third of the total in Alaska. These lands include grasslands, forests, high mountains, arctic tundra, and deserts. They contain diverse resources, including fuels and minerals; timber; forage; wild horses and burros; fish and wildlife habitat; recreation sites; wilderness areas; archaeological, paleontological, and historical sites; and other natural heritage assets. The agency also is responsible for approximately 700 million acres of federal subsurface mineral resources throughout the nation, and supervises the mineral operations on an estimated 56 million acres of Indian Trust lands. Another key BLM function is wildland fire management and suppression on approximately 370 million acres of DOI, other federal, and certain nonfederal lands.

Background

BLM was created in the Department of the Interior in 1946 by merging two agencies — the General Land Office and the U.S. Grazing Service. The General Land Office, created by Congress in 1812, helped convey lands to pioneers settling the western lands. The U.S. Grazing Service was established in 1934 to manage the public lands best suited for livestock grazing, in accordance with the Taylor Grazing Act of 1934. [41] This law sought to remedy the deteriorating condition of public rangelands due to their overuse as well as the drought of the 1920s and depression of the early 1930s.

The Taylor Grazing Act provided for the management of the public lands "pending [their] final disposal." This language expressed the view that the lands might still be transferred to private or state ownership, and that the federal government was serving only as custodian until that time. However, patenting of the more arid western lands had already slowed, and there was growing concern about the condition of resources on these lands. These factors, and a changing general attitude towards the public lands, contributed to their retention by the federal government.

For decades Congress debated whether to retain or dispose of the remaining public lands, and how best to coordinate their management. Studies throughout the 1960s culminated in the 1970 report of the Public Land Law Review Commission entitled *One-Third of the Nation's Land*. Three successive Congresses deliberated, and in 1976 Congress enacted a comprehensive public land law entitled the Federal Land Policy and Management Act of 1976 (FLPMA). [42]

FLPMA sometimes is called the BLM Organic Act because portions of it consolidated and articulated the agency's responsibilities. This law established, amended, or repealed many management authorities dealing with public land withdrawals, land exchanges and acquisitions, rights-of-way, advisory groups, range management, and the general organization and administration of BLM and the *public lands*, which were defined as the lands managed by BLM.

Congress also established in FLPMA the national policy that "the public lands be retained in federal ownership, unless as a result of the land use planning procedures provided for in this act, it is determined that disposal of a particular parcel will serve the national interest...." This retention policy contributed to a "revolt" during the late 1970s and early 1980s among some westerners who continued to hope that the federal presence in their states might be reduced through federal land transfers to private or state ownership. The resultant "Sagebrush Rebellion" —objecting to federal management decisions and in some cases to the federal presence itself— was directed primarily toward the BLM.

Since the 1780s, nearly 1.3 billion acres of federal land have been transferred to individuals, businesses, and states. This total includes approximately 287 million acres for homesteaders; 328 million acres to states for public schools, public transportation systems, and various public improvement projects; and 94 million acres for railroads.

The last large transfer of BLM land occurred in 1980 with passage of the Alaska National Interest Lands Conservation Act (ANILCA). [43] This act transferred approximately 80 million acres from BLM to the other federal land management agencies. BLM also is required by law (ANILCA, the Alaska Native Claims Settlement Act, and the Alaska Statehood Act) to transfer ownership of more than 155 million acres of federal lands to the state of Alaska and Alaska Natives. Approximately 127 million acres have been conveyed (or tentatively approved), and BLM continues to transfer land to Alaska and the Alaska Native corporations.

Organization

BLM headquarters in Washington, DC, is headed by the Director, a political appointee who reports to the Secretary of the Interior through the Assistant Secretary for Land and Minerals Management. There are 12 BLM state offices, each headed by a state director, and each BLM state office administers a geographic area that generally conforms to the boundary of one or more states. Under each state office there are field offices, each headed by a field manager responsible for "on the ground" implementation of BLM programs and policies. Line authority is from the director to state directors, terminating at the field manager level.

In addition, there are six national level support and service centers: the National Office of Fire and Aviation (Boise, ID); the National Training Center (Phoenix, AZ); the National Science and Technology Center (Denver, CO); the National Human Resources Management Center (Denver, CO); the National Business Center (Denver, CO); and the National Information Resources Management Center (Denver, CO). [44]

BLM maintains over 1 billion land and mineral records from the nation's history, including legal land descriptions, land and mineral ownership and entitlement records, and land withdrawal records. The agency conducts cadastral surveys to locate and mark the boundaries of federal and Indian lands. BLM's Public Land Survey System is the foundation of the nation's land tenure system. BLM is making its public lands and mineral records available on the Internet to improve public access to, and the quality of, the information. The survey records and land descriptions are being converted to digital, geospatial format. [45] BLM also is involved in a joint project with the Forest Service, states, counties, and private industry to develop a National Integrated Land System, a geospatial reference for lands throughout the nation regardless of ownership. A goal is to develop a common approach to compiling and making available the documents relating to the status of land so users can obtain all the attributes about a chosen parcel. [46]

Management

Overview

FLPMA set the framework for the current management of BLM lands. Among other important provisions, the law provides that:

- the national interest will be best realized if the public lands and their resources are periodically and systematically inventoried and their present and future use is projected through a land use planning process coordinated with other Federal and State planning efforts ...
- management be on the basis of multiple use and sustained yield unless otherwise specified by law ...

- the United States receive fair market value of the use of the public lands and their resources unless otherwise provided for by statute ...

- the public lands be managed in a manner that will protect the quality of scientific, scenic, historical, ecological, environmental, air and atmospheric, water resource, and archeological values; that, where appropriate, will preserve and protect certain public lands in their natural condition; that will provide food and habitat for fish and wildlife and domestic animals; and that will provide for outdoor recreation and human occupancy and use....

Thus, FLPMA established the BLM as a multiple-use, sustained-yield agency. However, some lands are withdrawn from one or more uses, or managed for a predominant use. The agency inventories its lands and resources and develops land use plans for its land units. All BLM lands (except some lands in Alaska), as well as the 700 million acres of mineral resources managed by BLM, are covered by a land use plan. Although plans are to be amended or revised as new issues arise or conditions change, a large number of land use plans were developed in the 1970s or 1980s and are in need of substantial revision or replacement to take account of changes during recent years. In FY2001, BLM began a multiyear effort to develop new land use plans and update existing ones, driven by such changes as increased demands for energy resources, a rise in use of off-highway vehicles and other types of recreation, additions to the National Landscape Conservation System, new listings of species under the Endangered Species Act, a buildup of biomass fuels on public lands, and a need to mitigate the effects of wildfires.

Rangelands

Livestock grazing is permitted on an estimated 162 million acres of BLM land. In some western states, more than half of all cattle graze on public rangelands during at least part of the year, although the forage consumed on federal lands is a small percentage of all forage consumed by beef cattle nationally. The grazing of cattle and sheep, and range management programs generally, are authorized by the Taylor Grazing Act, FLPMA, and the Public Rangelands Improvement Act of 1978 (PRIA). The Taylor Grazing Act converted the public rangelands from a system of common open grazing to one of exclusive permits to graze allotted lands. FLPMA set out overall public land management and policy objectives. PRIA reflected continuing concern over the condition and productivity of public rangelands and established more specific range management provisions for BLM. An example is a new grazing fee formula that was temporary but essentially has been continued under executive order.

BLM's range programs include management of wild horses and burros under the Wild, Free-Roaming Horses and Burros Act of 1971. [47] Currently there are about 60,000 wild horses and burros under BLM management — 36,000 on public land and 24,000 in long-term holding facilities. The herd size on the range is significantly more than the agency has determined is appropriate (ecologically sustainable) —approximately 26,400. BLM seeks to reduce animals on the range through adoption, fertility control, permanent or temporary holding facilities, and other means. In its FY2005 Budget Justification, BLM cites insufficient funds to remove animals from the range and care for those in holding facilities. For years, management of wild horses and burros has been controversial.

Energy and Minerals

BLM administers onshore federal energy and mineral resources. The agency is responsible for approximately 700 million acres of federal subsurface minerals, and supervises the mineral operations on about 56 million acres of Indian trust lands. An estimated 165 million of the 700 million acres have been withdrawn from mineral entry, leasing, and sale, except for valid existing rights. Lands in the National Park System (except National Recreation Areas), Wilderness System, and the Arctic National Wildlife Refuge (ANWR) are among those withdrawn. Mineral development on 182 million acres is subject to the approval of the surface management agency, and must not be in conflict with the land designation. Wildlife refuges (except ANWR), wilderness study areas, and identified roadless areas, among others, are in this category.

There are three approaches to development of federal mineral resources. One approach is locating and patenting mining claims for hard rock (locatable) minerals. A second approach is competitive and noncompetitive leasing of lands for leaseable minerals (oil, gas, coal, potash, geothermal energy, and certain other minerals). A third approach is the sale or free disposal of common mineral materials (e.g., sand and gravel) not subject to the mining or leasing laws.

In 2003, 42% of the coal, 11% of the natural gas, and 5% of the oil produced in the United States were derived from BLM managed resources. [48] These resources generate large revenues. For FY2003, the total on-shore mineral revenues (including royalties, rents, and bonus bids) were $2.2 billion, a substantial increase over recent years primarily due to higher oil and gas prices. The demand for energy from BLM managed lands continues to increase, and a goal of the Bush Administration is to augment energy supply from federal lands.

National Landscape Conservation System

In 2000, BLM created the National Landscape Conservation System, comprised of different types of units —national monuments, conservation areas, wilderness areas, wilderness study areas, wild and scenic rivers, and scenic and historic trails. Approximately 42 million acres currently are in the system (excluding trails and rivers), to give them greater recognition, management attention, and resources, according to BLM statements. Areas are managed based on their relevant authorities; for instance, the 6.5 million acres of designated wilderness are managed in accordance with FLPMA and the Wilderness Act. Another 15.6 million acres of wilderness study areas are to be managed by BLM to maintain their suitability for wilderness designation until legislation is enacted to determine their final status. (For more information on wilderness, see "The National Wilderness Preservation System," below.)

The agency's 15 national monuments and 17 national conservation areas are a particular focus of the system. BLM management emphasizes resource conservation overall and in general units are to serve outdoor recreationists. Other activities, such as grazing and hunting, may continue if they are compatible with the designation.

The proximity of BLM lands to many areas of population growth in the West has led to an increase in recreation on some agency lands. Recreational activities include hunting, fishing, visiting cultural and natural sites, birdwatching, hiking, picnicking, camping, boating, mountain biking, and off-highway vehicle driving. BLM collects money for permits for recreation on its lands, such as permits issued to hunting and fishing guide outfitters. The agency also charges entrance and use fees on some of its lands under the Recreational Fee

Demonstration Program authorized by Congress. The growing and diverse nature of recreation on BLM lands has increased the challenge of balancing different types of recreation, such as hiking and driving off-highway vehicles, and balancing recreation with other land uses.

Fire Management

Recent fire seasons have been among the most severe in decades due to long-term drought, build-up of fuels, and increased population in the wildland-urban interface. BLM carries out fire management on approximately 370 million acres of DOI, and certain other federal and nonfederal lands. [49] The Forest Service provides fire protection of the national forests. A focus of both agencies is implementation of the national fire plan, under a 10-year strategy developed jointly by the agencies and other partners. Goals of the strategy are to improve fire prevention and suppression, reduce fuels, restore fire- adapted ecosystems, and promote community assistance. Another focus of the agencies is implementation of the Healthy Forests Restoration Act of 2003 (P.L. 108-148), which sought to expedite fuel reduction on federal lands and authorized other forest protection programs.

Land Ownership

General

BLM lands often are intermingled with other federal or private lands. Many federal grants consisted of alternating sections of lands, often referred to as "checkerboard," resulting in a mixed ownership grid pattern. FLPMA consolidated procedures and clarified responsibilities regarding problems that arise because of this ownership pattern, including rights-of-way across public lands for roads, trails, pipelines, power lines, canals, reservoirs, etc. FLPMA also provided for land exchanges, acquisitions, disposals, and remedies for certain title problems.

Acquisition Authority [50]

BLM has rather broad, general authority to acquire lands principally under §205 of FLPMA. Specifically, the Secretary is authorized (43 U.S.C. §1715(a)):

> to acquire pursuant to this Act [FLPMA] by purchase, exchange, donation, or eminent domain, lands or interests therein: *Provided,* That with respect to the public lands, the Secretary may exercise the power of eminent domain only if necessary to secure access to public lands, and then only if the lands so acquired are confined to as narrow a corridor as is necessary to serve such purpose.

BLM may acquire land or interests in land, especially inholdings, to protect threatened natural and cultural resources, increase opportunities for public recreation, restore the health of the land, and improve management of these areas. The agency often acquires land by exchange, and completed 132 exchanges in FY2003. Although FLPMA and NFMA were amended in 1988 to "streamline ... and expedite" the process, exchanges may still be time consuming and costly because of problems related to land valuation, cultural and archaeological resources inventories, and other issues. Recent concerns about the BLM exchange program, including regarding the determination of fair market value and the extent of public benefit of exchanges undertaken, prompted BLM to change the requirements and procedures of the program. [51]

Disposal Authority

The BLM can dispose of public lands under several authorities. A primary means of disposal is through exchanges, just as a primary means of acquisition is through exchanges. Disposal authorities include sales under FLPMA, patents under the General Mining Law of 1872, transfers to other governmental units for public purposes, and other statutes. [52]

With regard to sales, §203 of FLPMA authorized the BLM to sell certain tracts of public land that meet specific criteria (43 U.S.C. §1713(a)):

1. such tract because of its location or other characteristics is difficult and uneconomic to manage as part of the public lands, and is not suitable for management by another Federal department or agency; or
2. such tract was acquired for a specific purpose and the tract is no longer required for that or any other Federal purpose; or
3. disposal of such tract will serve important public objectives, including but not limited to, expansion of communities and economic development, which cannot be achieved prudently or feasibly on land other than public land and which outweigh other public objectives and values, including, but not limited to, recreation and scenic values, which would be served by maintaining such tract in Federal ownership.

The size of the tracts for sale is to be determined by "the land use capabilities and development requirements." Proposals to sell tracts of more than 2,500 acres must first be submitted to Congress, and such sales may be made unless disapproved by Congress. [53] Tracts are to be sold at not less than their fair market value, generally through competitive bidding, although modified competition and non-competitive sales are allowed.

The General Mining Law of 1872 [54] allows access to certain minerals on federal lands that have not been withdrawn from entry. Minerals within a valid mining claim can be developed without obtaining full title to the land. However, with evidence of minerals and sufficient developmental effort, mining claims can be patented, with full title transferred to the claimant upon payment of the appropriate fee — $5.00 per acre for vein or lode claims (30 U.S.C. §29) or $2.50 per acre for placer claims (30 U.S.C. §37). Non-mineral lands used for associated milling or other processing operations can also be patented (30 U.S.C. §42). Patented lands may be used for purposes other than mineral development.

The Recreation and Public Purposes Act (43 U.S.C. §869) [55] authorizes the Secretary, upon application by a qualified applicant, to:

> dispose of any public lands to a State, Territory, county, municipality, or other State, Territorial, or Federal instrumentality or political subdivision for any public purposes, or to a nonprofit corporation or nonprofit association for any recreational or any public purpose consistent with its articles of incorporation or other creating authority.

The act specifies conditions, qualifications, and acreage limitations for transfer, and provides for restoring the lands to the public domain if conditions are not met.

BLM also conducts land disposals under two recent laws providing for land disposal and establishing funding sources for subsequent land acquisition. First, the Federal Land Transaction Facilitation Act (Title II, P.L. 106-248, 43 U.S.C. §2301) provides for the sale or exchange of land identified for disposal under BLM's land use plans "as in effect" at enactment. Land sales are being conducted under the provisions of FLPMA. The proceeds from the sale or exchange of public land are to be deposited into a separate Treasury account (the Federal Land Disposal Account). Funds in the account are available to both the Secretary of the Interior and the Secretary of Agriculture to acquire inholdings and other nonfederal lands (or interests therein) that are adjacent to federal lands and contain exceptional resources. However, the Secretary of the Interior can use not more than 20% of the funds in the account for administrative and other expenses of the program. Not less than 80% of the funds for acquiring land are to be used to purchase land in the same state in which the funds were generated, while the remaining funds may be used to purchase land in any state. The law's findings state that it would "allow for the reconfiguration of land ownership patterns to better facilitate resource management; contribute to administrative efficiency within Federal land management units; and allow for increased effectiveness of the allocation of fiscal and human resources within the Federal land management agencies..."

Second, the Southern Nevada Public Land Management Act (P.L. 105-623) allows the Secretary of the Interior, through the BLM, to sell or exchange certain land around Las Vegas. The Secretary, through the BLM, and the relevant local government unit jointly choose the lands offered for sale or exchange. State and local governments get priority to acquire lands under the Recreation and Public Purposes Act. Much of the money from the sales is deposited into a special account that may be used for purposes including the acquisition of environmentally sensitive lands in Nevada. Some of the proceeds of land sales are set aside for other purposes, such as the State of Nevada general education program.

Withdrawals [56]

FLPMA also mandated review of public land withdrawals in 11 western states to determine whether, and for how long, existing withdrawals should be continued. A withdrawal is an action that restricts the use or disposition of public lands; for instance, some lands are withdrawn from mining. The agency continues to review approximately 70 million withdrawn acres, giving priority to about 26 million acres that are expected to be returned by another agency to BLM, or, in the case of BLM withdrawals, made available for one or more uses. To date, BLM has completed reviewing approximately 8 million withdrawn acres, mostly BLM and Bureau of Reclamation land; the withdrawals on more than 7 million of these acres have been revoked. The review process is likely to continue over the next several years, in part because the lands must be considered in BLM's planning process and the withdrawals must be supported by documentation under the National Environmental Policy Act (NEPA).

Issues

The public continues to value and use BLM lands for their diverse attributes and opportunities — open spaces, cultural resources, recreational pursuits, energy development, livestock grazing, timber production, etc. Issues and conflicts arise from these diverse and often opposing interests, with energy issues being among the most contentious. The President is promoting an expanded role for federal lands in supplying energy, and Congress is debating the extent, type, and location of development on federal lands. BLM has adopted regulatory changes to increase access to energy resources, such as streamlining the permitting process for oil and gas exploration and development. The emphasis on expanded production has exacerbated old controversies over the balance of uses of federal lands.

The development and patenting of hardrock minerals on public lands continues to receive attention. A focus has been the effect of BLM's revised hardrock mining regulations on the environment and the level of mining activity. A perennial debate is whether to change the 1872 mining law, which allows claimants to develop the minerals within a claim without paying royalties, and to patent the lands and obtain full title to the land and its minerals for a small fee ($2.50 or $5.00 an acre). The amount of land withdrawn from mineral entry or development has long been controversial and the subject of many lawsuits. A recent Legal Opinion of the Solicitor of the Department of the Interior allowed for multiple millsites per mining claim, reversing a 1997 Opinion and continuing concerns over the environmental impact of mining and the availability of lands for mineral development.

Rangeland management presents an array of issues. They include recent proposed changes in grazing regulations that would allow shared title of range improvements and private acquisition of water rights, reduce requirements for public input into grazing decisions, and make other changes. Another issue involves the terms and renewal of expiring grazing permits and leases, with recent law authorizing the automatic renewal of permits and leasing expiring through FY2008. The restriction or elimination of grazing on federal land because of environmental and recreational concerns is being discussed, and the grazing fee that the federal government charges for private livestock grazing on federal lands has been controversial since its inception. Other range issues include the condition of federal rangelands, the spread of invasive plant species, consistency of BLM and FS grazing programs, the role of Resource Advisory Councils, access across private lands, and management of riparian areas. Concerns about the wild horse and burro program relate to the removal, adoption, and treatment of the animals and BLM's administration of the program. A focus is BLM's current efforts to achieve its identified optimal herd size on the range.

Recent, severe wildfires have challenged BLM's fire management program and prompted the adoption of the National Fire Plan and the Healthy Forests Restoration Act. One issue is reducing the risk of wildland fire on federal lands through fuels reductions and other treatments. A second issue is the sufficiency of funds and procedures for suppressing fires, and the effect of borrowing funds from other programs for fire fighting. A third issue is the effect of fire on resource conditions, a compounding factor in areas experiencing drought, invasive species, and other changes.

A number of preservation and recreation matters have come to the fore. These include whether to establish or restrict protective designations; the effect of protective designations on land uses; and the role of Congress, states, and the public in making designations. Congress is examining executive actions designating national monuments on BLM and other federal lands under the Antiquities Act of 1906, [57] and discussing whether to restrict the President's authority to create monuments. Conflicts over different types of recreation, especially high-impact (e.g., OHV use) versus low-impact uses (e.g., backpacking), appear to have become more prevalent. With dramatic population growth in the West in the vicinity of BLM lands, and the public value on federal lands for recreation, these conflicts can be expected to remain prevalent. Another issue is access to public lands, including restrictions such as limits on use of off-highway vehicles. Other issues are the impact minerals within a claim without paying royalties, and to patent the lands and obtain full title to the land and its minerals for a small fee ($2.50 or $5.00 an acre). The amount of land withdrawn from mineral entry or development has long been controversial and the subject of many lawsuits. A recent Legal Opinion of the Solicitor of the Department of the Interior allowed for multiple millsites per mining claim, reversing a 1997 Opinion and continuing concerns over the environmental impact of mining and the availability of lands for mineral development.

Rangeland management presents an array of issues. They include recent proposed changes in grazing regulations that would allow shared title of range improvements and private acquisition of water rights, reduce requirements for public input into grazing decisions, and make other changes. Another issue involves the terms and renewal of expiring grazing permits and leases, with recent law authorizing the automatic renewal of permits and leasing expiring through FY2008. The restriction or elimination of grazing on federal land because of environmental and recreational concerns is being discussed, and the grazing fee that the federal government charges for private livestock grazing on federal lands has been controversial since its inception. Other range issues include the condition of federal rangelands, the spread of invasive plant species, consistency of BLM and FS grazing programs, the role of Resource Advisory Councils, access across private lands, and management of riparian areas. Concerns about the wild horse and burro program relate to the removal, adoption, and treatment of the animals and BLM's administration of the program. A focus is BLM's current efforts to achieve its identified optimal herd size on the range.

Recent, severe wildfires have challenged BLM's fire management program and prompted the adoption of the National Fire Plan and the Healthy Forests Restoration Act. One issue is reducing the risk of wildland fire on federal lands through fuels reductions and other treatments. A second issue is the sufficiency of funds and procedures for suppressing fires, and the effect of borrowing funds from other programs for fire fighting. A third issue is the effect of fire on resource conditions, a compounding factor in areas experiencing drought, invasive species, and other changes.

A number of preservation and recreation matters have come to the fore. These include whether to establish or restrict protective designations; the effect of protective designations on land uses; and the role of Congress, states, and the public in making designations. Congress is examining executive actions designating national monuments on BLM and other federal lands under the Antiquities Act of 1906,[57] and discussing whether to restrict the President's authority to create monuments. Conflicts over different types of recreation, especially high-impact (e.g., OHV use) versus low-impact uses (e.g., backpacking), appear to have become more prevalent. With dramatic population growth in the West in the vicinity of BLM lands, and the public value on federal lands for recreation, these conflicts can be expected to remain prevalent. Another issue is access to public lands, including restrictions such as limits on use of off-highway vehicles. Other issues are the impact of recreation on resources and facilities and the collection of fees for recreation use, for example, under the Recreational Fee Demonstration Program.

Another key topic relates to the amount of land BLM owns and how the land is managed. Contemporary questions have centered on how much land should be acquired versus conveyed to state, local, or private ownership, and under what circumstances. Congress confronts concerns about acquisition of private land, the effectiveness of land exchange programs, and the effect of public ownership on state taxes and authorities. A related issue is whether to expand the non-federal role in managing federal lands.

Major Statutes

Alaska National Interest Lands Conservation Act of 1980: Act of Dec. 2, 1980; P.L. 96-487, 94 Stat. 2371. 16 U.S.C. §§3101, et seq.

Federal Land Exchange Facilitation Act of 1988: Act of Aug. 20, 1988; P.L. 100-409, 102 Stat. 1086. 43 U.S.C. §1716.

Federal Land Policy and Management Act of 1976: Act of Oct. 21, 1976; P.L. 94-579, 90 Stat. 2744. 43 U.S.C. §§1701, et seq.

Federal Land Transaction Facilitation Act: Act of July 25, 2000; P.L. 106-248, 114 Stat. 613. 43 U.S.C. §§2301, et seq.

General Mining Law of 1872: R.S. 2319, derived from Act of May 10, 1872; ch. 152, 17 Stat. 91. 30 U.S.C. §§22, et seq.

Materials Act of 1947: Act of July 31, 1947; ch. 406, 61 Stat. 681. 30 U.S.C. §§601, et seq.

Mineral Leasing Act for Acquired Lands: Act of Aug. 7, 1947; ch. 513, 61 Stat. 913. 30 U.S.C. §§351-359.

Mineral Leasing Act of 1920: Act of Feb. 25, 1920; ch. 85, 41 Stat. 437. 30 U.S.C. §§181, et seq.

Public Rangelands Improvement Act of 1978: Act of Oct. 25, 1978; P.L. 95-514, 92 Stat. 1803. 43 U.S.C. §§1901, et seq.

Southern Nevada Public Land Management Act of 1998: Act of Oct. 19, 1998; P.L. 105-263, 112 Stat. 2343. 31 U.S.C. §6901 note.

Taylor Grazing Act of 1934: Act of June 28, 1934; ch. 865, 48 Stat. 1269. 43 U.S.C. §§315, et seq.

Wild Horses and Burros Act of 1971: Act of Dec. 15, 1971; P.L. 92-195, 85 Stat. 649. 16 U.S.C. §§1331, et seq.

CRS Reports and Committee Prints [58]

CRS Issue Brief IB10076, *Bureau of Land Management (BLM) Lands and National Forests*, coordinated by Ross W. Gorte and Carol Hardy Vincent.

CRS Report RS21402, *Federal Lands, "Disclaimers of Interest," and R.S. 2477*, by Pamela Baldwin.

CRS Report RS21232, *Grazing Fees: An Overview and Current Issues*, by Carol Hardy Vincent.

CRS Report RL32244, *Grazing Regulations and Policies: Changes by the Bureau of Land Management*, by Carol Hardy Vincent.

CRS Report RL32142, *Highway Rights of Way on Public Lands: R.S. 2477 and Disclaimers of Interest*, by Pamela Baldwin.

CRS Issue Brief IB89130, *Mining on Federal Lands*, by Marc Humphries. CRS Report RS20902, *National Monument Issues*, by Carol Hardy Vincent. CRS Report RS21423,

Wild Horse and Burro Issues, by Carol Hardy Vincent. CRS Report RS21544, *Wildfire Protection Funding*, by Ross W. Gorte.

THE NATIONAL WILDLIFE REFUGE SYSTEM [59]

The National Wildlife Refuge System (NWRS) is dedicated primarily to the conservation of animals and plants. Other uses — hunting, fishing, recreation, timber harvest, grazing, etc. — are permitted only to the extent that they are compatible with the purposes for which the refuge was created. [60] In 1997, Congress established compatible wildlife-dependent recreation as a priority for the NWRS. Some have characterized the NWRS as intermediate in protection between the BLM and FS lands and NPS lands, but this is not entirely accurate. [61] The NWRS resembles the FS or BLM lands in allowing some commercial uses, but in certain cases, uses (e.g., public access) can be substantially more restricted than for NPS lands.

Background

The first national wildlife refuge was established at Pelican Island, FL, by executive order of President Theodore Roosevelt in 1903. By September 30, 2002, there were 540 refuges totaling 92.1 million acres in 50 states, the Pacific Territories, Puerto Rico, and the Virgin Islands (see Figures 4 and 5.) [62] The largest increase in acreage by far occurred with the addition of 53 million acres of refuge land under the Alaska National Interest Lands Conservation Act of 1980. Alaska now has 76.8 million acres of refuge lands — 80.5% of the system. Within 63 of the refuges are 78 designated wilderness areas, ranging from 2 acres at Green Bay National Wildlife Refuge (NWR) in Wisconsin to 8.0 million acres at Arctic NWR in Alaska. [63]

The NWRS includes two other categories of land besides refuges: (1) the 203 Waterfowl Production Area (WPA) districts, private lands managed in accordance with agreements between the farmers and ranchers who own the land and the FWS; and (2) 50 Wildlife Coordination Areas (WCAs), owned primarily by FWS, but also by other parties, including some federal agencies; they generally are managed by state agencies under agreements with the FWS. These bring the NWRS to 793 units. [64] These two additional categories bring the total land in the NWRS (counting refuges, WPAs, and WCAs) to 95.4 million acres. In approximately 1.7 million acres of the NWRS, FWS has secondary jurisdiction: the FWS has some influence over activities on these lands, but the lands are owned or managed principally by some other agency, subject to the mandates of that agency.

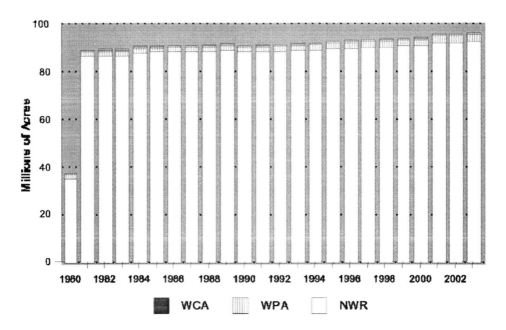

Figure 4. Acreage in the National Wildlife Refuge System (FY1980-FY2003).

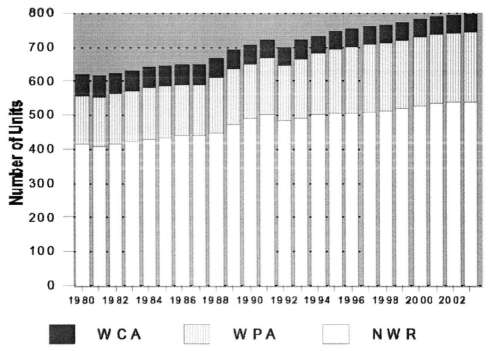

Source: Annual Report of Lands Under Control of the U.S. Fish and Wildlife Service, as of Sept. 30 of
    each fiscal year.
Notes: Major acreage was added to the system in December 1980 under ANILCA. ANILCA also
    consolidated a number of existing Alaskan refuges. In 1992, the number of units dropped due to
    consolidation of various refuges.

Figure 5. Number of Units in the National Wildlife Refuge System (FY1980-FY2003).

Organization and Management

The National Wildlife Refuge System Administration Act of 1966, [65] as amended, stated the purpose for establishing the system as consolidating the several authorities of the Secretary of the Interior over lands administered for the conservation and protection of fish and wildlife. Conservation of wildlife is the primary emphasis in the three types of areas in the NWRS, but the options for alternative resource use within the areas vary.

In the 105th Congress, the National Wildlife Refuge System Improvement Act of 1997 (P.L. 105-57) [66] addressed overarching refuge management controversies facing the FWS. This law clarified that the purpose of the NWRS is the "conservation, management and, where appropriate, restoration of the fish, wildlife and plant resources and their habitats." Another key provision of this law designated "compatible wildlife-dependent recreational uses involving hunting, fishing, wildlife observation and photography, and environmental education and interpretation as priority public uses of the Refuge System." It also required that priority public uses must "receive enhanced consideration over other general public uses in planning and management within the System." At the same time, the law continued the statutory policy that activities that are not wildlife-dependent (e.g., grazing, growing hay, etc.) may be permitted, provided they are compatible with wildlife. Some interest groups argued that the resulting regulations did not allow for sufficient public access for some forms of recreation, such as off-road vehicles or personal watercraft.

Wildlife refuges provide habitat for various plant and animal species, particularly emphasizing habitat for migratory waterfowl and for endangered species. Individual refuges may consist of single contiguous blocks or disjunct parcels scattered over a larger area. Research on wildlife conservation is carried out by the FWS on refuges (as well as on other areas). [67] Energy and mineral activities are permitted in certain refuges and under certain circumstances; any mineral rights owned by the United States are administered by BLM. Hunting, fishing, and other recreational uses frequently are permitted, but only to the extent that these activities are compatible with the major purposes for which a particular refuge was established. In refuges set aside for migratory birds, waterfowl hunting is limited to 40% of the refuge area unless the Secretary determines that hunting in a greater area is beneficial.

WPAs are managed primarily to provide breeding habitat for migratory waterfowl. [68] As of September 30, 2002, these areas totaled 2.9 million acres, of which 0.7 million acres were federally owned and 2.2 million acres were managed by the private landowners under leases, easements, or agreements with FWS. These areas are found mainly in the potholes and interior wetlands of the North Central states, a region sometimes called "North America's Duck Factory." In these areas, there is considerably less conflicting resource use, in part because the areas managed under lease are not subject to the federal mining and mineral leasing laws, and because the size of individual tracts is relatively small. However, the leased lands may be less secure as wildlife habitat because they may be converted later to agricultural use by the private owners. The WCAs (0.3 million acres) are owned primarily by FWS, but also by other parties, including some federal agencies; they are managed by state wildlife agencies under cooperative agreements with FWS.

The management of the NWRS is divided into three tiers: the 793 individual NWRS units under seven regional offices, and the national office in Washington, DC. Each of the seven regional offices is administered by a regional director who has considerable autonomy in operating the refuges within the region. FWS is headed by a director, a deputy director, and

11 assistant directors who head programs not only for the National Wildlife Refuge System, but also for Wildlife and Sport Fish Restoration; Migratory Birds; Fisheries and Habitat Conservation; Endangered Species; Law Enforcement (titled "Chief"); International Affairs; External Affairs; Budget, Planning, and Human Resources; Business Management and Operations; and Information Resources Technology Management.

Land Ownership

Growth of the NWRS may come about in a number of ways. Certain laws provide general authority to expand the NWRS, including primarily the Migratory Bird Treaty Act (MBTA) of 1929, [69] but also the Fish and Wildlife Coordination Act, the Fish and Wildlife Act of 1956, and the Endangered Species Act. [70] These general authorities allow the FWS to add lands to the Refuge System without specific congressional action.

Some units have been created by specific acts of Congress (e.g., Protection Island NWR, WA; Bayou Sauvage NWR, LA; or John Heinz NWR, PA). [71] Other units have been created by executive order. Also, FLPMA authorizes the Secretary of the Interior to withdraw lands from the public domain for additions to the NWRS, although all withdrawals exceeding 5,000 acres are subject to congressional approval procedures (43 U.S.C. §1714(c)). [72]

Acquisition Authority

The primary acquisition authority has been the MBTA. This act authorizes the Secretary to recommend areas "necessary for the conservation of migratory birds" [73] to the Migratory Bird Conservation Commission, after consulting with the relevant governor (or state agency) and appropriate local government officials (16 U.S.C. §715c). The Secretary may then purchase or rent areas approved by the Commission (§715d(1)), and "acquire, by gift or devise, any area or interest therein ..." (§715d(2)). [74]

New acquisitions result from transfers from the public domain or lands acquired from other owners. Nonfederal lands and interests in lands to create or add to specific NWRS units may be accepted as donations or purchased. Purchases may be made on a willing buyer/willing seller basis or under condemnation authorities. Condemnation authority was last used, under congressional direction contained in P.L. 99-333, for Protection Island NWR in 1986. [75] Purchases, regardless of authority or funding source, are rarely large. In FY2002, 68,014 acres were acquired (as opposed to transferred from other federal agencies), while $90.6 million was spent on acquisition. [76] As might be expected, refuges in western states tend to be formed from lands reserved from the public domain, while eastern refuges tend to be acquired lands.

The purchase of refuge lands is financed primarily through two funding sources: the Migratory Bird Conservation Fund (MBCF) and the Land and Water Conservation Fund (LWCF, see "Federal Lands Financing," above). [77] MBCF acquisitions have emphasized wetlands essential for migratory waterfowl, while LWCF acquisitions have encompassed the gamut of NWRS purposes. MBCF is supported from three sources (amounts in parentheses are FY2003 receipts deposited into the MBCF):

- the sale of hunting and conservation stamps (better known as *duck stamps*) purchased by hunters and certain visitors to refuges ($25.1 million); [78]

- import duties on arms and ammunition ($18.5 million); and
- 70% of certain refuge entrance fees ($0.15 million).

MBCF funds are permanently appropriated to the extent of these receipts and (after paying for engraving, printing, and distribution of the stamps) may be used for the "location, ascertainment, and acquisition of suitable areas for migratory bird refuges ... and administrative costs incurred in the acquisition" of the new acquisitions whose number varies from year to year (16 U.S.C. §718d(b)). However, the acquisition must be "approved by the Governor of the State or appropriate State agency" (§715k-5). The predictability of MBCF funding makes it assume special importance in the FWS budget. This contrasts with LWCF funding, which has fluctuated significantly from year to year. In FY2003, the MBCF received $43.8 million from its permanently appropriated sources, and Congress appropriated $72.9 million from the LWCF for FWS land acquisition.

Disposal Authority

With certain exceptions, NWRS lands can be disposed only by an act of Congress (16 U.S.C. §668dd(a)(6)). Also, for refuge lands reserved from the public domain, FLPMA prohibits the Secretary from modifying or revoking any withdrawal which added lands to the NWRS (43 U.S.C. §1714(j)). For acquired lands, disposal is allowed only if: (1) the disposal is part of an authorized land exchange (16 U.S.C. §668dd(a)(6) and (b)(3)); or (2) the Secretary determines the lands are no longer needed and the Migratory Bird Conservation Commission approves (§668dd(a)(5)). In the latter case, the disposal must recover the acquisition cost or be at the fair market value (whichever is higher).

Issues

The most enduring controversy concerning the NWRS has been that of conflicting uses, with some critics arguing that FWS has been too lenient in its decisions about commercial and extractive uses or developed recreation; others criticize its policies as too restrictive. Specific conflicts have arisen between such activities as grazing, energy extraction, power boat recreation, motorized access, and similar activities on the one hand, and the purposes for which refuges were designated on the other. [79]

In recent years, a controversy developed over the propriety of hunting (and, to a lesser extent, fishing) on refuge lands. The pro-hunting position is based largely on two arguments: (1) the purchase of migratory duck stamps by hunters has paid for a substantial portion of refuge land, mainly in areas suitable for waterfowl habitat; and (2) the animal population is the appropriate measure of conservation, and removal of individual animals for human use is not harmful, and may be beneficial as long as the population growth rate is maintained. The anti-hunting argument holds that no place can be considered a "refuge" if its major wildlife residents are regularly hunted. They contend further that since fewer people now hunt [80] and the enjoyment of this sport hinders use of the land by others (by restricting access for safety reasons), then hunting should be eliminated to allow fuller access by non-hunting users. While various bills have been introduced over the years to eliminate or restrict hunting on refuges, others have been introduced to support it.

Over the past several years, the backlog of unmet maintenance needs of the federal land management agencies has been an issue of focus of the Congress and the Administration. Although there is debate over the amount of FWS money that should be spent on the deferred maintenance backlog versus the acquisition of additional federal lands, there is broad consensus that maintenance of the NWRS has lagged. The funding for deferred maintenance projects in the NWRS increased from $48.1 million in FY2002 to $66.5 million in FY2004. The maintenance backlog is expected to figure in the debate over appropriations in future years.

One refuge — the Arctic National Wildlife Refuge — remains locked in a decades-long controversy regarding proposals for energy development in the biologically and geologically rich northern part of this refuge. This complex issue is covered extensively in CRS Report RL31278, *Arctic National Wildlife Refuge: Background and Issues*, coordinated by M. Lynne Corn, and in CRS Issue Brief IB10111, *Arctic National Wildlife Refuge (ANWR): Controversies for the 108th Congress*, by M. Lynne Corn, Bernard A. Gelb, and Pamela Baldwin.

Major Statutes

Alaska National Interest Lands Conservation Act of 1980: Act of December 2, 1980; P.L. 96-487, 94 Stat. 2371. 16 U.S.C. §3101, et seq.

Endangered Species Act of 1973: Act of Dec. 28, 1973; P.L. 93-205, 87 Stat. 884. 16 U.S.C. 1531-1544.

Fish and Wildlife Act of 1956: Act of August 8, 1956; ch. 1036, 70 Stat. 1120. 16 U.S.C. §742a, et seq.

Fish and Wildlife Coordination Act of 1934: Act of March 10, 1934; ch. 55, 48 Stat. 401. 16 U.S.C. §661-667e.

Migratory Bird Treaty Act of 1918: Act of July 13, 1918; ch. 128, 40 Stat. 755. 16 U.S.C. §703-712.

National Wildlife Refuge System Administration Act of 1966: Act of October 15, 1966; P.L. 90-404, 80 Stat. 927. 16 U.S.C. §668dd-668ee.

National Wildlife Refuge System Improvement Act of 1997: Act of October 9, 1997; P.L. 105-57. 16 U.S.C. §668dd.

San Francisco Bay National Wildlife Refuge: Act of June 30, 1972; P.L. 92-330, 86 Stat. 399. 16 U.S.C. §668dd note. (A typical statute establishing a refuge.)

CRS Reports and Committee Prints [81]

CRS Report RL31278, *Arctic National Wildlife Refuge: Background and Issues*, M. Lynne Corn, coordinator.

CRS Issue Brief IB10111, *Arctic National Wildlife Refuge (ANWR): Controversies for the 108th Congress*, by M. Lynne Corn, Bernard A. Gelb, and Pamela Baldwin.

CRS Report 90-192, *Fish and Wildlife Service: Compensation to Local Governments*, by M. Lynne Corn.

THE NATIONAL PARK SYSTEM [82]

Perhaps the federal land category best known to the public is the National Park System. The National Park Service (NPS) currently manages 388 system units, including 56 units formally entitled *national parks* (often referred to as the "crown jewels" of the system), as well as national monuments, battlefields, military parks, historical parks, historic sites, lakeshores, seashores, recreation areas, reserves, preserves, scenic rivers and trails, and other designations. The system has grown to a total of 84.4 million acres — 79.0 million acres of federal land, 1.2 million acres of other public land, and 4.2 million acres of private land — in 49 states, the District of Columbia, and U.S. territories. Passage of ANILCA in 1980 roughly doubled the acreage of the National Park System because of the large size of the new parks in Alaska. The acreage has been relatively stable in recent years, as new authorizations and land acquisitions have been modest. The NPS has the often contradictory mission of facilitating access and serving visitors while protecting and preserving the natural, historic, and cultural integrity of the lands and resources it manages.

Background

By the Act of March 1, 1872, Congress established Yellowstone National Park in the then-territories of Idaho, Montana, and Wyoming "as a public park or pleasuring ground for the benefit and enjoyment of the people." [83] The park was placed under the exclusive control of the Secretary of the Interior, who was responsible for developing regulations to "provide for the preservation, from injury or spoliation, of all timber, mineral deposits, natural curiosities, or wonders within said park, and their retention in their natural condition." [84] Other park functions were to include developing visitor accommodations, building roads and trails, removing trespassers (mostly poachers) from the park, and protecting "against wanton destruction of fish and game." [85]

When Yellowstone National Park was authorized, there was no concept or plan for the development of a system of such parks. The concept now firmly established as the National Park System, embracing a diversity of natural and cultural resources nationwide, evolved slowly over the years. This idea of a national park was an American invention of historic proportions, marking the start of a global conservation movement that today accounts for hundreds of national parks (or equivalent conservation preserves) throughout the world. The American National Park System continues to serve as an international model for preservation.

At the same time that interest was growing in preserving the scenic wonders of the American West, efforts were underway to protect the sites and structures associated with early Native American cultures, particularly in the Southwest. In 1906, Congress enacted the Antiquities Act to authorize the President "to declare by public proclamation [as national monuments] historic and prehistoric structures and other objects of historic or scientific interest." [86] In the years following the establishment of Yellowstone, national parks and monuments were authorized or proclaimed, principally from the public domain lands in the West, and were administered by the Department of the Interior (initially with help from the U.S. Army). However, no single agency provided unified management of the varied federal parklands.

On August 25, 1916, President Woodrow Wilson signed the act creating the National Park Service, a new federal agency in the Department of the Interior with the responsibility for protecting the national parks and many of the monuments then in existence and those yet to be established. This action reflected a developing national concern for preserving the nation's heritage. This "Organic Act" states that the National Park "Service then established shall promote and regulate the use of Federal areas known as national parks, monuments and reservations ... to conserve the scenery and the natural and historic objects and the wildlife therein and to provide for the enjoyment of the same in such manner and by such means as will leave them unimpaired for the enjoyment of future generations." [87] By executive order in 1933, President Franklin D. Roosevelt transferred 63 national monuments and military sites from the Forest Service and War Department to the National Park Service. This action was a major step in the development of a truly national system of parks. [88]

Of the four federal land management agencies, the NPS manages the most diverse collection of units. More than 20 different designations are used for park sites or areas, ranging from the traditional national park designation to scenic rivers and trails, memorials, battlefields, historic sites, historic parks, seashores, lakeshores, recreation areas, and monuments. Because of this variety of park unit designations and the public perception of lesser status for units lacking the *national park* designation, Congress sought to establish that all units in the system are to be considered of equal value. A 1970 law stated that all NPS units are part of "one national park system preserved and managed for the benefit and inspiration of all people of the United States...." [89] In 1978, Congress amended that law to reassert the system-wide standard of protection for all areas administered by the NPS. [90]

Organization and Management

The National Park Service manages the 388 units of the National Park System. The Director of the National Park Service, headquartered in Washington, DC, is the chief administrative officer of the Service, with an immediate staff of two deputy directors, five associate directors, and a number of policy and program office managers. Directly overseeing NPS operations is the Interior Department's Assistant Secretary for Fish, Wildlife, and Parks. In addition, the National Park Service Advisory Board, composed of private citizens with requisite experience and expertise, advises on management policies and on potential additions to the system. In 2001, the Advisory Board issued a report with recommendations on the future of the National Park System. [91]

The individual park units are arranged in seven regional offices, each headed by a regional director. The NPS had traditionally operated with 10 regional offices but eliminated three, while also forming a system of "park clusters." The reorganization, a part of the Clinton Administration's "reinvention" of government that involved downsizing and streamlining, was primarily designed to shift resources and personnel from central offices to field units. Regional offices and cluster support offices provide certain administrative functions and specialized staff services and expertise which were not believed to be practicable to have in each park unit. This shared assistance is particularly important to the smaller units. The individual units are overseen by a park superintendent, with staff generally commensurate with the size, public use, and significance of the unit. The park units in Alaska are an exception to this, with relatively few personnel in comparison to the large size of the holdings.

As stated, the basic NPS mission is twofold: (1) to conserve, preserve, protect, and interpret the natural, cultural, and historic resources of the nation for the public and (2) to provide for their enjoyment by the public. To a considerable extent, the NPS contributes to meeting the public demand for certain types of outdoor recreation. Scientific research is another activity encouraged in units of the Park System. Management direction is provided in the general statutes and in those that create and govern individual units. In general, activities which harvest or remove the resources within units of the system are not allowed. Mining, for instance, is generally prohibited, although in a limited number of national parks and monuments some mining is allowed, in accordance with the Mining in the Parks Act of 1976 (P.L. 94-429). Also, in authorizing certain additions to the system, Congress has specified that certain natural resource uses, such as oil and gas development or hunting, may — or shall — be permitted in specific units; examples include national preserves such as Big Cypress and national recreation areas such as Glen Canyon. Other uses are dealt with in specific enactments, such as the 1911 law dealing with rights-of-way through Park System units.

Land Ownership

Designation and Acquisition Authority

Most units of the National Park System have been created by Acts of Congress. In 1998, Congress amended existing law pertaining to the creation of new units to standardize procedures, improve information about potential additions, prioritize areas, focus attention on outstanding areas, and ensure congressional support for studies of possible additions. [92] The Secretary of the Interior is to investigate, study, and monitor nationally significant areas with potential for inclusion in the system. The Secretary is to submit annually to Congress a list of areas recommended for study for potential inclusion in the National Park System. The Secretary also is required to submit to Congress each year a list of previously studied areas that contain primarily historical resources, and a similar list of areas with natural resources, with areas ranked in order of priority for possible inclusion in the system. In practice, NPS performs these functions assigned to the Secretary.

In assessing whether to recommend a particular area, the NPS is required by law to consider: whether an area is nationally significant, and would be a suitable and feasible addition to the National Park System; whether an area represents or includes themes, sites, or resources "not adequately represented" in the system; and requests for studies in the form of public petitions and congressional resolutions. An actual study requires authorization by Congress, although the NPS may conduct certain preliminary assessment activities. In preparing studies, NPS must consider certain factors also established in law. After funds are made available, NPS must complete a study within three fiscal years.

Under the Antiquities Act of 1906, the President is authorized to proclaim national monuments on federal land, and to date about 120 monuments have been created by presidential proclamations. Many areas initially designated as national monuments were later converted into national parks by acts of Congress. Before 1940, Presidents used this authority frequently (for proclaiming 87 national monuments), but in 1978 President Carter set aside more land as national monuments (56 million acres in Alaska) than any other President. [93]

President Clinton used his authority under the Antiquities Act 22 times to proclaim 19 new monuments and enlarge 3 others. Other agencies also manage some national monuments, with the BLM managing many of the monuments created by President Clinton.

In addition to establishing a unit of the National Park System, an act of Congress may set the boundaries of the unit and authorize the NPS to acquire the nonfederal lands within those boundaries. The major funding source for such land acquisition has been the Land and Water Conservation Fund, described above in the section entitled "Federal Lands Financing." The Secretary is to include, in a report to Congress at least every three years, a "comprehensive listing of all authorized but unacquired lands within the exterior boundaries of each unit" (16 U.S.C. §1a-11(a)) and a "priority listing of all such unacquired parcels" (16 U.S.C. §1a-11(b)). Further, the general management plan for each unit is to include "indications of potential modifications to the external boundaries of the unit, and the reasons therefor" (16 U.S.C. §1a-7). The Secretary is to identify criteria to evaluate proposed boundary changes (16 U.S.C. §1a-12). Further, the Secretary is authorized to make minor boundary adjustments for "proper preservation, protection, interpretation, or management" and to acquire the nonfederal lands within the adjusted boundary (16 U.S.C. §460*l*-9(c)).

Disposal Authority

Units (and lands) of the National Park System established by acts of Congress can be disposed of only by acts of Congress. Non-NPS lands encompassed by minor boundary adjustments can be acquired through land exchanges, but, unlike for some of the other federal land management agencies, the Secretary may not convey property administered as part of the National Park System to acquire lands by exchange. [94] Finally, the Secretary cannot modify or revoke any withdrawal creating a national monument. [95] Thus, with minor exceptions, National Park System lands can be changed from that status or disposed of only by an act of Congress.

Issues

Striking a balance between appropriate public use of National Park System lands for recreation and protecting the integrity of park resources is a continuing challenge to the NPS and the congressional committees providing agency oversight. Motorized recreation in NPS units presents particular challenges, with debates over the economic and environmental impacts of, safety of, and level of access for such types of recreation and the adequacy of existing laws and regulations governing motorized use. Manufacturers and user groups fear that NPS limits would be economically damaging to communities and industries serving users, unfairly restrict access, and set a precedent for other federal land managers. Others, including environmentalists, fear that failure to adequately manage motorized use will damage resources and other park users, and increase pressure for additional forms of motorized access.

One focus of the motorized recreation debate is commercial air tours over NPS units. Currently, the NPS and the Federal Aviation Administration (FAA) are developing air tour management plans for park units to implement a law regulating commercial air tours over park units, and the FAA has proposed a rule providing safety standards for commercial air tours including over park units. Other issues relate to NPS regulation of the use of personal watercraft (PWC), such as jet skis, and snowmobiles in national parks. The NPS is developing regulations governing PWC for 16 park units to settle a successful lawsuit over unrestricted PWC use. Litigation and appeals continue over different versions of snowmobile regulations that would either restrict or allow snowmobile use in Yellowstone and Grand Teton National Parks and the John D. Rockefeller Jr. Memorial Parkway.

Over the years, Congress has added new units to the Park System as well as expanded the management responsibilities of the NPS. These new obligations, together with increased numbers of visitors, have stretched the Park System's operational capabilities and contributed to a multibillion dollar backlog of deferred maintenance. While overall NPS appropriations have increased annually in recent years, they have not kept pace with operational and maintenance needs. Increased priorities on security and protection also have affected park funding. The NPS claims to be implementing President Bush's initiative, begun in FY2002, to eliminate a then-estimated $4.9 billion maintenance backlog over five years; there is disagreement about whether the Administration is on track to eliminate the maintenance backlog. By the end of FY2004, the NPS expects to have completed a computerized inventory and assessment of every facility in the Park System, and by FY2006, estimated costs of repairing facilities and a list of maintenance priorities.

Congress authorized a Recreational Fee Demonstration Program to supplement NPS and other land management agency appropriations with higher entrance and recreation user fees (described above under "Federal Lands Financing"). There is controversy over whether to make the program permanent and if so in what form and for which agencies — for the NPS only, all four land management agencies currently participating in the program, or additional federal agencies (such as the federal water project agencies — the Bureau of Reclamation and the Corps of Engineers). The temporary program was initiated in the FY1996 Omnibus Consolidated Rescissions and Appropriations Act (P.L. 104-134, §315), and allows most of the higher fees charged by participating agencies to be retained at the sites where the money is collected, rather than returned to the U.S. Treasury. It continues to be tested by the agencies and has been extended in appropriations laws, most recently through December 2005 for fee collection to give the authorizing committees time to consider establishing a permanent program. Many citizens have objected to paying additional fees for previously free or low-cost recreation in the national forests, but have expressed few objections to higher fees for the National Park System. The Administration has asked Congress to make the program permanent for the four major federal land management agencies.

Over the last two decades, Congress has created two dozen National Heritage Areas (NHAs) to conserve, commemorate, and promote areas and their resources. There is disagreement over whether to enact generic legislation for the creation and management of NHAs, to continue allowing variety in their creation and operation, or to cease creating and funding these areas. For NHAs, the NPS assists communities in attaining the designation, and supports state and community efforts through seed money, recognition, and technical assistance. Proponents claim that heritage areas protect important resources and traditions; promote tourism and community revitalization; and help prevent new, and perhaps costly or inappropriate, additions to the Park System. Opponents fear that the heritage program is potentially costly and could be used to extend federal control over nonfederal lands.

Congressional leaders have at times packaged a large number of diverse park, public land, and recreation related bills into omnibus measures to expedite passage in the closing days of a Congress. Neither the 106th nor the 107th Congress enacted an omnibus parks bill.

Because of the growing number of park and recreation related bills that have passed in one chamber or been reported by an authorizing committee, some observers feel that prospects are favorable for development of an omnibus measure late in the 108th Congress.

Major Statutes [96]

Mining in National Parks: Act of Sept. 28, 1976; P.L. 94-429, 90 Stat. 1342. 16 U.S.C. §§1901-1912.

National Park Service General Authorities Act of 1970: Act of Aug. 18, 1970; P.L. 91-383, 84 Stat. 825. 16 U.S.C. §1a-1, §1c.

National Park Service Organic Act of 1916: Act of Aug. 25, 1916; ch. 408, 39 Stat. 535. 16 U.S.C. §§1-4.

National Parks Omnibus Management Act of 1998: Act of Nov. 13, 1998; P.L. 105-391, 112 Stat. 3497. 16 U.S.C. §5901, et seq.

Omnibus Parks and Public Lands Management Act of 1996: Act of Nov. 12, 1996; P.L. 104-333, 110 Stat. 4093. 16 U.S.C. §1, et seq.

Preservation of American Antiquities: Act of June 8, 1906; ch. 3060, 34 Stat. 225. 16 U.S.C. §§431-433.

Recreational Fee Demonstration Program: §315 of the Interior and Related Agencies Appropriations Act, 1996, §101(c) of the Omnibus Consolidated Rescissions and Appropriations Act, 1996, Act of Apr. 26, 1996; P.L. 104-134, 110 Stat. 1321-200. 16 U.S.C. §460*l* — 6a Note.

Yellowstone National Park Act: R.S. 2474, derived from Act of March 1, 1872; ch. 24, 17 Stat. 32. 16 U.S.C. §21, et seq.

CRS Reports and Committee Prints [97]

CRS Issue Brief IB10126, *Heritage Areas: Background, Proposals, and Current Issues*, by Carol Hardy Vincent and David Whiteman.

CRS Issue Brief IB10093, *National Park Management and Recreation*, coordinated by Carol Hardy Vincent.

CRS Report RS20158, *National Park System: Establishing New Units*, by Carol Hardy Vincent.

CRS Report RL31149, *Snowmobiles: Environmental Standards and Access to National Parks*, by James E. McCarthy.

SPECIAL SYSTEMS ON FEDERAL LANDS

There are currently three special management systems that include lands from more than one federal land management agency: the National Wilderness Preservation System, the National Wild and Scenic Rivers System, and the National Trails System. These systems were established by Congress to protect special features or characteristics on lands managed by the various agencies. Rather than establish new agencies for these systems, Congress directed the existing agencies to administer the designated lands within parameters set in statute.

THE NATIONAL WILDERNESS PRESERVATION SYSTEM [98]

The Wilderness Act defines wilderness as "undeveloped federal land ... without permanent improvements." Further, wilderness generally consists of *federal* land that is primarily affected by the forces of nature, relatively untouched by human activity, and primarily valued for solitude and primitive recreation. Lands eligible for inclusion in the system are areas that generally contain more than 5,000 acres or that can be managed to maintain their pristine character.

Background

The National Wilderness Preservation System was established in 1964 by the Wilderness Act. It was based on a FS system that was established administratively in 1924, but reserves to Congress the authority to include areas in the system. The Wilderness Act designated 9.1 million acres of national forest lands as wilderness, and required the FS, NPS, and FWS to review the wilderness potential of lands under their jurisdiction. These reviews were completed within the required 10 years, with wilderness recommendations presented to Congress. The FS also chose to expand its review to all NFS roadless areas (the first and second Roadless Area Review and Evaluation, RARE and RARE II), and presented wilderness recommendations in 1979. A comparable review of BLM lands was required by FLPMA in 1976, and the BLM finalized wilderness recommendations in 1991.

Organization and Management

Wilderness areas generally are managed to protect and preserve their natural conditions. Permanent improvements, such as buildings and roads, and activities which significantly alter existing natural conditions, such as timber harvesting, generally are prohibited. The Wilderness Act allowed mineral exploration and leasing for 20 years (through December 31, 1983), and directed that valid existing mineral rights be permitted to be developed under "reasonable regulations" to attempt to preserve the wilderness characteristics of the area. The Wilderness Act also specified that existing livestock grazing and motorboat or airstrip uses be allowed to continue. In addition, Congress has included exceptions to the act's management limitations in subsequent laws designating specific areas.

The National Wilderness Preservation System contains more than 105 million acres in 44 states, as shown in Table 5 (data column 1). This amounts to nearly one-sixth (16%) of all federal land. More than half of all wilderness acres are in Alaska (57 million, 55%); this accounts for about a quarter of the federal land in the state. Another 42 million acres of wilderness (41%) are in the 11 western states. In total, this wilderness acreage represents 12%

of the federal land in those states, ranging from 1% in Nevada to 38% in Washington. The remaining 4 million acres (4%) are in the other states (the Atlantic Coast through the Great Plains, plus Hawaii). This is 8% of the federal land in other states, ranging from 0% in six states to 52% in Florida.

No one agency manages the system. Rather, all four agencies currently manage wilderness areas. (See Table 5.) The FS manages nearly 35 million acres of designated wilderness. This comprises 18% of all NFS lands. Nearly 6 million acres of NFS wilderness land (16%) are in Alaska, and another 27 million acres (78%) are in the 11 western states. The FS also manages 2 million acres of wilderness in the other states (6%), and 26 (of those other 38) states have wilderness areas.

More than half of the NPS lands are designated wilderness (43 million acres, 56%). Approximately three-quarters of all NPS wilderness land is in Alaska (33 million acres, 76%), and significant NPS wilderness areas also are in California, Florida, and Washington.

The FWS manages nearly 21 million acres of wilderness. This represents 22% of FWS lands. Nearly 19 million acres of FWS wilderness areas (90%) are in Alaska, and significant FWS wilderness areas also are in Arizona. Overall, about half of the states have wilderness areas within the purview of the FWS.

The BLM currently manages more than 6 million acres of wilderness (as shown in Table 5), a small fraction of all BLM lands (2%). Approximately two-thirds of BLM wilderness is in the California desert, and another quarter is in Arizona. BLM also manages relatively small amounts of wilderness in several other states.

Designation

The Wilderness Act reserves to Congress the authority to designate wilderness areas as part of the National Wilderness System. Congress has designated many particular wilderness study areas, in addition to the broader agency reviews required under the Wilderness Act and FLPMA. How long study areas must be administered to preserve their wilderness character depends on the language of the law requiring the study; some areas are available for other uses when the agency recommends against designation, but others must be protected until Congress releases them.

Congress began expanding the system in 1968, four years after it was established. The most significant expansion was included in the Alaska National Interest Lands Conservation Act of 1980, which established 35 new wilderness areas in Alaska with more than 56 million acres. This action more than tripled the system at that time.

For the decade following the FS recommendations in 1979, Congress generally addressed possible wilderness designations for all FS lands within a state. Many statewide FS wilderness bills were introduced, but their enactment was held up in the early 1980s until a compromise over *release language [99]* broke the legislative stalemate. This compromise — which allowed but did not compel the FS to maintain wilderness attributes of released lands — led Congress to enact 21 wilderness laws designating 8.6 million acres of predominately NFS wilderness in 21 states. The 103rd Congress (1993-1994) also substantially expanded the system, with NFS wilderness areas in Colorado and BLM and NPS wilderness areas in the California Desert.

Congress continues to consider further expansion of the National Wilderness Preservation System. More than 29 million acres, mostly NPS lands in Alaska, have been recommended by the agencies to Congress for inclusion in the system. Numerous areas continue to be reviewed for their wilderness potential by the federal land management agencies.

Table 5. Federally Designated Wilderness Acreage, by State and Agency

State	Total Acreage	Forest Service	National Park Service	Fish and Wildlife Service	Bureau of Land Management
Alabama	41,367	41,367	0	0	0
Alaska	57,522,408	5,753,448	33,079,611	18,689,349	0
Arizona	4,528,973	1,345,008	444,055	1,343,444	1,396,466
Arkansas	153,655	116,578	34,933	2,144	0
California	14,154,062	4,430,849	6,122,045	9,172	3,591,996
Colorado	3,345,091	3,142,035	60,466	3,066	139,524
Connecticut	0	0	0	0	0
Delaware	0	0	0	0	0
Florida	1,426,327	74,495	1,300,580	51,252	0
Georgia	485,484	114,537	8,840	362,107	0
Hawaii	155,590	0	155,590	0	0
Idaho	4,005,712	3,961,667	43,243	0	802
Illinois	32,782	28,732	0	4,050	0
Indiana	12,945	12,945	0	0	0
Iowa	0	0	0	0	0
Kansas	0	0	0	0	0
Kentucky	18,097	18,097	0	0	0
Louisiana	17,025	8,679	0	8,346	0
Maine	19,392	12,000	0	7,392	0
Maryland	0	0	0	0	0
Massachusetts	2,420	0	0	2,420	0
Michigan	249,219	91,891	132,018	25,310	0
Minnesota	815,952	809,772	0	6,180	0
Mississippi	6,046	6,046	0	0	0
Missouri	71,113	63,383	0	7,730	0
Montana	3,443,038	3,372,503	0	64,535	6,000
Nebraska	12,429	7,794	0	4,635	0
Nevada	1,581,871	823,585	0	0	758,286
New Hampshire	102,932	102,932	0	0	0
New Jersey	10,341	0	0	10,341	0
New Mexico	1,625,117	1,388,262	56,392	39,908	140,555
New York	1,363	0	1,363	0	0
North Carolina	111,419	102,634	0	8,785	0
North Dakota	39,652	0	29,920	9,732	0
Ohio	77	0	0	77	0

Table 5. (Continued)

State	Total Acreage	Forest Service	National Park Service	Fish and Wildlife Service	Bureau of Land Management
Oklahoma	23,113	14,543	0	8,570	0
Oregon	2,274,152	2,086,504	0	925	186,723
Pennsylvania	9,031	9,031	0	0	0
Rhode Island	0	0	0	0	0
South Carolina	60,681	16,671	15,010	29,000	0
South Dakota	77,570	13,426	64,144	0	0
Tennessee	66,349	66,349	0	0	0
Texas	85,333	38,483	46,850	0	0
Utah	800,614	772,894	0	0	27,720
Vermont	59,421	59,421	0	0	0
Virginia	177,214	97,635	79,579	0	0
Washington	4,317,133	2,569,391	1,739,763	839	7,140
West Virginia	80,852	80,852	0	0	0
Wisconsin	42,323	42,294	0	29	0
Wyoming	3,111,232	3,111,232	0	0	0
Territories	0	0	0	0	0
Total	**105,176,917**	**34,807,965**	**43,414,402**	**20,699,338**	**6,255,212**

Sources: The sources for this table were generally the same as for Table 2, except NPS data are from their website at [http://wilderness.nps.gov/maplocator.cfm], visited February 24, 2004. Data in the table are updated by CRS to reflect laws enacted after the publication dates.

Issues

Wilderness designations continue to be controversial. Restrictions on the use and development of designated wilderness areas often conflict with the desires of some groups, while providing the values sought by others. In an attempt to find a balance between development and protection, Congress has enacted general standards and prohibitions for wilderness protection (e.g., no motorized access), and general and specific exemptions to those standards and prohibitions (e.g., *continued* motorboat use where such use was occurring prior to the designation). Exceptions often reflect agreements for specific areas, but widespread compromise between development and preservation interests generally remains elusive.

Several current issues surround the possible protection of the remaining FS and BLM roadless areas. One issue focuses on BLM lands in Utah, but has national relevance. Central to the controversy is whether BLM may currently designate *wilderness study areas* (WSAs) — areas that are statutorily entitled to automatic and continuing protections. Section 603 of FLPMA required the BLM to review the wilderness potential of its lands, and, by 1991, to recommend areas to the President, who then could recommend areas to the Congress for possible inclusion in the National Wilderness Preservation System. In response, BLM first inventoried its lands to determine which lands met the basic size criteria and might have wilderness characteristics, then conducted a more in-depth review of these lands to determine which among them possessed wilderness characteristics. Lands found to have wilderness character were designated as WSAs and studied further. Many WSA lands were then recommended to Congress in the early 1990s for inclusion in the National Wilderness Preservation System. These wilderness recommendations are still pending for Utah and many other western states. Approximately 1.9 million of 2.5 million acres of WSAs in Utah were recommended; some Utah wilderness bills before Congress have recommended more acreage, some less.

Section 603 also requires BLM to protect the wilderness characteristics of "such areas" until Congress directs otherwise. This *non-impairment* standard prevents most development, and BLM has applied it to all WSAs, including those that BLM did not recommend for wilderness designation. Therefore, whether BLM can designate new WSAs and whether the non-impairment standard can be applied to these or other lands are important issues both for those seeking to protect the lands and those seeking to develop them — either the automatic and continuing protections of the non-impairment standard apply, or protections may only be provided through the slower, less certain, and more changeable land use planning process under §202 of FLPMA.

Following debate over additional wilderness areas proposed in legislation, Secretary Babbitt in 1996 used the §201 FLPMA inventory authority to identify an additional 2.6 million acres in Utah as having wilderness qualities. Although the stated purpose of the inventory was only to ascertain which lands had wilderness characteristics and report on those, Utah filed suit alleging various flaws in this process, and alleging that the inventory was illegal, even under §201. The district court enjoined the inventory preliminarily, but the 10th Circuit remanded to the district court to dismiss (on various grounds) all but the claim that related to de facto wilderness management of the inventoried lands. (*Utah v. Babbitt* 137 F. 3d 1193 (10th Cir. 1998)). The Department of the Interior subsequently settled the case, and on September 29, 2003, issued new wilderness guidance (Instruction Memoranda No. 2003-274 and 2003-275). These directives apply to BLM lands nationwide, except for Alaska and certain categories of lands, and take the position that the §603 authority terminated in 1993, that BLM cannot administratively create more WSAs under §603 or other authority, and that the non-impairment standard cannot be applied to non-WSA lands. Rather, protective management of the remaining BLM roadless areas can occur only through the relevant land management plans. Others disagree with this interpretation, and the issues are currently in litigation. The importance of these issues is accentuated by the emphasis of the Bush Administration on energy development of the federal lands, and by the promulgation of new regulations on *disclaimers of interest* that may facilitate the validation of highway rights-of-ways in roadless areas, thereby disqualifying additional lands from further consideration.

Another Utah wilderness controversy has widespread implications for management of WSAs generally. A suit was filed to compel BLM to protect WSAs from impairment by increased off-road vehicle use. The Supreme Court ruled in *Norton v. Southern Utah Wilderness Alliance* that although the protection of WSAs was mandatory, it was a programmatic duty and not the type of discrete agency obligation that could be enforced under the APA. Also, the Court concluded that language contained in relevant FLPMA land use plans indicating that WSAs would be monitored constituted management goals that might be modified by agency priorities and available funding, and was not a basis for enforcement under the APA (5 U.S.C. §706(1)).

The management of the remaining FS roadless areas has also seen recent changes. FS roadless areas nationwide were administratively protected from most timber cutting and most roads under a Clinton Administration rule (66 Fed. Reg. 3244 (January 12, 2001)) that was subsequently enjoined, and would be replaced by a Bush Administration proposed rule (69 Fed. Reg. 42636 (July 16, 2004)). The proposed rule would provide interim management for the FS roadless areas and allow a period of time during which a state governor may petition for a special rule governing the management of roadless areas in a particular state. After this petition period, management of roadless areas would be governed by any special rules that are developed, or by the relevant forest plan.

Major Statutes

Alaska National Interest Lands Conservation Act of 1980: Act of Dec. 2, 1980; P.L. 96-487, 94 Stat. 2371.
California Desert Protection Act of 1994: Act of Oct. 31, 1994; P.L. 103-433, 108 Stat. 4471.
Wilderness Act: Act of Sept. 3, 1964; P.L. 88-577, 78 Stat. 890. 16 U.S.C. §§1131, et seq.

CRS Reports and Committee Prints [100]

CRS Report RL30647, *The National Forest System Roadless Areas Initiative*, by Pamela Baldwin.
CRS Report 98-848, Wilderness *Laws: Prohibited and Permitted Uses*, by Ross W. Gorte.
CRS Report RL31447, *Wilderness: Overview and Statistics*, by Ross W. Gorte.
CRS Report RS21917, *Bureau of Land Management (BLM) Wilderness Review Issues*, by Ross W. Gorte.
CRS Report RS21290, *Wilderness Water Rights: Language in Laws from the 103rd Congress to Date*, by Pamela Baldwin and Ross Gorte.

THE NATIONAL WILD AND SCENIC RIVERS SYSTEM [101]

Background

The National Wild and Scenic Rivers System was established in 1968 by the Wild and Scenic Rivers Act (P.L. 90-542, 16 U.S.C. §§1271-1287). The act established a policy of preserving selected free-flowing rivers for the benefit and enjoyment of present and future generations, to complement the then-current national policy of constructing dams and other structures (such as flood control works) along many rivers. Three classes of wild and scenic rivers were established under the act, reflecting the characteristics of the rivers at the time of designation, and affecting the type and amount of development that may be allowed thereafter. The classes of rivers are:

- *Wild* rivers are free from impoundments (dams, diversions, etc.) and generally inaccessible except by trail, where the watersheds (area surrounding the rivers and tributaries) are primitive and the shorelines are essentially undeveloped;
- *Scenic* rivers are free from impoundments in generally undeveloped areas but are accessible in places by roads;
- *Recreational* rivers are readily accessible by road, with some shoreline development, and possibly may have undergone some impoundment or diversion in the past.

Rivers may come into the system either by congressional designation or state nomination to the Secretary of the Interior. Congress initially designated 789 miles in 8 rivers as part of the National Wild and Scenic Rivers System. Congress began expanding the system in 1972, and made substantial additions in 1976 and in 1978 (413 miles in 3 rivers, and 688 miles in 8 rivers, respectively). The National Wild and Scenic Rivers System was more than doubled by designation of rivers in Alaska in ANILCA in 1980. In January 1981, Interior Secretary Cecil Andrus approved 5 rivers designated by the state of California, increasing the system mileage by another 20% (1,235 miles). The first additions under the Reagan Administration were enacted into law in 1984, with the addition of 5 rivers including more than 300 miles. The next large addition came in 1988, with the designation of more than 40 river segments in Oregon, adding 1,400 miles. In 1992, 14 Michigan river segments totaling 535 miles were added. The 106th and 107th Congresses added new designations to the system which now includes 163 river units with 11,302.9 miles in 38 states and Puerto Rico. [102] (See Table 6.)

Organization and Management

Land areas along rivers designated by Congress generally are managed by one of the four federal land management agencies, where federal land is dominant. However, land use restrictions and zoning decisions affecting private land in wild and scenic corridors generally are made by local jurisdictions (e.g., the relevant county, township, city, etc.) where appropriate. The boundaries of the areas along wild and scenic rivers are identified by either the Interior or Agriculture Secretary, depending on land ownership within the corridor. The area included may not exceed an average of 320 acres per mile of river designated (640 acres per mile in Alaska), an average of 1/4 mile width of land on each side of the river.

Where wild and scenic river corridor boundaries include state, county, or other public land, or private land, federal agencies have limited authority to purchase, condemn, exchange, or accept donations of state and private lands within the corridor boundaries. Additionally, federal agencies are directed to cooperate with state and local governments in developing corridor management plans.

In response to controversies associated with management of lands within wild and scenic river corridors, several recent designations have included language calling for creation of citizen advisory boards or other mechanisms to ensure local participation in developing management plans. Even without such direction, management plans for river corridors involving predominantly private lands usually are developed with input from local jurisdictions, prior to designation.

Table 6. Mileage of Rivers Classified as Wild, Scenic, and Recreational, by State and Territory, 2003

State	Wild	Scenic	Recreational	Total
Alabama	36.40	25.00	0.00	61.40
Alaska	2,955.00	227.00	28.00	3,210.00
Arizona	18.50	22.00	0.00	40.50
Arkansas	21.50	147.70	40.80	210.00
California	685.80	199.60	986.85	1,872.25
Colorado	30.00	0.00	46.00	76.00
Connecticut	0.00	0.00	14.00	14.00
Delaware[a]	0.00	15.60	79.00	94.60
Florida	32.65	7.85	8.60	49.10
Georgia[a]	39.80	2.50	14.60	56.90
Idaho[a]	321.90	34.40	217.70	574.00
Illinois	0.00	17.10	0.00	17.10
Kentucky	9.10	0.00	10.30	19.40
Louisiana	0.00	19.00	0.00	19.00
Maine	92.50	0.00	0.00	92.50
Massachusetts	0.00	33.80	38.50	72.30
Michigan	79.00	277.90	267.90	624.80
Minnesota[a]	0.00	193.00	59.00	252.00
Mississippi	0.00	21.00	0.00	21.00
Missouri	0.00	44.40	0.00	44.40
Montana	161.90	66.70	139.40	368.00
Nebraska[a]	0.00	76.00	126.00	202.00
New Hampshire	0.00	13.50	24.50	38.00
New Jersey[a]	0.00	119.90	146.80	266.70
New Mexico	90.75	20.10	10.00	120.85
New York[a]	0.00	25.10	50.30	75.40
North Carolina[a]	44.40	95.50	52.00	191.90
Ohio	0.00	136.90	76.00	212.90
Oregon[a]	635.65	381.40	798.05	1,815.10
Pennsylvania[a]	0.00	111.00	298.80	409.80
South Carolina[a]	39.80	2.50	14.60	56.90
South Dakota[a]	0.00	0.00	98.00	98.00
Tennessee	44.25	0.00	0.95	45.20
Texas	95.20	96.00	0.00	191.20
Washington	0.00	108.00	68.50	176.50
West Virginia	0.00	10.00	0.00	10.00
Wisconsin[a]	0.00	217.00	59.00	276.00
Wyoming	20.50	0.00	0.00	20.50
Puerto Rico	2.10	4.90	1.90	8.9
U.S. Total[b]	**5,350.60**	**2,457.20**	**3,495.10**	**11,302.90**

Source: U.S. Dept. of the Interior, National Park Service. *River Mileage Classifications for Components of the* National Wild and Scenic Rivers, Washington, DC: Jan. 2002, available on the NPS website at [http://www.nps.gov/rivers/wildriverstable.html], visited May 7, 2004.

Also, personal communication with John Haubert, Division of Park Planning and Special Studies, NPS, U.S. Dept. of the Interior, Washington, DC, on February 12, 2004.

a. This state shares mileage with some bordering states, where designated river segments are also state boundaries. Figures for each state reflect the total shared mileage, resulting in duplicate counting.

b. Figure totals represent the actual totals of classified mileage in the United States and do not reflect duplicate counting of mileage of rivers running between state borders. Because the figures for individual states do reflect the shared mileage, the sum of the figures in each column exceeds the indicated column total.

Management of lands within wild and scenic corridors generally is less restricted than in some protected areas, such as wilderness areas, although management varies with the class of the designated river and the values for which it was included in the system. Administration is intended to protect and enhance the values which led to the designation, but Congress also directed that other land uses not be limited unless they "substantially interfere with public use and enjoyment of these values" (§10(a) of the 1968 Act). Primary emphasis for management is directed toward protecting aesthetic, scenic, historic, archaeologic, and scientific features of the area. Road construction, hunting and fishing, and mining and mineral leasing may be permitted in some instances, as long as the activities are consistent with the values of the area being protected and with other state and federal laws.

Designation

Rivers may be added to the system either by an act of Congress, usually after a study by a federal agency, or by state nomination with the approval of the Secretary of the Interior. Congress has identified numerous rivers as potential additions to the system. The Secretaries of the Interior and Agriculture are required to report to the President on the suitability of these rivers for wild and scenic designation, who in turn submits recommendations to Congress.

State-nominated rivers may be added to the National Wild and Scenic Rivers System only if the river is designated for protection under state law, is approved by the Secretary of the Interior, and is permanently administered by a state agency (§2(a)(ii) of the 1968 Act). Management of these state-nominated rivers may be more complicated because of the diversity of land ownership in these areas. Fewer than 10% of the federal wild and scenic river designations have been made in this manner.

Issues

Concern over land management issues and private property rights have been the predominant issues associated with designation of wild and scenic rivers since the inception of the 1968 Act. Initially, the river designations involved land owned and managed primarily by the federal agencies; however, over the years, more and more segments have been designated that include private lands within the river corridors. The potential use of condemnation authority in particular has been quite controversial. Congress has addressed these issues in part by encouraging development of management plans during the wild and scenic river study phase, prior to designation, and by avoiding condemnation. According to the National Park Service, condemnation has "almost ceased to be used [since] the early 1980s." [103] Another issue that arises from time to time is the nature of state or federal projects allowed within a wild and scenic corridor, such as construction of major highway crossings, bridges, or other activities that might affect the flow or character of the designated river segment.

Major Statutes

Wild and Scenic Rivers Act: Act of Oct. 2, 1968; P.L. 90-542, 82 Stat. 906. 16 U.S.C. §1271, et seq.

CRS Reports and Committee Prints [104]

CRS Report RL30809, *The Wild and Scenic Rivers Act and Federal Water Rights*, by Pamela Baldwin.

NATIONAL TRAILS SYSTEM [105]

The National Trails System (NTS) was created in 1968 by the National Trails System Act (P.L. 90-543, 16 U.S.C. §§1241-1251). This act established the Appalachian and Pacific Crest National Scenic Trails, and authorized a national system of trails to provide additional outdoor recreation opportunities and to promote the preservation of access to the outdoor areas and historic resources of the nation. The system includes four classes of national trails:

- *National Scenic Trails (NST)* provide outdoor recreation and the conservation and enjoyment of significant scenic, historic, natural, or cultural qualities;
- *National Historic Trails (NHT)* follow travel routes of national historic significance;
- *National Recreation Trails (NRT)* are in, or reasonably accessible to, urban areas on federal, state, or private lands; and
- *Connecting or Side Trails* provide access to or among the other classes of trails.

Background

During the early history of the United States, trails served as routes for commerce and migration. Since the early 20th Century, trails have been constructed to provide access to scenic terrain. In 1921, the concept of the first interstate recreational trail, now known as the Appalachian National Scenic Trail, was introduced. In 1945, legislation to establish a "national system of foot trails" as an amendment to a highway funding bill, was considered but not reported by committee. [106]

As population expanded in the 1950s, the nation sought better opportunities to enjoy the outdoors. [107] In 1958, Congress established the Outdoor Recreation Resources Review Commission (ORRRC) to make a nationwide study of outdoor national recreation needs. A 1960 survey conducted for the ORRRC indicated that 90% of all Americans participated in some form of outdoor recreation and that walking for pleasure ranked second among all recreation activities. [108] On February 8, 1965, in his message to Congress on "Natural Beauty," President Lyndon B. Johnson called for the nation "to copy the great Appalachian Trail in all parts of our country, and make full use of rights-of-way and other public paths."

[109] Just three years later, Congress heeded the message by enacting the National Trail System Act. [110]

The National Trails System began in 1968 with only two scenic trails. One was the Appalachian National Scenic Trail, stretching 2,160 miles from Mount Katahdin, ME, to Springer Mountain, GA. The other was the Pacific Crest National Scenic Trail, covering 2,665 miles from Canada to Mexico along the mountains of Washington, Oregon, and California. The system was expanded a decade later when the National Parks and Recreation Act of 1978 designated four NHTs with more than 9,000 miles, and another NST, along the Continental Divide, with 3,100 miles. Today, the federal portion of the system consists of 23 national trails (8 scenic trails and 15 historic trails) covering almost 40,000 miles, more than 800 recreation trails, and 2 connecting and side trails. In addition, the act has authorized more than 1,100 rails-to-trails [111] conversions.

Organization and Management

Each of the 23 national trails is administered by either the Secretary of the Interior or the Secretary of Agriculture under the authority of the National Trails System Act. The NPS administers 16 of the 23 trails, the FS administers 4 trails, the BLM administers 1 trail, and the NPS and BLM jointly administer 2 NHTs. The Secretaries are to administer the federal lands, working cooperatively with agencies managing lands not under their jurisdiction. Management responsibilities vary depending on the type of trail.

National Scenic Trails

NSTs provide recreation, conservation, and enjoyment of significant scenic, historic, natural, or cultural qualities. The use of motorized vehicles on these long-distance trails is generally prohibited, except for the Continental Divide National Scenic Trail which allows: (1) access for emergencies; (2) reasonable access for adjacent landowners (including timber rights); and (3) landowner use on private lands in the right of way, in accordance with regulations established by the administering Secretary.

National Historic Trails

These trails follow travel routes of national historical significance. To qualify for designation as a NHT, the proposed trail must meet the following criteria: (1) the route must have historical significance as a result of its use and documented location; (2) there must be evidence of a trail's national significance with respect to American history; and (3) the trail must have significant potential for public recreational use or historical interest. These trails do not have to be continuous, and can include land and water segments, marked highways paralleling the route, and sites that together form a chain or network along the historic route. Examples include the Mormon Trail and the Oregon Pioneer Trail.

National Recreation Trails

The Park Service is responsible for the overall administration of the national recreation trails program, including coordination of nonfederal trails, although the FS administers NRTs within the national forests. NRTs are existing trails in, or reasonably accessible to, urban areas, and are managed by public and private agencies at the local, state, and national levels. Various NRTs provide recreation opportunities for the handicapped, hikers, bicyclists, cross country skiers, and horseback riders.

Connecting and Side Trails

These trails provide public access to nationally designated trails or connections between such trails. In 1990, the Secretary of the Interior designated: 1) the 18-mile Timm's Hill Trail, WI, which connects Timm's Hill to the Ice Age NST, and 2) the 186-mile Anvik Connector, AK, a spur of the Iditarod NHT which joins that trail to the village of Anvik on the Yukon River. Connecting and side trails are administered by the Secretary of the Interior, except that the Secretary of Agriculture administers trails on national forest lands.

Designation

As defined in the National Trails System Act, NSTs and NHTs are long distance trails and are designated by acts of Congress. NRTs and connecting and side trails may be designated by the Secretaries of the Interior and Agriculture with the consent of the federal agency, state, or political subdivision with jurisdiction over the lands involved. Of the 39 completed feasibility studies requested by Congress since 1968, 5 NSTs and 15 NHTs have been designated.

The Secretaries are permitted to acquire lands or interest in lands for the Trails System by written cooperative agreements, through donations, by purchase with donated or appropriated funds, by exchange, and, within limits, by condemnation. The Secretaries are directed to cooperate with and encourage states to administer the nonfederal lands through cooperative agreements with landowners and private organizations for the rights-of-way or through states or local governments acquiring such lands or interests.

Issues [112]

The level of funding continues to be a major issue. With the exception of the Appalachian and the Pacific Crest NSTs, the National Trails System Act does not provide for sustained funding of designated trails operations, maintenance, and development, nor does it authorize dedicated funds for land acquisition. The Federal Surface Transportation Program is a major funding source for trails, shared use paths, and related projects in the United States. Prior to 1991, highway funds were to be used only for highway projects and selected bicycle transportation facilities. With the passage of the Intermodal Surface Transportation Efficiency Act of 1991 (ISTEA, P.L. 102-240) and subsequently reauthorized as the Transportation Equity Act for the 21st Century (TEA-21, P.L. 105-178), many trail projects paths became eligible to receive federal highway program funds. Program funding increases are being considered in the reauthorization of TEA-21. [113]

One of the weaknesses of the system, according to critics, is that "a poor definition exists of which kinds of trails should be part of the system (except for NHT criteria)." [114] While it is relatively easy to add new trails, it has proven more difficult to provide them with adequate staffing and partnership resources.

Another issue is whether the federal government should be given authority to acquire land for existing trails, and the extent of any such authority. Between 1978 and 1986, Congress authorized nine national scenic and historic trails but prohibited federal authority for land acquisition. The trails are the Oregon, Mormon Pioneer, Lewis and Clark, Iditarod, and Nez Perce National Historic Trails, and the Continental Divide, Ice Age, North Country, and Potomac Heritage National Scenic Trails. Legislation to authorize federal land management agencies to purchase land from willing sellers was considered, but not enacted, by the 106th and 107th Congresses. Willing seller legislation has been reintroduced in the 108th Congress. Trails authorized since 1986 typically have included land acquisition authority.

Finally, some trails supporters have advocated a nationwide promotion to inform the public about the National Trails System. They assert that most Americans are unaware of the system and the breathtaking scenes and journeys into the past which can be experienced along the national scenic and historic trails. However, there is concern that a significant increase in the number of trails users could overwhelm present staffing and resources.

Major Statutes [115]

National Parks and Recreation Act of 1978: Act of Nov. 10, 1978; P.L. 95-625, 92 Stat. 3467.
National Trails System Act: Act of Oct. 2, 1968; P.L. 90-543, 82 Stat. 919. 16 U.S.C. §1241, et seq.
Outdoor Recreation Act of 1963: Act of May 28, 1968; P.L. 88-29. 16 U.S.C. §4601.

APPENDIX 1. MAJOR ACRONYMS USED IN THIS ARTICLE

ACEC:	Area of Critical Environmental Concern
ANILCA:	Alaska National Interest Lands Conservation Act
ANWR:	Alaska National Wildlife Refuge
BLM:	Bureau of Land Management
DOD:	Department of Defense
DOI:	Department of the Interior
EIS:	Environmental Impact Statement
FAA:	Federal Aviation Administration
FLPMA:	Federal Land Policy and Management Act of 1976
FS:	Forest Service
FWS:	Fish and Wildlife Service
ISTEA:	Intermodal Surface Transportation Efficiency Act of 1991
LWCF:	Land and Water Conservation Fund
MBCF:	Migratory Bird Conservation Fund
MUSYA:	Multiple-Use Sustained-Yield Act of 1960
NEPA:	National Environmental Policy Act of 1969

NFMA:	National Forest Management Act of 1976
NFS:	National Forest System
NHA:	National Heritage Area
NHT:	National Historic Trails
NPS:	National Park Service
NRT:	National Recreation Trails
NST:	National Scenic Trails
NWR/NWRS:	National Wildlife Refuge/National Wildlife Refuge System
O and C:	Oregon and California (grant lands)
OCS:	Outer Continental Shelf
ORRRC:	Outdoor Recreation Resources Review Commission
PILT:	Payments in Lieu of Taxes (Act and Program)
PRIA:	Public Rangelands Improvement Act of 1978
PWC:	Personal Watercraft
RPA:	Forest and Rangeland Renewable Resources Planning Act of 1974
RTP:	Recreational Trails Program
TEA-21:	Transportation Equity Act for the 21st Century
USDA:	United States Department of Agriculture
WCAs:	Wildlife Coordination Areas
WPAs:	Waterfowl Production Areas

APPENDIX 2. DEFINITION OF SELECTED TERMS

Acquired lands: land obtained by the federal government from a state or individual, by exchange, or through purchase (with or without condemnation) or gift. One category of federal lands.

Entry: occupation of public land as first step to acquiring title; can also mean application to acquire title.

Federal land: any land owned or managed by the federal government, regardless of its mode of acquisition or managing agency.

Homesteading: the process of occupying and improving public lands to obtain title. Almost all homesteading laws were repealed in 1976 (extended to 1986 in Alaska).

Impoundment: man-made impediment to the free flow of rivers or streams, such as a dam or diversion.

Inholdings: state or private land inside the designated boundaries of lands owned by the federal government, such as national forests or national parks.

Land and Water Conservation Fund: the primary source of federal funds to acquire new lands for recreation and wildlife purposes to be administered by federal land management agencies. The fund is derived largely from receipts from the sale of offshore oil and gas (16 U.S.C. 460*l*), but funds must be appropriated annually.

Land withdrawal: an action that restricts the use or disposition of public lands, e.g., for mineral leasing.

Leaseable minerals: minerals that can be developed under federal leasing systems, including oil, gas, coal, potash, phosphates, and geothermal energy.

Lease: contractual authorization of possession and use of public land for a period of time.

Mining claim: a mineral entry and appropriation of public land that authorizes possession and the development of the minerals and may lead to title.

Multiple use land: federal lands which Congress has directed be used for a variety of purposes.

Patent: a document that provides evidence of a grant from the government —usually conveying legal title to public lands.

Payments in Lieu of Taxes: a program administered by BLM which provides payments to local governments which have eligible federal lands within their boundaries.

Public domain land: One category of federal lands consisting of lands ceded by the original states or obtained from a foreign sovereign, through purchase, treaty, or other means. By contrast, "acquired lands" are obtained from an individual or state.

Public land: various meanings. Traditionally has meant the public domain lands subject to the public land disposal laws. Defined in FLPMA to refer to the lands and interests in land owned by the United States that are managed by the BLM, whether public domain or acquired lands. Also, commonly used to mean all federal, state, and local government-owned land.

Rangeland: land with a plant cover primarily of grasses, forbs, grasslike plants, and shrubs. Most federal rangeland is managed by the BLM and the FS and is leased (or used under permit) for private grazing use.

Release language: congressional direction on the timing and extent of future wilderness considerations, and on the management of roadless areas pending future wilderness reviews, if any.

Reservation: public land withdrawn from general access for a specific public purpose or program.

Right-of-way: a permit or easement that authorizes the use of lands for specific purposes, such as construction of a forest access road, installation of a pipeline, or placement of a reservoir.

Subsurface mineral estate: typically refers to a property interest in mineral resources below ground.

Surface estate: typically refers to a property interest in surface lands and the above-ground resources.

Sustained yield: a high level of resource outputs maintained in perpetuity, but without impairing the productivity of the land.

Water right: right to use or control water. Such rights typically are granted by the states, although the United States may have federal water rights as well.

Wetlands: areas predominantly of soils that are situated in water-saturated conditions during part or all of the year, and support water-loving plants, called hydrophytic vegetation. They are transitional between terrestrial and aquatic systems, and are found where the water table generally is at or near the surface.

Wilderness: undeveloped federal land, usually 5,000 acres or more and without permanent improvements, that is managed (either administratively or by statute) to protect and preserve natural conditions.

Wildlife refuge: land administered by the FWS for the conservation and protection of fish and wildlife. (Hunting, fishing, and other forms of wildlife-related recreation typically are allowed, consistent with the purposes of the refuge.)

ENDNONTES

[1] This section was prepared by Carol Hardy Vincent.

[2] U.S. General Services Administration, *Overview of the United States Government's Owned and Leased Real Property: Federal Real Property Profile as of September 30, 2003.* See Table 16 of the report on the agency's website at *[http://www.gsa.gov/ gsa/cm_attachments/ GSA_DOCUMENT/Annual%20Report%20%20FY2003-R4_R2M-n11_0Z5RDZ-i34K-pR. pdf]*, visited March 8, 2004.

[3] In this article, the term *federal land* refers to any land owned or managed by the federal government, regardless of its mode of acquisition or managing agency. *Public domain land* is used when the historical distinction regarding mode of land acquisition is relevant, i.e., when a law specifically applies to those lands that originally were ceded by the original states or obtained from foreign sovereigns (including Indian tribes) as opposed to being acquired from individuals or states. *Public land* refers to lands managed by the Bureau of Land Management, consistent with §103(e) of the Federal Land Policy and Management Act of 1976 (FLPMA, P.L. 94-579; 43 U.S.C. §§1701, et seq.).

[4] Several other agencies manage some of the remaining 43.4 million acres (6.5%) of federal land. The Department of Defense (DOD), including the Army Corps of Engineers, is the fifth largest federal land manager. Because land management is not DOD's primary mission, these lands are not discussed in this article. Nonetheless, military lands often are noteworthy for their size, which can provide important open space, and for their historic, cultural, and biological resources. Moreover, because access is sometimes severely restricted, these lands may contain ecological resources in nearly pristine condition. In addition, the General Services Administration owns or rents lands and buildings to house federal agencies and also administers the excess/surplus system of property disposal.

[5] For the text of the law and other information, see the Indiana Historical Bureau, *Land Ordinance of 1785*, at *[http://www.statelib .lib.in.us/www/ihb/resources/docldord. html]*, visited April 1, 2004.

[6] For the text of the law and other information, see: *[http://www.ourdocuments.gov/ doc.php?doc=8]*, visited April 1, 2004.

[7] These major land acquisitions gave rise to a distinction in the laws between *public domain lands*, which essentially are those ceded by the original states or obtained from a foreign sovereign (via purchase, treaty, or other means), and *acquired lands*, which are those obtained from a state or individual by exchange, purchase, or gift. (Some 601.5 million acres, 89.5% of all federal lands, are public domain lands, while the other 70.3 million acres, 10.5% of federal lands, are acquired lands.) Many laws were passed that related only to the vast new public domain lands. Even though the distinction has lost most of its underlying significance today, different laws may still apply depending on the original nature of the lands involved. The lessening of the historical significance of land designations was recognized in the FLPMA, which defines *public lands* as those managed by BLM, regardless of whether they were derived from the public domain or were acquired. For more information on the Louisiana Purchase, see [http://www.ourdocuments.gov/ doc.php?doc=18], and on the 1848 Treaty with Mexico

see [http://www.ourdocuments.gov/ doc.php?doc=26], both visited April 1, 2004. For more information on the Oregon Compromise, see the Center for Columbia River History, *The Oregon Treaty, 1846*, at *[http://www.ccrh.org/comm/ river/docs/ ortreaty.htm]*, visited April 1, 2004.

[8]   For more information, see the Act of May 20, 1862; ch. 75, 12 Stat. 392 and *[http://www.ourdocuments.gov/doc.php?doc=31]*, visited April 1, 2004.

[9]   U.S. Dept. of the Interior, Bureau of Land Management, *Public Land Statistics, 2002*, Table 1-2 (Washington, DC: GPO, April, 2003). Available on the BLM website at *[http://www.blm.gov/natacq/pls02/]*, visited April 1, 2004.

[10] U.S. Dept. of Commerce, Bureau of the Census, *Historical Statistics of the United States, Colonial Times to 1970* (Washington, DC: GPO, 1976), H. Doc. No. 93-78 (93rd Congress, 1st Session), pp. 428-429. FLPMA, enacted in 1976, repealed the Homestead Laws; however, homesteading was allowed to continue in Alaska for 10 years. For the text of FLPMA and other information on the law, see the BLM website at *[http://www.blm.gov/ flpma]*, visited April 1, 2004.

[11] For more information, see *[http://www.ourdocuments.gov/ doc.php?doc=45]*, visited April 1, 2004.

[12] "Yo-Semite" was established by an act of Congress in 1864, to protect Yosemite Valley from development, and was transferred to the State of California to administer. In 1890, surrounding lands were designated as Yosemite National Park, and in 1905, Yosemite Valley was returned to federal jurisdiction and incorporated into the park. For the text of the law, see the NPS website at *[http://www.cr.nps.gov/ history/online_ books/anps/anps_1a.htm]*, visited April 1, 2004. Still earlier is the 1832 establishment in Arkansas of Hot Springs Reservation, which was dedicated to public use in 1880 and as Hot Springs National Park in 1921.

[13] For the text of the law establishing the system, see the National Park Service website at *[http://www.cr.nps.gov/history/online_books/anps/ anps_1i.htm]*, visited April 1, 2004.

[14] For more information, see the BLM website at *[http://www.blm.gov/ flpma/organic. htm]*, visited April 1, 2004.

[15] For more information, see 30 U.S.C. §§ 181, et seq. and the BLM website at *[http://www.ca.blm.gov/caso/1920act.html]*, visited February 12, 2004.

[16] 43 U.S.C. §§ 315, et seq.

[17] FLPMA also established a comprehensive system of management for the remainder of the western public lands, and a definitive mission and policy statement for the BLM.

[18] The most current copies of CRS products are available at *[http://www.crs.gov/]*.

[19] This section was prepared by Ross W. Gorte.

[20] This is also known as the Dingell-Johnson Act and the Wallop-Breaux Act.

[21] This is also known as the Pittman-Robertson Act.

[22] Since FY1998, this account has been available for forest health improvement activities, as well as for building and repairing roads and trails.

[23] Funding for land acquisition under SNPLMA is excluded from FY2004 and FY2005 figures because funds are released after (1) monies from federal lands sales have been collected, and (2) lands have been nominated for acquisition. For FY2004, the SNPLMA budget for lands nominated for acquisition is $110.6 million, but not all nominated lands will be acquired. Nominations for FY2005 will not be completed until after the end of FY2004.

[24] For national forests that contain northern spotted owl habitat, which led to lower timber sale levels, payments were set at 85% of the FY1986-FY1990 average for FY1994, and declining by 3 percentage points annually, to 58% in FY2003.

[25] A third of the county payment (i.e., 25% of the total) is returned to the General Treasury to cover appropriations for access roads and reforestation; thus, the counties actually receive 50% of the revenues.

[26] The most current copies of CRS products are available at *[http://www.crs.gov/]*.

[27] This section was prepared by Ross W. Gorte.

[28] For more information, see the Forest History Society, *U.S. Forest Service History*, at *[http://www.lib.duke.edu/forest/usfscoll/]*, visited February 20, 2004.

[29] The second principal FS program continues the original role of the Bureau of Forestry: to provide forestry assistance to states and to nonindustrial private forest owners. The authorities for assistance programs were consolidated and clarified in the Cooperative Forestry Assistance Act of 1978. Forestry research is the third principal FS program. Congress first authorized forestry research in 1928 "to insure adequate supplies of timber and other forest products"; the research authorities were streamlined by the Forest and Rangeland Renewable Resources Research Act of 1978.

[30] Congress enacted the limitation in response to Roosevelt's 1906 reservations. Roosevelt needed the funds provided in the 1907 act, but proclaimed additional reserves after it was enacted, but before he signed it into law.

[31] U.S. Dept. of Agriculture, Forest Service, *Land Areas of the National Forest System, as of September 30, 2004*, Table 1 at *[http://www.fs.fed.us/land/staff/lar/LAR03/]*, visited Feb. 20, 2004.

[32] See U.S. Congress, Office of Technology Assessment, *Forest Service Planning: Setting Strategic Direction Under RPA*, OTA-F-441 (Washington, DC: U.S. Govt. Print. Off., July 1990). Available on the Princeton University website, at *[http://www.wws. princeton.edu/~ota/ disk2/1990/9019_n.html]*, visited February 12, 2004.

[33] Since 1997, provisions in the Interior Appropriations Acts have prohibited the FS from completing the overdue 1995 and 2000 RPA Programs, because, it has been asserted, the Government Performance and Results Act (GPRA) planning and reporting requirements have replaced the RPA Program. A Presidential Statement of Policy accompanied the first (1976) RPA Program, and Congress enacted a second Statement of Policy (1980), but no subsequent Statements of Policy have been issued. The *Report of the Forest Service* is printed annually, although no report was published for FY1999 or FY2000, and the reports typically are published several months later than required by law. They are required to be presented to Congress with the annual budget justifications. The Assessments continue to be prepared.

[34] See U.S. Congress, Office of Technology Assessment, *Forest Service Planning: Accommodating Uses, Producing Outputs and Sustaining Ecosystems*, OTA-F-505 (Washington, DC: U.S. Govt. Print. Off., Feb. 1992). Available on the Princeton University website, at *[http://www.wws.princeton.edu/~ota/disk1/1992/9216_n.html]*, visited February 12, 2004.

[35] Available on the Forest Service website at *[http://www.fs.fed.us/emc/ nfma/includes/ cosreport/Committee%20of%20Scientists%20 Report.htm]*, visited February 12, 2004.

[36] U.S. Dept. of Agriculture, Forest Service, *Land Areas of the National Forest System, as of September 30, 2003*, Tables 10-12 and 15-26, at *[http://www.fs.fed.us/land/staff/lar/LAR03/]*, visited February 20, 2004.

[37] The 1891 authority was repealed by §704(a) of FLPMA. The following day, in §9 of NFMA, Congress also prohibited the return of any NFS lands to the public domain without an act of Congress.

[38] The President can still create new national forests from lands acquired under the Weeks Law of 1911 (16 U.S.C. §521).

[39] The most current copies of CRS products are available at [http://www.crs.gov/]. Also, for further information on the Forest Service, see its website at *[http://www.fs.fed.us]*, visited February 12, 2004.

[40] This section was prepared by Carol Hardy Vincent.

[41] For more information, see 43 U.S.C. §§315, et seq. and the website of the University of New Mexico School of Law at *[http://ipl.unm.edu/cwl/fedbook/taylorgr.html]*, visited April 1, 2004.

[42] P.L. 94-579; 90 Stat. 2744, 43 U.S.C. §§ 1701, *et seq.*

[43] For the text of the law, see the FWS website at *[http://www.r7.fws.gov/asm/anilca/toc.html]*, visited April 1, 2004.

[44] For information on the six support and service centers, see the BLM website at *[http://www.blm.gov/nhp/directory/index.htm]*, visited April 1, 2004.

[45] The system, the Geographic Coordinate Data Base, is available on the BLM website at *[http://www.blm.gov/gcdb/]*, visited March 16, 2004.

[46] More information on the National Integrated Land System is available on the BLM website at *[http://www.blm.gov/nils/]*, visited March 16, 2004.

[47] For more information, see 16 U.S.C. §§1331, et seq. and the BLM website at *[http://www.wildhorseandburro.blm.gov/theact.htm]*, visited April 1, 2004.

[48] Fifty percent of the revenues collected from on-shore leasing are returned to the states (except Alaska which receives 90%) in which the lands are located (30 U.S.C. §191).

[49] For BLM wildland fire statistics, see the agency's website at *[http://www.fire.blm.gov/stats/]*, visited April 1, 2004.

[50] Under Title II of P.L. 106-248, the Federal Land Transaction Facilitation Act (43 U.S.C. §2301), the Secretary of the Interior and the Secretary of Agriculture may use funds from the disposal of certain BLM lands to acquire inholdings and other nonfederal lands. Also, the Southern Nevada Public Land Management Act of 1998 (P.L. 105-263) provides for the disposal, by sale or exchange, of lands in Nevada. The proceeds are used to acquire environmentally sensitive lands in Nevada, among other purposes. A description of these funding sources is provided under "disposal authority." The Land and Water Conservation Fund, addressed in the chapter on "Federal Lands Financing," is a primary means of funding BLM land acquisition.

[51] Other authorities provide for acquisitions in particular areas.

[52] Desert lands can be disposed under other laws. The Carey Act (43 U.S.C. §641) authorizes transfers to a state, upon application and meeting certain requirements, while the Desert Land Entry Act (43 U.S.C. §321) allows citizens to reclaim and patent 320 acres of desert public land. These latter provisions are seldom used, however, because the lands must be classified as available and sufficient water rights must be obtained. Other authorities provide for land sales in particular areas. The Homestead Act and

many other authorities for disposing of the public lands were repealed by FLPMA in 1976, with a 10-year extension in Alaska. The General Services Administration has the authority to dispose of surplus federal property under the Federal Property and Administrative Services Act of 1949; however, that act generally excludes the public domain, mineral lands, and lands previously withdrawn or reserved from the public domain (40 U.S.C. §472(d)(1)).

[53] 43 U.S.C. §1713 (c). This procedure and certain other provisions of FLPMA may be unconstitutional under *Immigration and Naturalization Service (INS)* v. *Chadha*, 462 U.S. 919 (1983).

[54] For a description of the law, see the BLM website at *[http://www.blm.gov/nhp/300/wo320/minlaw.htm]*, visited April 1, 2004.

[55] For a description of the law, see the BLM website at *[http://www.blm.gov/nhp/what/lands/realty/rppa.htm]*, visited April 1, 2004.

[56] For a table identifying public land withdrawals 1942-2003, see the BLM website at *[http://www.blm.gov/nhp/what/plo/plo7394.htm]*, visited April 1, 2004.

[57] For the text of the law, see the NPS website at *[http://www.cr.nps.gov/local-law/anti1906.htm]*, visited April 1, 2004.

[58] The most current copies of CRS products are available at [http://www.crs.gov/]. Also, for further information on BLM, including on many of the programs and responsibilities addressed in this section, see the agency's website at *[http://www.blm.gov]*, visited April 1, 2004.

[59] This section was prepared by M. Lynne Corn.

[60] Distinct pre-existing rights (e.g., to develop minerals, easements, etc.) are rarely acquired along with the land. Where they exist and their ownership is considered essential, these rights must be purchased from the landowners, who are otherwise able to exercise them.

[61] For example, some refuges (especially island refuges for nesting seabirds) may be closed to the public — an unlikely restriction for an NPS area, given the NPS mandate to provide for public enjoyment of park resources.

[62] In FY1992, there was a consolidation of units of the Refuge System. The drop in numbers of units shown in Figure 5 in that year is due to this change.

[63] There is also one wilderness area at an FWS National Fish Hatchery in Colorado.

[64] The 482 administrative sites and 69 fish hatcheries administered by FWS are not part of the system, and total only 22,671 acres.

[65] For the text of the law and other information, see the FWS website at *[http://refuges.fws.gov/policyMakers/mandates/index.html]*, visited Feb. 13, 2004.

[66] For the text of the law and other information, see the FWS website at *[http://refuges.fws.gov/policyMakers/mandates/HR1420/index.html]*, visited Feb. 13, 2004.

[67] Most of the research function was administratively transferred to the U.S. Geological Survey (in the Department of the Interior) in FY1996.

[68] This program is distinct from USDA programs to conserve wetlands.

[69] For the text of the law and other information, see the FWS website at *[http://migratorybirds.fws.gov/intrnltr/treatlaw.html]*, visited Feb. 13, 2004.

[70] For the text of the law and other information, see the FWS website at *[http://migratorybirds.fws.gov/intrnltr/treatlaw.html]*, visited Feb. 13, 2004.

[71] Of the 540 refuges, 34 (6.3%) were created under specific laws naming those particular refuges.

[72] These procedures result in congressional termination of executive actions other than by statute, and thus may be unconstitutional in light of *INS* v. *Chadha*, 462 U.S. 919 (1983).

[73] While the MBTA definition of "migratory bird" includes, potentially, almost all species of birds, in practice, the focus of acquisition has been on game birds (e.g., certain ducks, geese, etc.). Non-game species tend to benefit secondarily, though areas without game birds are rarely acquired with MBTA funds.

[74] This authority (and its related funding mechanism) is so commonly used that the distribution of refuges is a good approximation of the four major flyways for migratory waterfowl.

[75] Personal communication from FWS Realty Office, Feb. 9, 2004. Not counted are 11 instances of so-called "friendly condemnations," in which FWS, in cooperation with a willing seller, used the courts to achieve favorable tax treatment, or to settle questions of fair market value, clouded title, or similar problems. Some critics of condemnation authority have suggested that the existence of so-called "hostile" condemnation authority has affected some land sales, to the extent that some sellers feel intimidated — that they have little real choice in the decision to sell, even if condemnation authority was not formally used. If such intimidation exists, its extent is unclear, but legislation was introduced in the 105th Congress to restrict FWS land acquisitions without specific congressional approval. Ultimately, a provision was added in P.L. 105-277 forbidding the use of "any of the funds appropriated in this Act for the purchase of lands or interests in lands to be used in the establishment of any new unit of the National Wildlife Refuge System unless the purchase is approved in advance by the House and Senate Committees on Appropriations in compliance with the reprogramming procedures contained in Senate Report 105-56." This or a similar provision has been incorporated in subsequent appropriations acts. Because the Migratory Bird Conservation Fund and the Southern Nevada Public Lands Management Act funds are not appropriated in annual appropriations acts, purchases from those funds are unaffected by such provisions.

[76] The dollars spent were not necessarily spent on those particular 68,014 acres, due to a lag between payments and transfers of title, completion of paperwork, and other factors.

[77] See "Land Ownership" in BLM chapter, above, for information on a funding source created under the Southern Nevada Public Land Management Act. Funds obtained under this act from federal land sales may be used to acquire environmentally sensitive lands in Nevada, among other purposes. Some of these Nevada acquisitions have become additions to the National Wildlife Refuge System.

[78] For information on how "duck stamp" money is spent, see the FWS website at *[http://duckstamps.fws.gov/Conservation/conservation.htm]* visited February 13, 2004.

[79] U.S. General Accounting Office, *National Wildlife Refuges: Continuing Problems with Incompatible Uses Call for Bold Action*, GAO/RCED 89-196 (Washington, DC: GPO, Sept. 1989), 84 p.

[80] U.S. Dept. of the Interior, Fish and Wildlife Service, *2001 National Survey of Fishing, Hunting, and Wildlife-Associated Recreation* (Washington, DC: 2001). The survey is available on the FWS website at *[http://fa.r9.fws.gov/surveys/surveys.html]*, visited

Feb. 13, 2004. The number of hunters did not decline significantly from the previous surveys, but as a percent of the total U.S. population, there has been a general downward trend over approximately 30 years.

[81] The most current copies of CRS products are available at *[http://www.crs.gov/]*. Also, for further information on the National Wildlife Refuge System, including on many of the programs and responsibilities addressed in this chapter, see the FWS website at *[http://www.fws.gov]*, visited February 13, 2004.

[82] This section was prepared by David Whiteman.

[83] 16 U.S.C. §21.

[84] 16 U.S.C. §22. In the early years, the Interior Department relied on the U.S. Army for enforcement of the regulations and protection of the park units.

[85] For more information on the establishment of Yellowstone National Park, see Aubrey L. Haines, *Yellowstone National Park: Its Exploration and Establishment* (Washington, DC: 1974), available on the NPS website at *[http://www.cr.nps.gov/history/online_books/ haines1/]*, visited Mar. 8, 2004.

[86] 16 U.S.C. §431.

[87] 16 U.S.C. §1.

[88] For more information, see U.S. Dept. of the Interior, *History of the National Park Service*, available on the NPS website at *[http://www.cr.nps.gov/history/hisnps/NPShistory.htm]*, visited Mar. 8, 2004.

[89] National Park System General Authorities Act of 1970, P.L. 91-383; 16 U.S.C. §1a-1, §1c.

[90] Redwood National Park Expansion Act, P.L. 95-250; 16 U.S.C. §1a-1.

[91] *Rethinking the National Parks for the 21st Century*, National Park Service Advisory Board Report 2001, available on the NPS website at *[http://www.nps.gov/policy/futurereport.htm]*, visited Mar. 8, 2004.

[92] National Parks Omnibus Management Act of 1998, P.L. 105-391; 16 U.S.C. §1a-5.

[93] Congress rescinded these withdrawals and reestablished most of the lands as national monuments, national parks, or national preserves in ANILCA.

[94] 16 U.S.C. §460l-9(c).

[95] 43 U.S.C. §1714(j). While Presidents may modify monument boundaries, it is not certain that a President can revoke a national monument. (See CRS Report RS20647, *Authority of a President to Modify or Eliminate a National Monument*, by Pamela Baldwin.)

[96] There are hundreds of laws establishing or modifying specific units of the National Park System, in addition to the few general laws listed here.

[97] The most current copies of CRS products are available at *[http://www.crs.gov/]*. Also, for further information on the National Park System, see the NPS website at *[http://www.nps.gov]*, visited Mar. 8, 2004.

[98] This section was prepared by Ross W. Gorte.

[99] Release language provides congressional direction on the timing and extent of future wilderness considerations (i.e., when the land would be reviewed for possible wilderness), and on the interim management of roadless areas, pending any future wilderness reviews. See CRS Report RS21917, *Bureau of Land Management (BLM) Wilderness Review Issues*, by Ross W. Gorte.

[100] The most current copies of CRS products are available at *[http://www.crs.gov/]*.

[101] This section was prepared by Sandra L. Johnson.

[102] U.S. Dept. of the Interior, National Park Service, *River Mileage Classifications for Components of the National Wild and Scenic Rivers System* (Washington, DC: Jan. 2002). Available on the NPS website at *[http://www.nps.gov/rivers/wildriverstable. html]*, visited May 7, 2004.

[103] U.S. Dept. of the Interior, National Park Service, *Wild and Scenic Rivers and the Use of Eminent Domain* (Washington, DC: Nov. 1998). Available on the NPS website at *[http://www.nps.gov/rivers/ publications/eminent-domain.pdf]*, visited Feb. 13, 2004. Condemnation and subsequent acquisition of land by the federal government (in fee title, or fee-simple) has been used along 4 rivers since 1968: the Rio Grande, the Eleven Point River, the St. Croix, and the Obed, resulting in the acquisition of 1,413 acres. Condemnation of land for easements has occurred on 8 rivers amounting to 6,339.7 acres. The FWS is the only agency that has never used condemnation to acquire land or an easement for a wild and scenic river corridor.

[104] The most current copies of CRS products are available at *[http://www.crs.gov/]*. Also, for more information on the National Wild and Scenic Rivers System, see the NPS website at *[http://www.nps.gov/rivers]*, visited Feb. 13, 2004.

[105] This section was prepared by Sandra L. Johnson.

[106] Donald D. Jackson, "The Long Way 'Round," *Wilderness*, vol. 51, no. 181 (summer, 1998): 19-20.

[107] Outdoor Recreation Resources Review Commission, *Outdoor Recreation for America* (Washington, DC: Jan. 1962), p. 34.

[108] Ibid., p. 1.

[109] *Congressional Record*, vol. 111 (Feb. 8, 1965): 2087.

[110] The act is available on the NPS website at *[http://www.nps.gov/ncrc/ programs/nts/ legislation.html]*, visited Feb.13, 2004.

[111] 16 U.S.C. §1247(d); 49 C.F.R. §1152.29.

[112] For information on current legislation related to trails, see CRS Issue Brief IB10093, *National Park Management and Recreation*, coordinated by Carol Hardy Vincent.

[113] ISTEA also established the National Recreational Trails Funding Program, renamed the Recreational Trails Program (RTP) under TEA-21. RTP is not part of the National Trails System. Rather, RTP is a state-administered, federal-aid grant program which provides funds to local governments. The fund is administered by the Department of Transportation in consultation with the Department of the Interior. RTP provides funds to the states to develop and maintain recreational trails and trail-related facilities for both nonmotorized and motorized recreational trail uses. Trail uses include bicycling, hiking, in-line skating, crosscountry skiing, snowmobiling, off-road motorcycling, all-terrain vehicle riding, four-wheel driving, or using other off-road motorized vehicles.

[114] Steven Elkinton, "How the National Trails System Has Changed Since 1968," *Pathways Across America*, (Spring 1998): 10.

[115] For further information on the National Trails System, see the NPS website at *[http://www.nps.gov/ncrc/programs/nts/index.html]*, visited Feb. 13, 2004.

In: Progress in Environmental Research
Editor: Irma C. Willis, pp. 75-126

ISBN 978-1-60021-618-3
© 2007 Nova Science Publishers, Inc.

Chapter 2

ENVIRONMENTAL PERTURBATION AND COASTAL BENTHIC BIODIVERSITY IN URUGUAY

Ernesto Brugnoli, Pablo Muniz, Natalia Venturini and Leticia Burone

Sección Oceanología, Facultad de Ciencias, Universidad de la República Oriental del Uruguay (UdelaR), Iguá 4225, Montevideo, PC 11400, Uruguay

ABSTRACT

The Uruguayan coastal zone is bathed by the waters of the Southwest Atlantic Ocean (SWAO, 230 km) and the Río de la Plata Estuary (450 km), one of the largest estuaries in the world. The main tributaries of this estuary are the Paraná-Paraguay and Uruguay rivers, which drain the second largest basin in South America and provide the major source of freshwater runoff to the SWAO. Typical coastal ecosystems are sandy beaches with rocky points, sub-estuaries flowing along the Río de la Plata one and coastal lagoons in the Atlantic region. The estuarine portion is characterised by muddy sediments while sandy-shell debris are the dominant sediment type in the Atlantic portion. One of the most relevant features in this coastal zone is the interaction between the SWAO and the Río de la Plata waters, being that salinity is of primary importance in regulating the benthic biodiversity. In general, autochthonous fauna of the Uruguayan coast is characteristic of the temperate zone with temperate-cold and temperate-warm components, which correspond to the Patagonic biogeographic province, and show a break in the region of the Río de la Plata Estuary influence. Several studies had demonstrated that the Uruguayan coastal zone is under the effects of different kinds of human related stressors. Urbanisation, harbour, shipping and industrial activities are the main perturbation factors for the Río de la Plata portion, while, agricultural and tourism affect preferentially the Atlantic one. In addition, these studies had shown that petroleum hydrocarbons, heavy metals and the organic enrichment of bottom sediments have a direct influence on biodiversity patterns. Furthermore, some morphological anomalies have been detected in benthic foraminifera, which seem to be related to the heavy metal and organic content of the sediments. Despite the existence of a clear salinity gradient in this estuarine area; the utilization of different approaches, together with the integration of physical, chemical and biological data, demonstrated the occurrence of an environmental quality gradient with the improvement of conditions from the inner stations of Montevideo Bay to the outer coastal ones. Recent studies warned about the presence of non indigenous invasive species in both zones of the Uruguayan littoral, however, the degree of incidence seems to be greater in the estuarine portion. Their introduction would be related to the discharge of ballast water, and their distribution determined by salinity patterns. The available information about marine biodiversity and environmental perturbation in the Uruguayan coastal zone was improved during the last decades; however, it is still restricted to isolated areas along it and to some aspects of aquatic ecosystems. The implementation and development of integrative baseline studies on these topics are highly relevant, in order to contribute to the conservation of benthic biodiversity in the coastal zone of Uruguay.

INTRODUCTION

Coastal areas are the most dynamic portions of the global ecosystem, and also, the most subjected to population concentration. Natural and anthropogenic impacts, poor planning and management of the conflictive activities developed in coastal areas have major influence on the resident biota, functioning and productivity of all coastal ecosystems, which in turn, affect directly the global biodiversity and the whole ecosystem health. The consciousness about the consequences of man-induced impacts on marine ecosystems has increased during the last years, and so, the need of applying effective measures to prevent or stop them (Howarth et al., 2000). Regarding this necessity, the development of research to identify and to attenuate the impact is fundamental, aiming the prevention of socio-economical prejudices and health risks (Constanza et al., 1997). Furthermore, the search of solutions for the conflictive uses of an environment should lay down on a solid scientific base that permits to obtain information about how it works, which are its responses to external impacts and its capacity of recovery, in order to establish priorities and make the best decisions for its management and conservation.

Nowadays, an important challenge for the human beings is the maintenance of natural coastal systems that provide goods and services. Within the coastal area, estuaries are highly dynamic systems with an intrinsic strong natural variability and consequently, a high level of stress for the inhabitant organisms. Estuaries are environments where the meeting of water masses with contrasting physical and chemical properties occurs, promoting the establishment of strong horizontal and vertical gradients. Hydrographical regimes are very variable among estuaries and also, temporally, within a particular one. Then, populations and species associated with these relatively complex environments should develop physiological adaptations to deal with this high natural variability (Day et al., 1989; Lalli & Parsons, 1997). In addition, they are zones of transition between the marine and the terrestrial environments that perform essential ecological functions, including nutrient degradation and regeneration, as well as, the control of nutrient, water, particles and organisms fluxes, from and to the continental margins, rivers and oceans. The ecological and economical importance of estuaries and coastal zones in general, is worldwide recognized. Besides, they are the most complex, diverse and productive areas on earth, they are the most vulnerable ones, due to the competing demands between the natural and the socioeconomic systems. The former, is represented by the physical, chemical and biological components, whereas, the latter is represented by anthropogenic activities and the infrastructure needed to develop them (Constanza et al., 1997).

Within aquatic ecosystems, the benthic environment has an important function as an efficient natural trap for several substances, and also, as a natural regulator of sedimentary biogeochemical processes. Bottom sediments constitute a source of nutrients for the water column above them, leading to benthic-pelagic coupling and influencing primary productivity (Jorgensen, 1996). In addition, sediments accumulate natural and anthropogenic products from the overlying water, being that, they may act both, as a sink or a source of contaminants. Furthermore, marine sediments could act as a non-point source of contaminants causing adverse effects to organisms and to human health through trophic transfer. Heavy metals, hydrocarbons and other kind of pollutants derived from anthropogenic activities, produce perturbations in an ecosystem changing its abiotic conditions and affecting its biota. In these sense, the analysis of the structure of the benthic communities is an important tool to describe changes in space (with applications on point source pollution monitoring) and time (with applications on the description of changes concerning the state of the marine system) (Heip, 1992).

Benthic organisms have been used extensively as indicators of environmental status and trends. Numerous studies had demonstrated that they respond predictably to many kinds of natural and human induced stressors (López-Jamar, 1985; Ritter & Montagna, 1999; Borja et al., 2000). Benthic macrofauna can be suitable used to describe changes in a particular environment because the organisms are relatively sedentary and have comparatively long life spans (Thouzeau et al., 1991). In addition, the macrofauna consists of different species that exhibit different tolerances to stress (Dauer, 1993), allowing the monitoring of the environmental quality. Exposure to hypoxia is typically great in near-bottom waters and anthropogenic contaminants often accumulate in sediments where the benthos lives. The limited mobility of most adult macrobenthic organisms has advantages in environmental assessment, due to, unlike most pelagic fauna; their assemblages reflect local environmental conditions (Gray, 1979). The macrobenthos can also be employed to understand the incidence of certain ecological factors, such as predation and competition for space or food, which are responsible for the structure and productivity of benthic communities (Saiz-Salinas, 1997). In benthic ecosystems, the variability of environmental conditions has effects on species composition, which are well established. The distribution of invertebrate species is strongly influenced by the physico-chemical environment over a wide range of scales (Hall et al., 1994). Regionally, the species changes over geographic gradients, both latitudinal and longitudinally (Hillebrand, 2004; Heip, 1992). The most important variables identified, in the literature, as primary influencing factors are depth, temperature, water movement patterns and sediment type (Rees et al., 1999; Sanvicente-Anorve et al., 2002). Sediment mobility together with the amount and quality of the organic matter presented, have also been evocated as influential factors at the regional scale (Heip et al., 1992). Elsewhere, salinity can be a determinant factor of large-scale invertebrate composition (Giberto et al., 2004). On a minor scale, sediment composition and other related factors seems to be of variable importance in determining benthic composition assemblages (Brown et al., 2002; Thrush et al., 2003). At a local scale, seabed morphology is also important for structuring benthic patterns (Giménez et al., 2006), although, there may be multiple variables involved in complex species-environment relationships at this scale.

The reviews developed by Masello & Menafra (1996) and Calliari et al. (2003) provide detail information about characteristics, components and dynamics of the benthic communities in the Uruguayan coastal zone. In general, these authors coincided in the fact that there are a limited number of investigations developed, and they focused mainly on intertidal sandy beaches. Nevertheless, some recent works had analysed several aspects of the intertidal and subtidal macrobenthic communities in the Río de la Plata and the Atlantic coast providing some new knowledge, which permit recognize some distributional patterns, scales associated with this patterns and several environmental variables that regulate the structure and dynamics of these benthic communities (Rodríguez- Capítulo et al. 2003; Giménez et al. 2005, 2006; Defeo & McLachlan, 2005; Carranza et al., in press; Lercari & Defeo, 2006).

This chapter summarised the main natural and anthropogenic stressors for macrobenthic fauna in the Uruguayan coastal zone and its effects at the population and community levels. We focused specifically, on the shallow sublittoral benthic fauna of the coastal area of Montevideo, which is the most studied in relation to anthropogenic impacts. We also commented some published and original data about biological invasions in the Uruguayan coastal zone.

URUGUAY AND THE URUGUAYAN COASTAL ZONE

Uruguay is a small country situated in South America (30°-35° S and 53°- 58° W), between Argentina and Brazil (Figure 1). In Table 1, are presented the most important territorial, social, and economic indicators of Uruguay. The weather is subtropical temperate with an annual mean temperature of 16° C and 1000 mm of precipitation. It is situated in the lower portion of the Río de la Plata Basin, the second most important of South America after the Amazonian Basin. The Río de la Plata Basin, with an area of 3.100.000 km^2, includes five countries (Brazil, Argentina, Paraguay, Uruguay and Bolivia) and a population of ca. 120.000 million inhabitants. It is the most industrialised area of South America. The most important rivers are the Paraná (ca. 4000 km) and the Uruguay (ca.1.600 km), which flowed in the Río de la Plata Estuary, with an annual river flow of 16.000 m^{3}s^{-1} and 3.900 m^{3}s^{-1}, respectively (Tundisi et al., 1999). With a total extension of 680 km, the Uruguayan coast has 450 km lie on the Río de la Plata and 230 km on the Atlantic Ocean (Figure 2). One of the most important aspects in this coastal ecosystem is the interaction between the freshwater from the Río de la Plata Estuary and the Atlantic Ocean, which promotes a saline gradient along the coast. These coastal zone can be grossly divided in three different sub-environments according to the prevalent haline conditions: freshwater (between Punta Gorda and Montevideo corresponding to the upper and middle Río de la Plata, with a salinity between 1 and 10), estuarine (between Montevideo and Punta del Este in the lower Río de la Plata, with a highly salinity variation, between 1 and 33) and oceanic (between Punta del Este and Barra del Chuy in the Atlantic Ocean) (Figure 2) (Guerrero et al., 1997; Nagy et al., 1997, 2002).

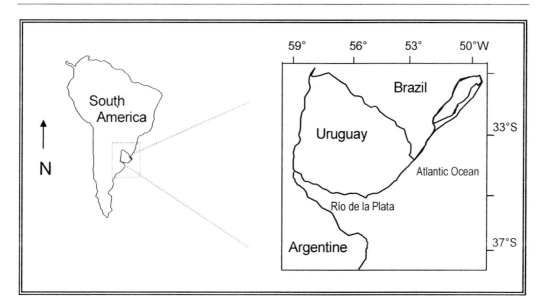

Figure 1.

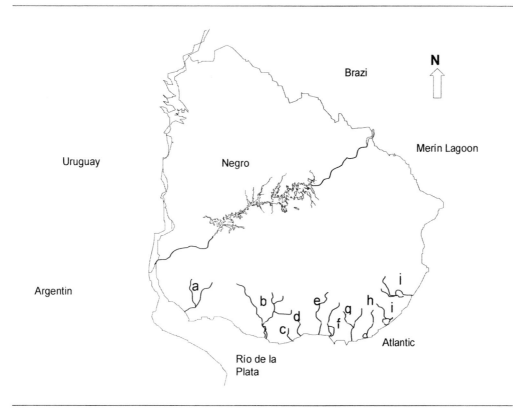

Figure 2.

Table 1. Main territorial and social-economical characteristics of Uruguay

Location	30° to 35 ° S – 53° to 58° W
Area*	176.000 km² (Terrestrial surface)
	137.567 km² (jurisdictional waters)
Borders	Argentine (W) and Brazil (E, NE).
Extension coastal zone**	680 km
Population***	3.305.723 inhabitants (2005)
Pop. distribution at coastal zone	73% live in coastal zones
Pop. density in the country****	19 inhabitants km²
Pop. density in the costal zone****	74 inhabitants km²
Pop. density in Montevideo city****	2600 inhabitants km²
PBI*****	12.329
PBI distribution	Fish and agriculture (9%), Industry
	(17%), Services (58%), Others (16%)

* Include terrestrial surface and jurisdictional waters;
** Include Río de la Plata and Atlantic coast;
*** www.ine.org.uy;
**** Populations projections to 2001 (www.ine.org.uy);
**** millions of US$ (year 2003).

The Río de la Plata ($35°00'$-$35°10'$ S and $55°00'$-$58°10'$ W) may be defined as a funnel-type coastal plain tidal river with a semi-enclosed shelf sea at the mouth and a surface area ca. 36.000 km². The Paraná and Uruguay Rivers feed freshwater into the Río de la Plata Estuary with a seasonal and interannual discharge variation between 22.000-28.000 m³ s^{-1}, and with extreme values during El Niño (> 30.000 m³ s^{-1}) and La Niña (< 20.000 m³ s^{-1}) (Nagy et al., 2002). Tides are semidiurnal with amplitude of about 40 cm on the Uruguayan coast. Features such as salinity, depth of the halocline and vertical mixing vary with astronomic tidal oscillation on an hourly basis, while axial winds influence water height and salinity variations on a daily basis. The river flow governs monthly to interannual variations. The mean water temperature (1981-1987) is 15 °C (CARP, 1990). Fine grained sediments are confined to the upper and middle Río de la Plata, while sand covers almost the entire outer Río de la Plata and the adjacent continental shelf (López-Laborde & Nagy, 1999). In this area, the dominant coastal features consist of sandy beaches and estuaries (López-Laborde et al., 2000). At least 67% of the Uruguayan population lives in this area, so, the main industrial and urban activities are concentrated there. Several polluted sub-estuaries are located in it and consequently habitat degradation is growing with an accelerated rate. Sedimentation and final deposition of dredging sediments are other important problems recently perceived. Among the main pollutants introduced in the aquatic environment we should mentioned heavy metals, petroleum hydrocarbons, surfactants, DDT's and PCB's (Cranston et al., 2002; Muniz et al., 2002, 2004a, b, 2005a; Viana et al., 2005; Moyano et al., 1993, Moresco & Dol, 1996). In addition, this region shows conditions of moderate eutrophication (Nagy et al., 2002). Recently, the increment of harbour activities and the transoceanic transport have called the attention for another problem, the introduction of exotic species with ballast water (Clemente & Brugnoli, 2002; Brugnoli et al., 2005; Brugnoli et al., in press; Muniz et al., 2005b).

The most relevant characteristics of the Atlantic zone are the large sandy beaches and the presence of coastal lagoons. Coastal lagoons in Uruguay are usually shallow water bodies receiving variable amounts of fresh water. Due to their geomorphological and hydrological characteristics, environmental conditions in the lagoons undergo frequent fluctuations on a daily and seasonal basis. This instability causes changes in the distribution of benthic species and the structure of communities, which sometimes are accentuated by anthropogenic influences. These ecosystems are important sites, functioning as nursery and feeding areas for several aquatic species of economic value. In this area, is localized the "Bañados del Este" Biosphere Reserve (MAB, UNESCO), the most important natural protected area of the country. Recently, a significant demographic growth was detected in the basins of these lagoons; therefore, some environmental problems were reported such as eutrophication, habitat degradation, the occurrence of exotic species and the increment in agriculture activities (Nión et al., 1979; Rodríguez-Gallego et al., 2003; Muniz et al., 2005b; Borthagaray et al., 2006). Sandy beaches located on the Atlantic coastal zone are interrupted by rocky points or freshwater systems, both, natural and artificial ones. The mainly artificial freshwater system that flows directly into the Atlantic Ocean is the Andreoni Channel, situated 800 m south of La Coronilla (Defeo & De Alava, 1995). It discharges pesticides, high concentration of suspended solids (Mendez & Anciaux, 1995) and freshwater from rice plantations that affect the benthic intertidal community adjacent to its mouth. For a review of these effects and detailed information of the effects of the channel over the intertidal fauna see Calliari et al., 2003; Defeo & de Alava, 1995, 2002; Defeo et al., 1996; Lercari & Defeo, 1999; Lercari & Defeo, 2003; Lercari et al., 2002.

SOME BIOGEOGRAPHICAL ASPECTS OF THE URUGUAYAN REGION

According to the influence of temperature, latitudinal gradients, local circulation patterns and water properties combined with the spatial arrangement of the continents and oceans, is possible to divide the oceans into a series of provinces or biogeographic regions with characteristic biological assemblages. These regions show high degree of endemism of its flora and fauna (Balech, 1954, 1964; Boschi, 1976, 2000) being characterised by specific organism assemblages that are based on qualitative data about the distribution of plant and/or animal species, primary, production, biomass or trophic relationship. The latitudinal gradient of species diversity is the most robust pattern in biogeography. Specifically, the gradient in species richness, a negative correlation between the number of species and latitude (Pianka, 1966; Rhode, 1992; Brown & Lomolino, 1998) is seen in most of taxonomic groups: vascular plants, algae, birds, mammals, reptiles, amphibians, fishes, arthropods, fungi and many others (e.g. Fischer 1959; Gaines & Lubchenco, 1982; Willig & Selcer, 1989; Currie, 1991; Rex et al., 1994; Blakburn & Gaston, 1996).

From Trinidad (~ 10°N) to Cape Horn (~ 55°S) the South American margin changes from tropical to cool temperate. It includes a great range of ecosystems including the enormous Amazonian delta, the coral reef of Fernando de Noronha Island, mangrove swamps and tidal marshes, hypersaline and brackish lagoons as well as narrow and wide continental shelves. Among the first studies about zoogeography in South America we can mentioned those developed by d'Orbigny (1835-43) and Dana (1853), who realised the occurrence of faunistic regions, centres of dispersion and speciation, and also the relationship between the regional marine fauna and different ocean temperature zones. Dana (1853), include within zoogeographic regions associated to the Uruguayan coastal zone, the Uruguayan Province (temperate, located 30° S northwards from la Plata Cape) and the Platense Province (sub-temperate, located in the mouth of the Río de la Plata Estuary), indicating a essential faunistic change in the Río de la Plata region. Moreover, Balech (1954), recognize 4 zoogeographic regions in the Southwest Atlantic (Antillana, Sur de Brasil, Argentina, Magallánica). Boltovskoy (1970) published the major biogeographic divisions in the Southwestern Atlantic included four domains in the neritic zone: a tropical domain (from equator to 20° S), subtropical (20° to 30 - 35° S), transition (30 - 35° S to 46 - 48° S) and subantartic (46 - 48° S to 55 -60° S). Recently, Sealey & Bustamante (1999), mentioned the warm-temperate Southwestern Atlantic Province, defined to the south by the Valdes Peninsula (41°S) and to the north by Cabo Frío (Brazil 23° S), comprising a coast line of 8.154 km length. This province has a warm temperate climate that constitutes a transition between the cold temperate South America Province and the Tropical Southwestern Atlantic Province. The transitional, that is ascribed to the Argentina Biogeographical Province (and its boundaries with subtropical and subantartic zones) is the most dynamic, and their limits vary widely seasonally and multiannually. This zone is also defined as an area of mixing of subtropical and subantartic fauna (Boltovskoy et al., 1999), being characterised by a high diversity of fishes and invertebrates, numerous colonies of sea mammals and birds that belong to South Brazilian and Sub Antarctic regions, so, the expected degree of endemism is very low.

The marine domain of Uruguay is made up by the Río de la Plata and the adjacent shelf and shares ecosystems with Brazil and Argentina. According to several authors, the Uruguayan coastal zone belongs to the transitional zoobiogeographic division, located between subtropical and subantartic zones (Balech, 1954; Palacio, 1982; Boltovskoy, 1970) (Figure 3). In the classification of Sealey & Bustamante (1999), the Uruguayan coastal zone includes the Uruguayan - Buenos Aires coastal shelf and the Río de la Plata coastal biogeographic province constituted by a warm-temperate biota, rather than a cold-temperate one. The Río de la Plata represents a major geographic feature of the Argentine Biogeographical Province, located approximately in its centre. Its presence imposes a not completely proved role of biogeographical barrier. According to Boltovskoy & Wright (1976), considering the benthic foraminifera communities, the Argentine Province is an area of temperate water characterized by the dominance of *Buccella peruviana*. This species also occurs along the Pacific coast of South America but its true domain is in the Argentina Province. Other species also characteristics of this province are *Bolivina compacta*, *Bulimina patagonica*, *Buliminella elegantissima*, *Elphidium spp. Epistominella exigua*, *Milolinella subrotunda*, *Quinqueloculina seminulum*, *Pyrgo rigens* and others, but the dominant species is *Bucella peruviana*. Murray (1991) divided the estuary of the Rio de la Plata in three zones according to the occurrence of living foraminifera: fluvial (salinity < 10) with *Haynesina germanica*, fluvial marine (salinity 10 - 30) with *Ammonia beccarii* and marine (salinity > 30) with *Buliminella elegantissima* (Boltovskoy & Lena, 1974).

Furthermore, the great majority of the intertidal fauna species have in the Uruguayan coastal zone their dispersion limits, restricting their latitudinal distribution. This characteristic is principally due to the water mean temperature, with presence of temperate cold and temperate warm faunal limits, and the existence of the Río de la Plata Estuary (salinity variation) that appears to act as an ecological barrier (Maytia & Scarabino, 1979; Escofet et al., 1979). Costal communities off Río de la Plata are characterised by the absence of the typical cold-temperate south Atlantic forms, the absence of significant invertebrate predators from the rocky intertidal zone, as well as the presence of a community dominated by the yellow clam (*Mesodesma mactroides*) in exposed sandy beaches (Sealey & Bustamante, 1999). Some authors ascribe these attributes to the river while others consider that this barrier does not exist. Recent information suggest that rivers act as an "intermittent barrier" for the distribution of flora and fauna (see Mianzan et al., 2001), such information must be taken into consideration when analysing species richness spatial patterns. The subtidal macrofauna of the Río de la Plata Estuary was less diverse than that of adjacent marine areas but shows higher densities and biomass. Bottom type, salinity and the presence of a turbidity front are the main physical variables structuring benthic communities of this estuary and coastal adjacent zone (Giberto et al., 2004). Table 2 synthesise the most representative taxa (abundance, biomass, with ecological or economic relevance) of the macrofaunal benthic community (intertidal and subtidal) of the Uruguayan coastal zone, according to studies developed in the last 30 years.

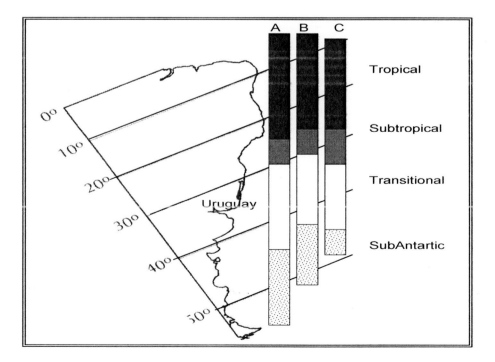

Figure 3.

Table 2. Benthic taxa most representative of coastal Uruguayan zone

Species	Intertidal (*)		Subtidal (**)	Río de la Plata				Atlantic coast		
	SS	HS		E	SE	RP	SB	CL	RP	SB
Plantae										
Juncus acutus	X				X		X			
Ruppia maritima	X							X		
Spartina longispina	X				X		X			
Spartina montevidensis	X				X		X			
Algae										
Codium sp.		X							X	
Enteromorpha sp.		X				X			X	
Hypnea musciformis		X							X	
Polysiphonia sp		X				X			X	
Porphyra sp.		X							X	
Ulva lactuca		X							X	
Artrophoda										
Crustacea										
Artemesia longinaris	X		X							
Chtamalus bisinuatus		X				X			X	
Chasmagnathus granulata	X				X			X		
Cyrtograpsus angulatus		X			X	X		X		
Eubalanus amphitrite		X							X	
Eubalanus improvisus		X				X		X	X	
Emerita brasiliensis	X									X
Excirolana armata	X						X			X
Excirolana brasiliensis	X									X
Lygia exotica (Esp.)		X		X						
Neomysis americana (Csp.)			X	X	X					
Ocypode cuadrata	X									X
Panaeus paulensis								X		
Seriolis marplatensis	X		X							
Annelida (Polychaeta)										
Ficopmatus enigmaticus (Esp.)		X	X		X			X		
Heteromastus similis	X				X			X		
Hemipodus olivieri	X									X
Laeonereis acuta	X				X					
Neanthes succinea	X				X		X		X	X
Nephtys fluviatille	X				X			X		
Onophis tenuis			X							
Mollusca										
Adelomenon brasiliana (Fsp.)			X					X		
Brachidontes darwinianus		X		X		X			X	
Brachidontes rodriguezi		X		X		X				

Table 2. (Continued)

Species	Intertidal (*)		Subtid al (**)	Río de la Plata				Atlantic coast		
	SS	HS		E	SE	RP	SB	CL	RP	SB
Buccinanops cochildium			X	X						
Corbicula fluminea (Esp.)	X		X	X						
Corbula patagonica			X							
Donax hanleyanus (Fsp.)	X									X
Erodona mactroides	X		X		X		X	X		
Heleobia australis	X		X	X	X			X		
Limnoperna fortunei (Esp.)		X		X	X					
Littoridina australis		X				X				
Mactra isabelleana	X		X	X	X		X			
Mesodesma mactroides (Fsp.)	X									X
Mytella charruana		X		X	X	X				
Mytilus edulis platensis (Fsp.)		X		X		X			X	
Olivancillaria vesica auricularia	X				X					X
Perna perna		X							X	
Rapana venosa (Esp.)		X	X	X		X				
Siphonaria lessoni		X								
Tagelus plebeius	X				X		X	X		
Echinodermata										
Encope emarginata	X		X							

* supra, meso and infralitoral;

** under infralittoral zone (Río de la Plata and adjacent shelf waters);

SS: Soft substrate; HS: Hard Substrate, E: Estuary; SE: Sub-estuary; RP: Rocky point; SB: Sandy beach; CL: Coastal Lagoon.; Fsp: Fisheries species; Esp: Exotic species; Csp. Criptogenic species. From Scarabino et al. (1975), Escofet et al. (1979), Maytia & Scarabino (1979), Nión (1979), Boschi (1988), Riestra et al. (1992), Masello & Menafra (1996), Muniz & Venturini (2001), Orensanz et al. (2002), Giberto et al. (2004), Calliari et al. (2001); Brugnoli et al. (2005), Giménez et al. (2005), Carranza et al. (in press).

THE MONTEVIDEO COASTAL ZONE:
ITS BIOTA AND MAIN STRESSORS

Montevideo Bay (MB) and the adjacent coastal zone (AZC) are located in the middle portion of the Río de la Plata Basin. Data about surface water temperature indicate the existence of a hot season from December to March and a cold season from June to September with a temperature difference between them of approximately 10 °C (Guerrero et al., 1997). Mean surface salinity of the Río de la Plata waters near Montevideo varies between 5 and 10. Hourly variations in salinity, halocline depth and vertical mixing, are coupled with tidal oscillations, which are semi-diurnal. The river discharge governs monthly and interannual variations of salinity and turbidity, being that the maximum of turbidity is generally associated with the limit of saline intrusion. Furthermore, it is related to gravitational circulation and flocculation of clay particles, and has a high suspended organic matter load (López-Laborde & Nagy, 1999). The MB with an area of 10 km^2 has a mean depth of 5 m, excepting in the navigation channels. In relation to geological features, it is characterized by the presence of Precambric outcrops with more recent material derived from them in some regions (Cardellino & Ferrando, 1969). Modern sediments consist of silt and clay (Muniz et al., 2002). Water circulation within the MB is clockwise, being determined mainly by winds from NE and W-SW. This area has a great economic importance for Uruguay as a navigation and commercial route. Its waters are used for sport activities and the extraction of artesian and industrial fish resources, among other demands.

The BM and its ACZ are under the influence of different anthropogenic impacts such as the input of urban and industrial effluents, oil refining processes and maritime traffic. Water quality of Montevideo Bay is highly deteriorated due to several point and non-point sources and harbour activities. This subsystem, of 10 km^2 of area, receives most of its water from the Río de la Plata, with minor contribution from three little creeks and several sewage outlets heavily charged in heavy metals, nutrients and BOD (Muniz et al., 2002, 2004a, b). Within the bay is the ANCAP oil refinery, La Teja dock, the Central Batlle electric power plant (UTE) and the Montevideo Harbour. Untreated or partially treated municipal effluents, produced by one and a half million people are discharged towards the east portion of BM, in the Punta carretas zone, where the largest sewage pipe of the country is located. The standing stock of particulates generally attains its peak concentration in the late summer and early fall, reflecting a trend towards a unimodal seasonal pattern in the development of autotrophic biomass, i.e. chlorophyll-a concentration. In summer, water conditions are clearly hypertrophic with high concentrations of ammonium (up to 120 micro M), chlorophyll a (> 100 μg/g chl a l^{-1}), and hypoxia (0-20 % of oxygen saturation) as a consequence of domestic effluent discharges and the relative high residence time of the water mass. Depth in the navigation channels that permit the access to Montevideo Harbour can reach between 9 and 11 m because they are frequently dredged. Is estimated that 700000 m^3 of sediment are removed every year, which are deposited in the Río de la Plata some kilometers off the bay. There is no information about the behavior of these sediments after their deposition (Muniz et al., 2004a).

The knowledge about local pollution levels and its effects is limited, mainly in relation to the biota. The first studies concerning heavy metal and hydrocarbon pollution in the Montevideo coastal zone were published by Moyano et al. (1993) and Moresco & Dol (1996). These authors found near the mouth of the Pantanoso Stream sediments severely contaminated by chromium derived from leather fabrics. Altogether, these activities promote the input of different contaminant classes that can modified environmental conditions and have detrimental effects on the ecosystem biota. These effects should be identified and quantified. In view of this background information, the following section presents results generated during 1997 and 1999 under a series of research projects developed by our group. The projects were developed to evaluate the degree of contamination and its effects on the ecosystem health through the study of the benthic communities and its environment in a portion of the Montevideo coastal zone, specifically between Punta Carretas and Punta Yeguas, including Montevideo Bay and Montevideo Harbour (Figure 4). The main goal was to create scientific-technical basis for the correct instauration of an environmental monitoring plan for the Montevideo coastal zone. With this objective we collected field information in a series of sampling stations concerning environmental and biotic variables.

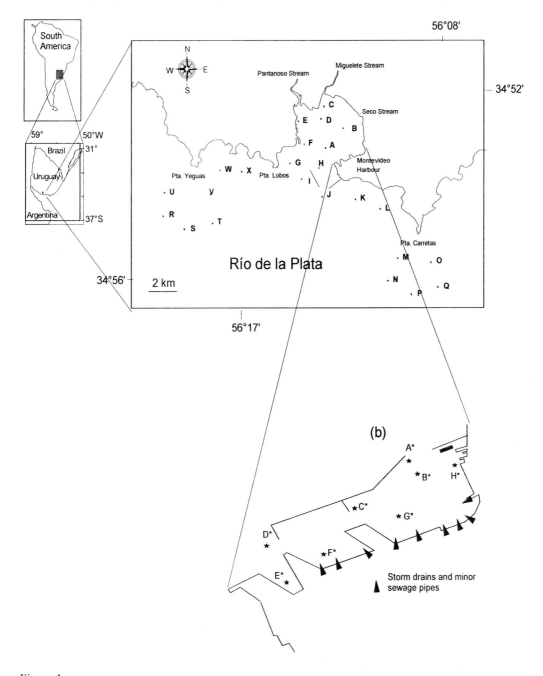

Figure 4.

In the Montevideo Bay (MB) and the adjacent coastal zone (ACZ) samples were collected monthly, in 24 stations during one year. Stations A to E were localised in the most inner portion of the bay, near the harbour, the electric power plant (Central Batlle), the oil refinery (ANCAP) and the mouths of Miguelete and Patanoso Streams (Figure 4). The west portion of the bay included stations F, G and I and the most external one included stations H and J. Stations K to Q were placed toward the east of the bay and in Punta Carretas zone, where the main submarine outfall of Uruguay is located. Stations R to X were positioned in the western part of the bay, in Punta Yeguas zone. The rationale of this sampling design was to evaluate and compare the environmental characteristics of these three zones, related with the influence of the urban effluents discharged by the Punta Carretas submarine sewage pipe and the two streams that flow into the bay. In the harbour area, samples were collected in winter and summer in 8 stations distributed to cover different portions inside the harbour area (stations A* to H*; Figure 4). Surveys were carried out onboard vessels provided by the Uruguayan Army and the Government of Montevideo City (IMM).

Salinity, temperature and pH of the water column were measured *in situ* using a YSI® multiparameter. Water was collected with HYDRO-BIOS water samplers to analyse oxygen dissolved concentration according to Winkler titration method (Grasshoff, 1983). At each station, 11 sediment samples were collected with manual Kajac corers of 4.5 cm internal diameter. Granulometric analysis was performed on three corers, the upper centimetre of three other corers, stored in a PE vial at 20°C, served for the quantification of organic matter and photosynthetic pigments. Sediments contained in three metallic corers were transferred to a tin foil and frozen in aluminium bottles until the analysis of hydrocarbons. Two acrylic corers were stored in vertical position in the cold; upon arrival at the laboratory, one of them was used for heavy metal analysis and the other one was used to measure redox potential. Danulat et al. (2002) and Muniz et al. (2002, 2004a) describe in detail the laboratory methods for granulometric, organic matter, chlorophyll *a*, heavy metals and hydrocarbons determinations, as well as redox measurements.

Biological samples for the study of macrobenthic communities' structure were collected in triplicate with an Ekman grab (0,053 m^2). They were sieved through a 0.4 mm mesh and preserved in 4% formol. After their separation from the sediment, macrobenthic organisms were preserved in 70% ethanol, counted and identified in most of the cases to the species level. To study the foraminiferal fauna, samples were collected with a Kajak corer (5 cm internal diameter) and the uppermost 3 cm layer of the sediment was taken at each station forming a volume of about 60 cm^3 per sample. Immediately after sampling, the material was stained with buffered rose Bengal dye (1 g of rose Bengal in 1000 ml of alcohol) for 48 h to differentiate between living and dead foraminifera (Walton, 1952). In the laboratory, the wet samples were carefully washed through 0.500 mm, 0.250 mm and 0.062 mm sieves to separate the size fractions. All living specimens in each sample were picked and identified following the generic classification of Loeblich & Tappan (1988).

RELATIONSHIPS BETWEEN MACROFAUNA AND MAIN STRESSORS IN THE MONTEVIDEO COASTAL ZONE

Montevideo Harbour

In the Montevideo Harbour area, in summer, only four benthic species were found, the gastropod *Heleobia australis*, the polychates *Alitta succinea* and *Nephtys fluviatilis* and the bivalve *Erodona mactroides*. Macrobenthic fauna was absent in the most inner part of the harbour (St. H*). *H. australis* was the only zoobenthic species found at the two sites in the inner harbour (Sts. B* and G*) and contributed to more than 94% of total abundance (Table 3). Very small organisms (< 2mm total length) made up a significant portion (ca. 21%) of the total number of specimens. In contrast, in winter the community was composed only by two species, the dominant *H. australis* and *N. fluviatilis*. The very small size of the *H. australis* individuals and the highly variable percentage of small specimens could be indicating that the recolonisation, settlement of juveniles, and their further survival are greatly altered, probably due to dredging and frequent sediment resuspension. Larvae of the polychaete *Heteromastus similis*, a small opportunistic species adapted to variable environmental conditions, poor oxygenation and high concentrations of organic matter of the sediment (Pearson & Rosenberg, 1978), were recorded in plankton samples (Danulat et al., 2002). However, no adults were found colonising the sediments. These polychaete species that inhabit the harbour area are considered as tolerant to variable environmental conditions (Méndez et al., 1998; Muniz & Pires, 2000), and in Brazil, *Alitta succinea* is commonly found in eutrophic coastal areas (Amaral et al., 1998).

Table 3. Spatial distribution of four macrozoobenthos species (individuals m^2) in Montevideo Harbour, in March compared to July of 1998.

Location	< 2 mm	Heleobia 2-4 mm	6 mm	total	Nephthys	Neanthes	Erodona
March 1998							
A*	14500	14187	438	29125	0	63	0
B*	3750	8000	1125	12875	0	0	0
C*	5063	13250	11000	29313	0	0	63
D*	4313	16375	15437	36125	63	125	13
E*	187	938	312	1437	63	63	0
F*	2063	11625	15375	29063	63	0	0
G*	1687	6313	5875	13875	0	0	0
H*	0	0	0	0	0	0	0
July 1998							
A*	125	1250	125	1500	0	0	0
B*	250	562	125	937	0	0	0
C*	187	17000	34750	51937	0	0	0
D*	6250	169937	90000	266187	125	0	0
E*	250	813	250	1313	0	0	0
F*	375	2250	2750	5375	63	0	0
G*	312	1500	438	2250	0	0	0
H*	0	0	0	0	0	0	0

For the gastropod *Heleobia cf. australis australis*, total number of specimens as well as abundance of three size classes is presented, while only the total number of individuals is indicated for the polychaetes *Nephthys fluviatilis* and *Neanthes succinea*, and the pelecipod *Erodona mactroides*. Results represent mean values based on three replicate sediment samples retrieved by grab and five obtained by corer.

Related with the abiotic benthic environment, sediments showed little variation of the grain size. The predominant fraction was silt (up to 85%) and the only area where sand reached 15% was that near the fluvial dock (St. E*). Organic matter content of the sediments was high, with a clear spatial gradient from the inner area (St. H* = 16.5 %) to the outer one of the harbour (St. G = 9.6%). Accordingly, redox potential evidenced the lack of oxygen in surface sediments at most of the stations. Reduced conditions were detected at 1cm depth within the sediment column with values ranging between – 90 and + 100 mV, only at St. D* the top 1 cm of the sediment tend to be oxygenated (+210 mV) (Figure 5).

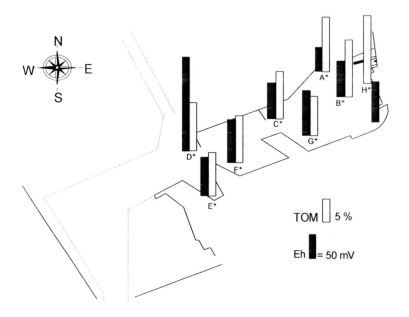

Figure 5.

Table 4 present data of heavy metals in different marine and estuarine environments of Latin America, including the harbour area, MB and the adjacent coastal zone. The comparison of the results with those reported for other regions of Latin America indicates that levels and ranges of variation of our data are similar to those reported from sites with high anthropogenic disturbance. These data and those derived from the analysis of hydrocarbons were evaluated for potential adverse effects on biological organisms using available sediment quality values and sediment quality guidelines (Muniz et al., 2004a). Concerning heavy metals and considering that toxicity rarely occurs below the TEL and frequently above PEL (MacDonald et al., 1996), the authors concluded that: i) the inner region of the harbour showed heavy metal concentrations that may cause major adverse biological effects, except for Ni and Hg; and almost all the metals were above PEL level; the exceptions were Cd, Ni and Ag that were between TEL and PEL; ii) stations D*, E* and F* presented Pb and Cr levels between TEL and PEL; iii) only at St. H* Ag would cause major adverse biological effects on benthic fauna. Related to PAHs, at least one of the analysed contaminants was present in excess of the TEL level, and most of them (ca. 81%) had at least one compound in excess of PEL. This situation is more severe in summer, when all compounds listed by the U.S. Environment Protection Agency were above PEL (Muniz et al., 2004a). These results were coincident with those observed for the benthic fauna abundance patterns (Danulat et al., 2002).

Table 4. Range of variation of the heavy metals of sediments in different environments of Latin America n.a = not analyzed, concentrations are all in mg kg^{-1}, (bl) = background level

Location	Cd	Zn	Cu	Cr	Ni	Pb	Ag	Hg	Reference
Uruguay									
Montevideo Harbour	<1.0-1.6	183-491	58-135	79-253	26-34	44-128	<1.0-2.3	0.3-1.3	Muniz et al. 2004a
Montev. Coastal Zone	41-231	2.4-105	1.3-4.0	n.a	n.a	40-148	n.a	n.a	Moyano et al. 1993
Montev. Bay	0.1-0.2	n.a	10-112	n.a	n.a	40-148	n.a	n.a	Moresco & Dol 1996
Montev. Bay	109	300	150	300	n.a	81	n.a	n.a	Cranston et al. 2002
Montev. Bay	n.a	n.a	n.a	68-1062	n.a	99-365	n.a	n.a	Muniz et al. 2002
Montev. Coastal Zone	n.a	n.a	n.a	37-50	n.a	38-56	n.a	n.a	Muniz et al. 2002
Carrasco Creek				10-807		17-73			Lacerda et al. 1998
Brazil									
Patos Lagoon	0.1-20	20-214	0.8-20	8-337	n.a	8-267	n.a	n.a	Baisch et al.1988
Patos lagoon	n.a	n.a	n.a	n.a	n.a	n.a	n.a	0.06 (bl)	Mirlaen et al. 2001
Coastline of RJ State	n.a	14-795	11-166	18-121	11.2	10-83	n.a	n.a	Lacerda et al. 1982
Jurujuba Sound	n.a	15-337	5-213	10-223	15-79	5-123	n.a	n.a	Baptista Neto et al. 2000
Mangroves in RJ	n.a	26-610	18-80	n.a	6-12	20-130	n.a	n.a	Machado et al. 2002
Argentine									
Bahia Blanca	0.44	35.5	7.3	n.a	n.a	17	n.a	n.a	Villa et al. 1988
Venezuela									
Coral reef sediments	n.a	36-77	6-40	18-32	n.a	17-36	n.a	0.2-0.33	Bastidas et al. 1999
Chile									
South Fjords	0.1-0.5	91-122	16-22	49-82	24-31	26-29	n.a	n.a	Ahumada &Contreras 1999
Mexico									
Baja California	n.a	n.a	n.a	n.a	n.a	n.a			Carreón-Martínez et al. 2001

The biological benthic diversity of the Montevideo Harbour is extremely poor. Macrobenthic fauna consists of only four species of small size and is dominated strongly by the gastropod *H. australis*, an opportunistic species that feed on the surface sediment. This general pattern of low diversity and high abundance of a single species is common to estuaries world-wide, including those in the same geographic region (Benvenuti, 1997; Ieno & Bastida, 1998; Muniz & Venturini, 2001; Giménez et al., 2005, 2006). Diversity and species richness, however, are substantially lower in Montevideo Harbour than at the locations in the immediate vicinity. Scarabino et al. (1975) found seven macrobenthic species just outside the harbour area that were considered typical for the entire Montevideo coastal zone. In addition to the dominant *H. australis* the authors found three bivalves (*E. mactroides*, *Tageleus plebius* and *Mactra isabelleana*) and one species of Tubificidae (oligochaete). In the contiguous Montevideo Bay Muniz et al. (2000, 2005a, b) and Venturini et al. (2004) recorded ten species, nevertheless, the dominant was *H. australis*. Further than the fact that low biodiversity found in estuaries can be merely the result of highly variable salinity and unstable sediments (Tenore, 1972; Wilson, 1994), it has also been established that chronic contamination produces and additional reduction of macrofaunal diversity (Heip, 1995). Data analysed for the Montevideo Harbour clearly suggest the latter case.

Montevideo Bay

In Montevideo Bay, considering the ten sampling stations (Figure 4) in an annual cycle, the total mean abundance determined for macrobenthic species was of 30,118 individuals belonging to the Phyla Arthropoda, Nematoda, Mollusca and Annelida. As in the neighbouring area of Montevideo Harbour, the most frequent and dominant species was *Heleobia australis*, following by *Nephtys fluviatilis*, *Erodona mactroides*, *Heteromastus similis*, *Neanthes succinea*, *Goniadides sp.*, *Glycera sp.*, *Sigambra grubii*, an unidentified isopod as well as unidentified Nematoda and Ostracoda.

Density was very variable at each station on a month scale. Over all stations, the highest values occurred in May, June and July 1997 and the lowest in March 1998 (Figure 6a). In general, stations B, C and D presented a lower number of individuals per unit area than the remaining stations. Ignoring unidentified Nematodes, Ostracods and Barnacles a total of 10 species were recorded. Maximum species richness occurred between April and August 1997 (Figure 6b). Stations F and G showed the highest species richness, and the lowest values were found at stations B, C and D. Species richness was low throughout the study. A higher number of individuals did not always correspond to a higher number of species. Shannon diversity was also very low, reflecting the high dominance of *Heleobia australis* that occurs frequently in very high abundance in several regions of the bay. The maximum diversity was registered in September 1997 at station G (1.63) and the minimum at station B (0) where only Nematodes were present. Between April and August 1997 the highest diversity values were registered in the majority of the sampling stations. In November of the same year diversity decreased notably and in February 1998 it increased again (Figure6b). As for abundance, biomass was also variable in all stations during the period of study. The highest biomass values of the dominant species *Heleobia australis* were occasionally exceeded by those corresponding to the second most abundant species, the bivalve *Erodona mactroides*. The highest biomass was recorded in July 1997 and the lowest in March 1998 (Figure 6a).

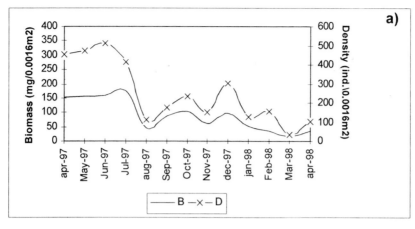

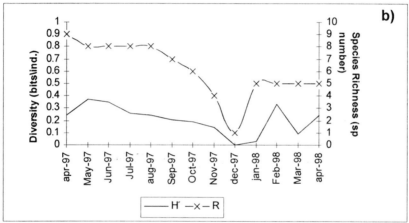

Figure 6.

The decline in the number of *H. australis* during some months of the studied period is reflecting that it is a short-lived species. The relation among the rise of organic matter concentration, the reduction in species number, diversity and the increment of the abundance of one or two species of small size have been well reported in previous studies (see for example Pearson & Rosenberg, 1978; Méndez et al., 1998; Oug et al., 1998; Sánchez-Mata et al., 1999). These species are generally considered as indicators of organically enriched sediments. In such communities, perturbed by organic contamination, the frequency of disturbance is higher than the recovery rate. Thus, opportunistic species of small size and short lifetime will be favoured and could colonise such habitats with any type of biological competition. Then, such species can be adapted to a high frequency of continuous disturbance. Even though, *Heleobia australis* was the most abundant (80% of the total abundance) and the dominant macrobenthic species, many of the other species, especially the polychaetes *Nephtys fluviatilis, Allita succinea, Heteromastus similis* and *Goniadides sp.*, have been also reported in environments with a high organic load elsewhere (e.g. Dauer & Conner, 1980; Amaral et al., 1998). The high frequency of occurrence of these species, in addition to the presence of large-bodied nematodes, which were retained in the 0.4 mm sieve, would be related to the high organic content of the sediments. Cluster analysis of abundance and biomass data (annual arithmetic mean of pooled data) showed two groups of stations at near 60% of similarity (Figure 7). One group was composed by the most-inner stations B, C and D, whereas the remaining stations constituted the other. The same two groups appeared in the MDS ordinations (Figure 7).

Although the bay was a very variable system, it was possible to differentiate, by means of the cluster analyses and the MDS ordination, discrete faunal associations, in regions with particular environmental characteristics. The cluster formed by Sts B, C and D that showed less abundance and biomass of benthic organisms, corresponds to the inner part of the bay where environmental conditions are very unfavourable. Water circulation is limited; there is a high percentage of organic matter, chromium, lead and PHAs in the sediments and a tendency to the presence of reduced sediments (Muniz et al., 2002). The other cluster formed by the remaining stations A, E, F, G, H, I and J corresponds to regions of the bay, which are heterogeneous, but in general have more favourable environmental conditions than the inner region. At Sts F and G particularly, the high water circulation and oxygenation of the sediment column, together with the small percentage of organic matter, may be responsible for the great abundance and biomass of benthic organisms recorded.

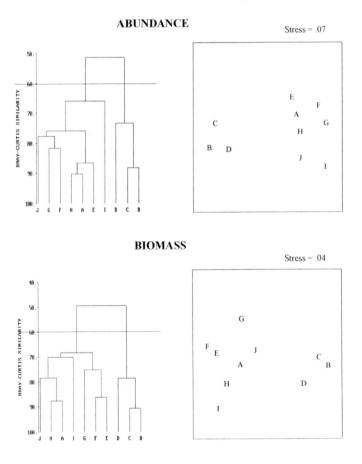

Figure 7.

Montevideo bay and the Adjacent Coastal Zone: Their Degree of Perturbation and Ecological Status

When considered Montevideo Bay and the adjacent coastal zone (Stations A to X, Figure 4) the main results using multivariate statistical techniques showed that the region can be divided in three zones with different abiotic (Figure 8) and macrofaunal patterns (Figure 9). The inner portion is the most heterogeneous according to sedimentological composition. It has a high organic load and is highly contaminated by chromium, lead and oil derived hydrocarbons. The outer portion of the bay and the ACZ showed a moderate contamination level. Despite the dominance of *Heleobia australis* in the whole area, the difference in environmental quality among the three regions was reflected in their macrobenthic community structure. In the inner portion of MB, benthic communities showed a very simple structure, being dominated by nematodes, organisms that belong to the meiofauna and generally are associated to organically enriched environments (Warwick 1986) and also some individuals of *Heleobia* (Muniz et al., 2000; Venturini & Muniz, 2001; Venturini et al., 2004). In the other two zones, benthic communities showed a more complex structure with a higher number of species and diversity. Moreover, in the inner portion of the bay the individuals of the dominant species presented epibiontic parasites (Ciliophora of the family Vortecellidae: *Zoothanium elegans*), a smaller size and thinner shells than those individuals of the outer portion and the ACZ. Mean biomass values for Gastropoda in the group 1 of stations, were 20 times lower than those in group 3, and 6 times lower than in group 2.

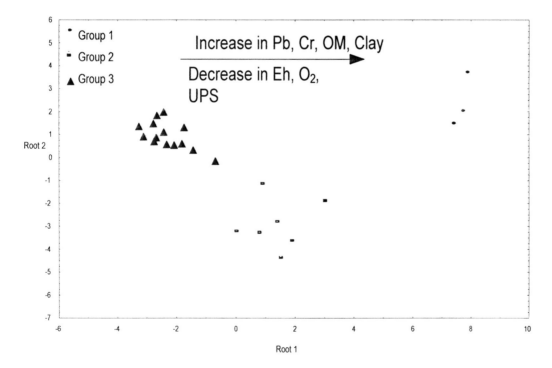

Figure 8.

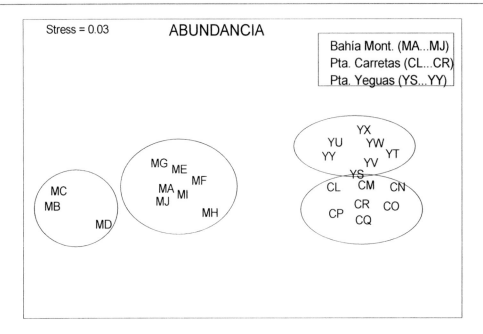

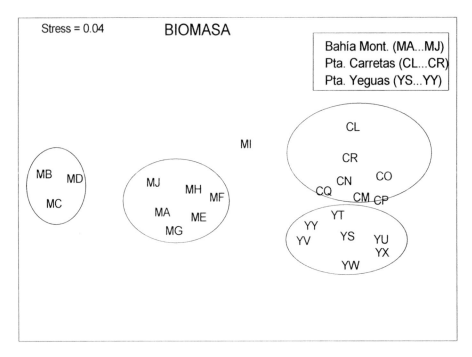

Figure 9.

The environmental variables that best explained the macrofauna distributional pattern were lead and polycyclic aromatic hydrocarbon concentrations and salinity (Venturini et al., 2004). It is known that coastal regions and estuaries are very dynamic environments characterised by great variations in their abiotic parameters and subjected to continuous natural disturbances. This natural variability can represent the main cause of stress for organisms (Turner et al., 1995), however, the input of nutrients, organic matter and human-derived contaminants can alter environmental conditions in a different manner from that expected by natural causes alone (Pearson & Rosenberg, 1978; Mucha et al., 2002).

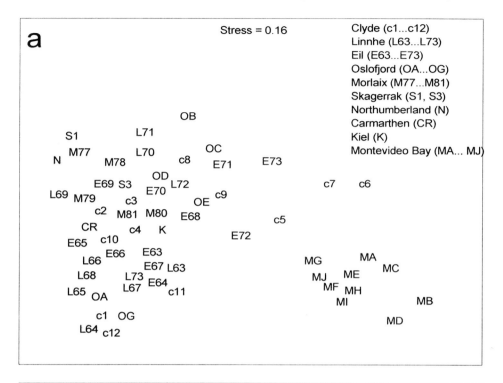

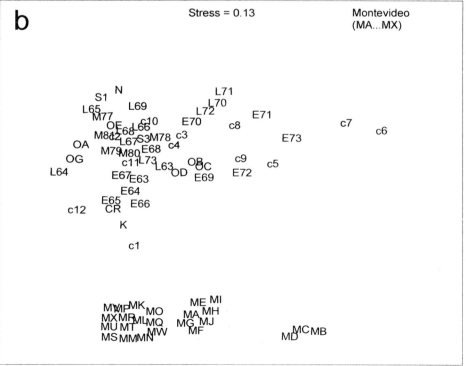

Figure 10.

Warwick & Clarke (1993) created a method denominated "Phylum-level Meta-analysis", which using abundance and biomass data at the phylum level, allows the evaluation of the degree of perturbation of a particular benthic community in a global scale of anthropogenic impact. Applying this method to data obtained in the Montevideo coastal zone, it was observed that the inner stations of MB were located in the right side of the diagram (Figure 10), indicating a high degree of perturbation of these communities (Warwick & Clarke, 1993). The other two groups formed by the stations of the outer portion of MB and the ACZ, were located in a gradient from lower to higher impact, respectively. The position of stations B, C and D in the right side of the ordination diagram is related to presence of large size nematodes, coupled to the great organic load of the sediments. The vertical separation of Montevideo stations from the others is the result of the high dominance of the gastropoda *H. australis*, which would tolerate the instability of environmental conditions, as well as, the elevated organic and heavy metal loads. Within this framework, stations B, C and D can be classified as highly contaminated, station A and stations E to J as contaminated and the others (K to X) as moderately contaminated (Venturini et al., 2004).

All previous studies indicated the occurrence of a vertical separation between the samples studied and the NE Atlantic samples. This separation was attributed to the alteration of the balance between echinoderms and crustaceans due to estuarine characteristics of the coast of Trinidad (Agard et al., 1993), to the higher proportion of annelids due to oxygen-deficient conditions in central Chile (Tam & Carrasco, 1997) and to the higher proportion of crustaceans and annelids and the lower proportion of molluscs and echinoderms in the Gulf of Cádiz (Drake et al., 1999). Non-mined samples from southern Africa exhibited a lower proportion of molluscs and echinoderms but a larger proportion of crustaceans than the NE Atlantic samples (Savage et al., 2001). In the Montevideo coastal zone data, part of the separation could be explained by the absence of echinoderms due to the salinity gradient that would prevent the presence of these osmoconformer organisms (Agard et al., 1993). It could also be attributed to the lower proportion of annelids and the higher proportion of crustaceans in the Montevideo coastal zone samples with regard to those from the NE Atlantic. However, the prominent separation of the Montevideo coastal zone samples seems to be mainly due to the high proportion of molluscs. Similar results were obtained in mined samples from the southern African coast, which were ascribed to the capacity of some gastropods to withstand the physical disturbance caused by the mining process (Savage et al., 2001). In this case the high proportion of molluscs is the consequence of the high dominance of the gastropod *Heleobia australis* that seems to tolerate the environmental instability typical of estuarine areas and the high organic and inorganic loads existing in the Montevideo coastal zone. According to Rakocinski et al. (2000), hydrobiid gastropods are among many opportunistic and/or tolerant estuarine taxa associated with sites moderate or high contaminated with both metals and organic chemicals.

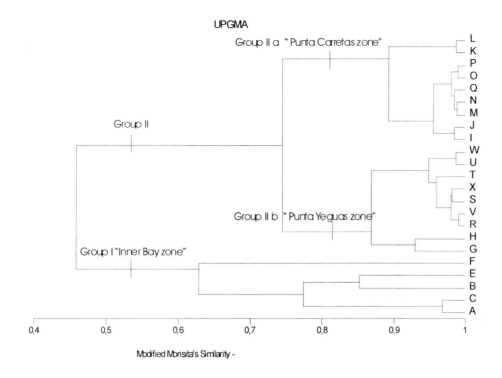

Figure 11.

A conceptual framework, which has been the basic reference in the literature concerning the effects of organic enrichment on benthic communities, has been established by Pearson and Rosenberg (1978). Within this context, several investigators have developed biotic indices to estimate macrobenthic community disturbance level and to establish the ecological status of soft-bottom benthos (Hilly et al., 1986; Grall & Glémarec, 1997). All such studies have emphasised the importance of biological indicators, to measure the ecological quality of a marine environment (Engle et al., 1994; Grall & Glémarec, 1997; Weisberg et al., 1997). Recent approaches have developed a biocriteria-based predefined reference condition and, upon this, several deviations (disturbance classes) were established (Dauer & Alden, 1995; Weisberg et al., 1997; Van Dolah et al., 1999). Borja et al. (2000) have proposed the adoption of AZTI´s Marine Biotic Index (AMBI), using macrobenthic organisms as bio-indicators. These authors have explored the response of soft-bottom communities, to natural and man-induced changes in water quality. Such approach has integrated the long-term environmental conditions in several European estuarine and coastal environments. This index is based essentially upon the distribution of 5 ecological groups, of soft-bottom macrofauna (Grall & Glémarec, 1997); these are in relation to their sensitivity to an increasing stress gradient. Such an approach has the advantage of being simple, in terms of calculation, compared to those adopted previously; it is based upon a formula which permits the derivation of a coefficient (AMBI, Borja et al., 2003, 2004), allowing statistical analysis of the results. Further, Borja et al. (2003) have established that benthic samples subjected to different impact sources e.g. organic enrichment, physical alterations of the habitat, heavy metal inputs, etc., along the European coast, were classified correctly according to the Marine Biotic Index (see Muxika et al., 2005 and Borja et al., 2006, for a review of the increasing use of this index, within

Europe). In this context, Muniz et al. (2005a) applied AMBI to the data set of the Montevideo coastal zone, including Montevideo Bay area. For Montevideo Bay and the adjacent coastal zone, Muniz et al. (2002) and Venturini et al. (2004) had classified the innermost stations of the bay (B, C and D) as grossly polluted, Stations A and E to J as polluted and the remainder as moderately polluted. The inner part of Montevideo Bay: (a) was associated with high concentrations of chromium, lead and polycyclic aromatic hydrocarbons in the sediments; (b) presented anoxic conditions, with negative values of redox potential; and (c) the benthic communities were dominated by large-sized nematodes, which were not considered in the AMBI calculations (see discussion below).

Although, the dominance of the second-order opportunistic species *Heleobia australis* was apparent over the whole area, in most of the outer stations of Montevideo Bay and in the adjacent coastal zone, the benthic communities were richer, more diverse and the bottom conditions were less severe. This general trend was observed clearly with high AMBI values in the innermost stations of Montevideo Bay; these decreased throughout the outermost part of the bay and the adjacent coastal zone (Table 5). Based upon the AMBI (without considering the nematodes), the innermost stations were classified as moderately disturbed (Table 5). These stations were dominated, both in terms of abundance and biomass, by large nematodes (retained in 0.4 mm sieve mesh). For this reason, in a previous study when data were analysed with the phylum-level meta-analysis approach, they presented the status of grossly polluted (Venturini et al., 2004). According to the ecological group assignment followed to the calculation of AMBI and BI values, nematodes belong to EG III (Borja et al., 2000); this will result in the final classification of these inner stations as being slightly disturbed, instead of grossly polluted, sites. Perhaps, these large nematodes, probably oncholaimid, should be assigned to ecological group V, since their dominance could be a symptom of polluted conditions (Warwick, 1986). Conversely, Stations K and L, which were classified previously as moderately polluted (based upon chemical data, Muniz et al., 2002, 2004b), revealed AMBI values which are similar to those presented by the innermost stations of Montevideo Bay. Along this sector of the coastline, there is an important sewage outfall that carries also the city runoff and other untreated industrial effluents (Moyano et al., 1993); this promotes, probably, adverse effects on the benthic fauna, reflected in AMBI values. Venturini et al. (2004) did not identify any difference in the macrobenthic communities between the two coastal zones adjacent to Montevideo Bay. However, these investigators argued that Punta Carretas should be more perturbed than Punta Yeguas, due to the effluents released by the largest and most important sewage outfall of Uruguay, situated in this area. With the use of the AMBI and BI, it was possible to detect differences between these two zones (Table 5). Punta Yeguas was classified as an area with an unbalanced benthic community health (slightly disturbed) and Punta Carretas as moderately disturbed, with a transitional to pollution benthic community health. Although the dominance of *Heleobia australis* was evident in the two zones, the crustacean species in Punta Yeguas (mainly EG I) had an important contribution, in terms of abundance (between 25 and 49 % of the total abundance); this suggests that the AMBI appeared to be more sensitive in this respect, than the meta-analysis approach previously applied to this data set.

Table 5. Values of AMBI, BI, total abundante (ind/0.053 m^2), total organic matter content (%TOM), chlorophyll *a* content (ug/g), chromium (mg/kg) and lead (mg/kg) content of the sediments, benthic community health (BCH) and site disturbance classification (SDC) in the 24 stations of the Montevideo Coastal Zone

St	AMBI	BI	Abund	TOM	Chl *a*	Cr	Pb	BCH	SDC
A	4.4	4	396	6.6	6.2	131.5	215.1	Polluted	Moderately disturbed
B	4.5	4	3	11.3	11.6	81.1	246.7	Polluted	Moderately disturbed
C	4.5	4	1	8.3	3.5	91.5	369.6	Polluted	Moderately disturbed
D	4.5	4	1	12.0	8.8	657.1	352.2	Polluted	Moderately disturbed
E	4.3	3	134	12.8	0.8	368.1	64.9	Polluted	Moderately disturbed
F	4.2	3	144	3.5	0.5	43.7	44.7	Transitional to pollution	Moderately disturbed
G	3.9	3	32	7.5	0.3	30.9	39.1	Transitional to pollution	Moderately disturbed
H	4.2	3	65	9.4	0.5	83.7	65.4	Transitional to pollution	Moderately disturbed
I	3.6	3	17	6.2	0.2	42.1	38.5	Transitional to pollution	Moderately disturbed
J	4.1	3	17	9.5	0.3	56.2	41.7	Transitional to pollution	Moderately disturbed
K	4.4	4	3091	6.8	0.1	38.9	56.4	Polluted	Moderately disturbed
L	4.3	3	3375	6.9	0.5	42.5	57.9	Polluted	Moderately disturbed
M	3.2	2	388	4.8	0.5	40.1	58.5	Unbalanced	Slightly disturbed
N	3.8	3	153	4.5	0.5	40.3	58.9	Transitional to pollution	Moderately disturbed
O	3.9	4	867	6.1	0.8	39.3	57.9	Transitional to pollution	Moderately disturbed
P	3.7	3	338	4.6	0.4	36.7	55.1	Transitional to pollution	Moderately disturbed
Q	3.8	4	305	5.9	0.2	39.3	55.2	Transitional to pollution	Moderately disturbed
R	2.7	2	2412	6.1	0.3	38.8	54.5	Unbalanced	Slightly disturbed
S	2.3	2	1804	5.6	0.3	36.9	55.4	Unbalanced	Slightly disturbed
T	2.6	2	1679	5.2	0.2	37.4	55.1	Unbalanced	Slightly disturbed
U	3.1	2	537	5.5	0.4	38.3	56.2	Unbalanced	Slightly disturbed
V	2.9	2	2267	6.6	0.6	38.9	56.7	Unbalanced	Slightly disturbed
W	3.2	2	49	6.5	0.1	42.1	54.8	Unbalanced	Slightly disturbed
X	3.2	2	322	5.4	0.3	38.1	54.9	Unbalanced	Slightly disturbed

Table 6. Density of foraminifera species present in Montevideo Coastal Zone at the 24 stations

Species/Stations	A	B	C	D	E	F	G	H	I	J	K	L	M	N	O	P	Q	R	S	T	U	V	W	X
Ammobaculites exiguus							25			2			1		21			6	17	9	4	2	1	
Ammonia parkinsoniana							152	51			66	42	11	15	48	2	1	11	42	65		31		56
Ammonia rolshauseni																								5
Ammonia tépida	3	8	12		12	2	241	159	112	117	308	137	911	1120	6390	1230	1000	157	665	600	379	242	329	530
Ammotium salsum							38	12			2		3	1	3		1				1		8	
Bolivina pulchella															6									
Brizalina striatula													17	8	1	4	16		12		9			5
Bulimina marginata															3		1							
Buliminella elegantissima	7	1	5	8																5				

Table 6. (Continued)

Species/Stations	A	B	C	D	E	F	G	H	I	J	K	L	M	N	O	P	Q	R	S	T	U	V	W	X
Cibicides variabilis												1	1	1	6	1	5							
Discorbis williamnsoni														2	4	3								
Elphidium excavatum					5	2	16	7	8	5	10	3	9	25	12	1	3	1	11	30	1		3	2
Milammina fusca						8	287	116			2		2	6	34	5	24	93	600	222	191	174	202	415
Pararotalia cananeiaensis								7	1	3	5	1	11		21	1	1							
Pseudononion atlanticum								8					4	2	15	8	12		5					
Psammosphera sp.	3	2	5	7																				
Rosalina sp.														1	1									
Trochammina sp.						1	5													1				
Total	13	11	22	32		13	764	360	121	127	393	184	970	1181	6565	1255	1064	268	1352	939	585	449	543	1026

RELATIONSHIPS BETWEEN FORAMINIFERAL FAUNA AND MAIN STRESSORS IN THE MONTEVIDEO COASTAL ZONE

Studies carried out by Burone et al. (2004) and Burone et al. (2006) in the Montevideo coastal zone to study the foraminifera responses to polluted sediments allowed the identification of a total of 18 species and 18,341 individuals of benthic foraminifera at the 24 stations sampled (see Figure 4 and Table 6) belonging to the suborders Rotaliina (13 species) and Textulariina (five species). According to the results of these works, three different sub-environments based on foraminiferal assemblages and population parameters can be distinguished (Figures 11, 12 and 13).

In one hand, Montevideo Bay, the inner portion in particular, showed an extremely poor foraminiferal fauna, including a totally azoic station, evidencing the high degree of local contamination. *Psammosphera* sp. and *Buliminella elegantissima* were the representative spices recorded in this region, however, other species like *Ammonia tepida* and *Elphidium excavatum* were also observed but in a very low density. In addition, high percentages of abnormal test were observed, being that most of them were classified as complex (Figure 14). Intermediate values of diversity (between 0.760 and 1.337) and high evenness, with values approaching 1 were registered. This sub-environment is under stressed environmental conditions due to the different pollutants that come from different sources and strongly affect this zone, such as the high concentration of organic matter and heavy metals (Muniz et al., 2002, 2004a, b).

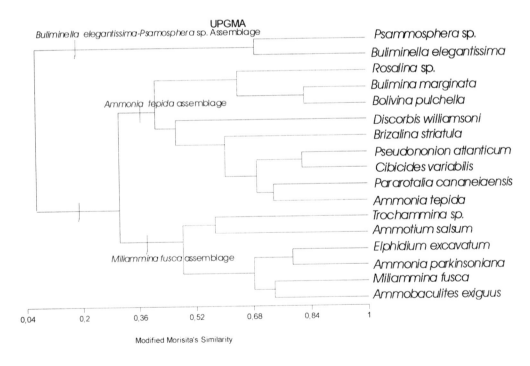

Figure 12.

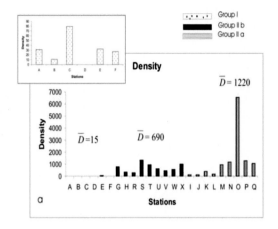
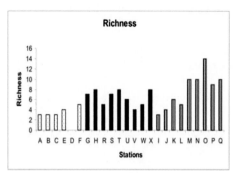
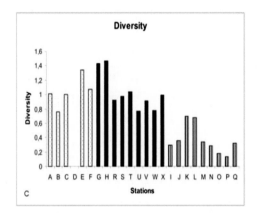
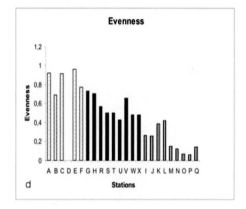

Figure 13.

On the other hand, Punta Carretas region is characterised by *Ammonia tepida* assemblage that consisted of nine hyaline species (Figure 12) with *A. tepida* represented 99.1 % of the total assemblage. This sub-environment showed the highest mean density (D = 1220) and the diversity was low ranged between 0.134 and 0.918. The species richness showed the highest values of the studied area with a maximum of 14 species. A positive effect on density of foraminifera, especially on *A. tepida* in this local was noted, which seems to be related to the discharges of the major sewage pipe of Uruguay that lies in this region, therefore, with a more pure organic contamination.

The third sub environment identified corresponding to Punta Yeguas zone was represented by *Miliammina fusca* assemblage (73.3 %) that was constituted by six species (two hyaline and four agglutinated). This zone showed values of pH and Cr slightly lower than those recorded in Punta Carretas zone (Burone et al., 2006), presented intermediate values of mean density (D = 600), as well as intermediate species richness and evenness, which ranged between 4 and 8, and from 0.428 to 0.734, respectively.

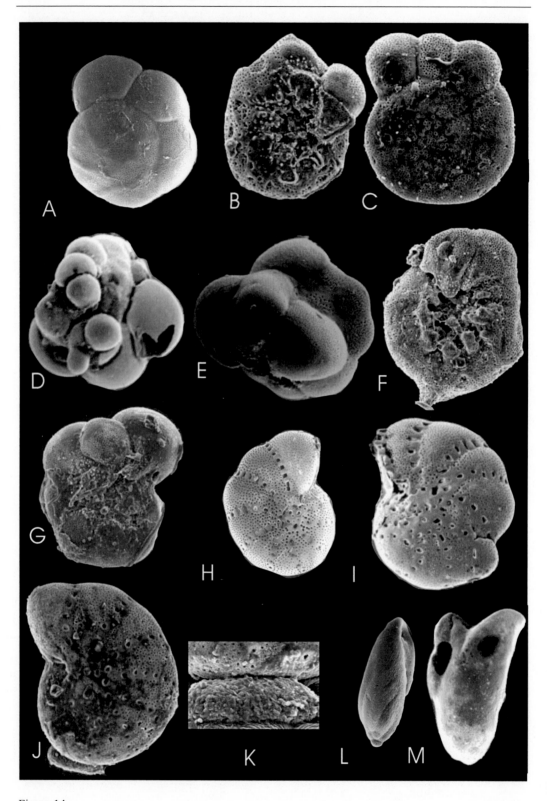

Figure 14.

A high number of agglutinates species was recorded in the stations of Punta Yeguas, while calcareous species dominated in Punta Carretas zone. This pattern can be related to the great marine influence on the latter region as indicated by the presence of *Pararotalia cananeiaensis*. This is a small marine species easy to be transported by currents and used as an indicator of marine influence (Debenay et al., 2001). Moreover, the presence of *P. cananeiaensis* in this area can be the response to the entrance of more saline waters, considering that the mean upstream limit of the saline intrusion is located just at the transverse section of Punta Yeguas (Nagy et al., 2002).

Out of the 18 species found in the study area, three of them exhibited morphological abnormalities (*A. tepida*, *B. elegantissima* and *Elphidium excavatum*). Basically, all the hyaline specimens observed in the inner portion of MB showed at least one type of abnormality. Besides, more than 58% of the hyaline specimens showed morphological deformities in this region. When it is consider the total population at each sampling station, this number reaches 72.7 %. Morphological abnormalities in this zone were manifested as protuberances of one chamber (Figure 14 F), aberrant chamber shape and size (Figure 14 B and I), additional chambers (Figures 14 J and K), overdeveloped chambers (Figure 14 C) double apertures (Figure 14 M) and Siamese twins (Figure 14 E). Some specimens accumulate more than one type of deformation and were named as complex deformities (Figure 14 D). In the outermost regions (Punta Carreta and Punta Yeguas zones), the percentages of deformed specimens were much lower (between 0.03 % and 0.08%). In these zones, the morphological deformities were manifested basically by aberrant chamber size (Figure 14 B).

A strong relationship between organic matter, oxygen and heavy metal concentrations, as well as redox potencial and pH values with the mean density of each sub-environment was detected. The diverse pollution sources and the complex mixture of different contaminants in the sediments make difficult to identify the effect of a single stressor on benthic foraminifera, even more, in high variable environments such as estuaries. Based in the results obtained it was concluded that such an extreme harmful condition in the Montevideo Bay is a consequence of the combined action of all polluted factors presented in the bay. Moreover, the high percentages of abnormal tests in Montevideo Bay seem to be related with the high contamination level.

Summarising, through the integration of abiotic and biotic data, it is possible to classify the coastal area of Uruguay near Montevideo, in at least three zones with different environmental quality and degree of anthropogenic impact (Figure 15). The first one corresponds to the inner portion of Montevideo Bay and includes the Montevideo Harbour. This zone is the most impacted by heavy metals and oil derived hydrocarbons with several compounds present in concentrations potentially harmful for benthic organisms, has the highest organic load and sometimes showed evidences of lack of oxygen. As a result of these conditions, benthic diversity was very poor and species richness very low with the dominance of only one opportunistic species and nematodes, and also with high percentages of abnormal foraminifera. The second zone corresponds to the outer portion of Montevideo Bay, which is more heterogeneous. There, environmental conditions seem to be more favorable for the establishment and development of benthic organisms, due to higher hydrodynamic energy, oxygenation of the sediments, lower organic matter and contaminant concentrations than in the inner part. The greater abundance, biomass and diversity of benthic organisms recorded confirm this trend. In a global scale, the less impacted is the adjacent coastal zone of

Montevideo, as was shown by the "Phylum-level Meta-analyses" approach. However, according to the AMBI results we can distinguish between Punta Yeguas and Punta Carretas, being that the former can be classified as slightly disturbed and the latter as moderately disturbed, probably, as a consequence of the sewage input through the submarine pipe. Is relevant to remark that despite a clear salinity gradient exists from the inner stations of Montevideo Bay to the outer coastal stations; through the utilization of different approaches, the occurrence of an environmental quality gradient (with the same direction) was effectively established.

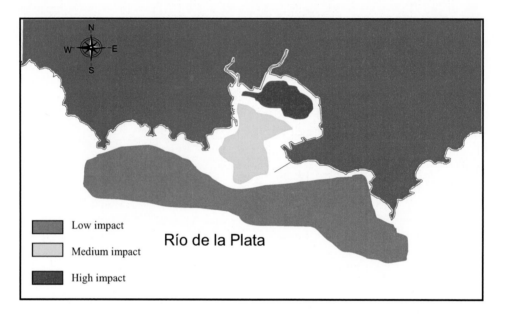

Figure 15.

BIOLOGICAL CONTAMINATION BY INVASIVE EXOTIC SPECIES

The introduction of species into a new ecosystem represents an economic and environmental serious risk. Under favourable environmental conditions, without predators, parasites and/or natural competitors these new species can reach high densities and are difficult to eliminate once established (Carlton, 1989). After their introduction into the aquatic environment invasive species may promote several changes. They can, alter the local hydrological regime, produce loss of biodiversity with the elimination of native species, changes in the trophic web, habitat modification, and also cause negative economic impacts to human populations (de Poorter 1999; Darrigran, 2002; Silva et al., 2004). In the last two centuries, introduced species have caused considerable changes in the biogeography of coastal areas, with an important decline in local biodiversity (Raffaelli & Hawkins, 1997).

Table 7. Aquatic exotic species reported in the coastal zone of Uruguay
(after Orensanz et al 2002, and Brugnoli et al. in press)

Species	Biogeographic Origen
Artrophoda	
Crustacea, Amphipoda	
Monocorophium insidiosum	N Atlantic
Crustacea, Cirripedia	
Amphibalanus amphitrite	Cosmopolitan
Crustacea, Isopoda	
Lygia exotica	Cosmopolitan
Synidotea laevidorsalis	Japan y China
Annelida	
Polychaeta	
Boccardiella ligerica	W Europe
Ficopomatus enigmaticus	Cosmopolitan
Mollusca	
Bivalvia	
Corbicula fluminea	SE Asia
Corbicula largillierti	SE Asia
Limnoperna fortunei	SE Asia
Gastropoda	
Myosotella myosotis	Europe
Rapana venosa	Japan
Chordata	
Ascidiacea	
Styela plicata	Asia
Piscies	
Cyprinus carpio	Europe-Asia
C.carpio var. especularis	

Exotic species are aloctone organisms that can be considered as biological contamination (Ricciardi & Atkinson, 2004). According to the introduction manner they can be divided in intentional and accidental. Important vectors are intercontinental shipping and the commercial transport of aquaculture and aquarium products (Carlton, 1996; Ruiz et al., 2000; Naylor et al., 2001; Semmens et al., 2004). The intentional introduction includes aquaculture and aquarium human activities, while fouling and ballast waters are the principal vector for the accidental introductions. Shipping ballast water is known as the highest risk vector for international introductions, causing a coastal community homogenization. At any given moment it is estimated that 10,000 different species are being transported between biogeographic regions in ballast tanks alone (Carlton, 1999).

As in other areas of the world, the South-Western Atlantic contains many exotic aquatic species (Schwindt, 2001; Orensanz et al., 2002; Silva & Souza, 2004) and Uruguayan coastal ecosystems are not an exception (Maytia & Scarabino, 1979; Nión et al., 1979; Scarabino & Verde, 1995; Amestoy et al., 1998; Brugnoli, et al., 2005, Muniz et al., 2005b; Brugnoli et al., in press). According to Brugnoli et al. (in press), Uruguay has 12 exotic species (Table 7) accidentally introduced by ballast water. These benthic species have in common an estuarine or marine distribution in the Río de la Plata or Atlantic coastal ecosystems that is related with their salinity tolerance (Brugnoli et al., 2005; Muniz et al., 2005b; Brugnoli et al. in press). There are reports concerning this kind of organisms in Uruguay from the beginning of this century, however, this ecological problem and its perception by the society has been increased during the last years (Masello & Menafra 1996; Orensanz et al., 2002; Muniz et al., 2005b; Brugnoli et al., in press). Some studies about three exotic organisms that occur in the coastal zone of Uruguay are following described.

Annelida, Polychaeta, Serpulidae

Ficopomatus enigmaticus (Fauvel, 1923) is an exotic reef-building polychaete distributed in most brackish waters in temperate zones throughout the world (Ten Hove & Weerdenburg, 1978) and originates in Australia (Allen, 1953). *Ficopomatus* have different sexes, external fecundation and trochophore larvae that once settled in hard substrata produces a calcareous tube that is secreted by the collar glands (Obenat & Pezzani, 1994; Obenat, 2001). Tube size and shape varies according to environmental variables, but in general the reefs are circular structures that can reach 2.5 m in diameter and grow with an estimated rate of 1.6 cm/month (Schwindt et al., 2004a). Calcareous reefs can produce large extensions in shallow water and low energy environments. These reefs can act as efficient traps for sediments, generating topographic heterogeneity that promotes changes in the abundance and distribution of the benthic associated communities (Schwindt & Iribarne, 2000). As a species producing habitat creation and modification, it is known as an ''ecosystem engineer'' (Schwindt et al., 2004b).

In Uruguay, *F. enigmaticus* has been recorded in Las Brujas stream (Monro, 1938), the outlet of Coronilla, El Bagre, Pando and Pantanoso Streams (Scarabino et al., 1975, Brugnoli et al., in press and refers), Montevideo Bay (Muniz et al., 2005b), Castillos, Garzón and José Ignacio coastal lagoons (Nión, 1979; Orensanz et al., 2002; Brugnoli et al., in press), in the most inner region of the Solís Grande Stream estuary (Muniz & Venturini, 2001) (Figure 16). More recently, in the innermost region of Rocha coastal lagoon, calcareous tubes were recorded (Borthagaray et al., 2006). In the Solís Grande Stream estuary the reefs are of small size (approx. 20 cm) and are generally attached to dead and live valves of the native mollusc *Tagelus plebeius* (Muniz & Venturini, 2001). The species has a negative effect, making the benthic community health classified as heavily perturbed (Muniz et al., 2005a). In the Rocha Lagoon, tubes of this species were found only in the shallow innermost region, where the concentration of suspended matter is high and the current speed low (Conde et al., 1999; Borthagaray et al., 2006). Borthagaray et al. (2006) developed a prediction analysis of the potential invasion rate of *Ficopomatus* in Rocha coastal lagoon, identifying ecological characteristics of different coastal ecosystems in Uruguay that promote it presence and colonisation. This species also causes negative economic impacts in Uruguay (Brugnoli et al., in press). Recently, large tubes (twice in size to those found in natural environments) were recorded obstructing the cooling system of the ANCAP oil refinery situated in the Montevideo Bay (Muniz et al., 2005b).

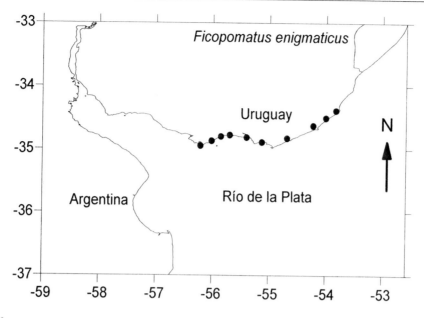

Figure 16.

Mollusca, Bivalvia, Mytilidae

Limnoperna fortunei Dunker 1857, also known as golden mussel, is a mytilid invasive species of the Río de la Plata Basin that is native from freshwater systems of China, Southeast Asia. The golden mussel was first found in South America in the coast of the Río de la Plata, Buenos Aires province (Pastorino et al., 1993). It was introduced accidentally in the region in 1991 with ballast waters (Darrigran & Pastorino, 1995). It has an epifaunal aggregate behaviour (Cataldo & Boltovskoy, 2000; Darrigran & Ezcurra de Drago, 2000) and occurs in fresh and brackish water systems, until a salinity of 3 (Darrigran, 2002). It is a dioeciously species with external fecundity and a swimming larvae phase (Cataldo & Boltovskoy, 2000). Since its arrival to the Río de la Plata Basin, this species had been found in several kinds of hard substrates, natural or artificial, showing an increase in its population abundance and changing the benthic community composition. It has been recorded in the main hydrologic freshwater systems of the region (Paraguay, Paraná, Salado and Uruguay rivers) (Darrigran & Ezcurra de Drago, 2000; Darrigran, 2002; Brugnoli et al., 2005), Los Patos Lagoon (Mansur et al., 1999, 2003) and in coastal areas of the Río de la Plata (Scarabino & Verde, 1995; Darrigran et al., 1998; Brugnoli et al., 2005).

In Uruguay, it has been recorded in five of the six main hydrographical basins: Río de la Plata, Santa Lucía, Negro and Uruguay rivers (Scarabino & Verde, 1995; Brugnoli et al., 2005) and recently in the Merín Lagoon (Langone, 2005) (Figure 17). According to Brugnoli et al. (2005) and Dabezies et al. (2005) salinity could be the most important environmental variable in determining its distributional limit in the Uruguayan costal zone. Although there are any records of this mussel in the Atlantic Basin, Brugnoli et al. (2005) suggested that it could access this basin through the San Gonzalo Channel that connects Los Patos and Merín lagoon systems. After their settlement larvae would be disperse by different vectors (ships or birds) to other aquatic ecosystems in the Atlantic Basin. Langone (2005) recently reported the occurrence of the golden mussel in the Merín Lagoon and Río Branco River and mentioned its potential dispersion by tourism boats.

The golden mussel had also been reported as a new item in the diet of native fishes in the Río de la Plata Basin (López-Armengol & Casciotta, 1998; Darrigran & Ezcurra de Drago, 2000; Montalto et al., 1999; Penchaszadeh et al., 2000), having a potential impact on the native freshwater malacofauna of Uruguay with special reference to endemic species (Scarabino et al., 2004) and causing problems of macrofouling in hydraulic installations (Clemente & Brugnoli, 2002; Mansur et al., 2003; Muniz et al., 2005b; Brugnoli et al., in press).

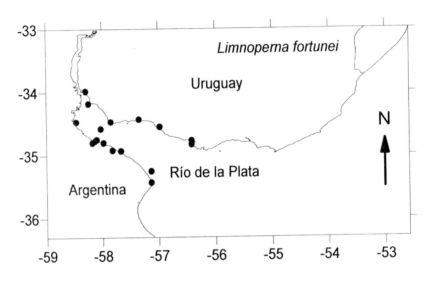

Figure 17.

Mollusca, Gastropoda, Muricidae

Rapana venosa (Valenciennes, 1846) (rapa whelk) is a predatory mollusc native of the Sea of Japan, Yellow Sea, Bohai Sea, and from the East China Sea to Taiwan (ICES, 2004). Nowadays, it has a distribution out of its native biogeographic range, being that ballast water transport was claimed as the main vector for the introduction of this species (ICES, 2004). This species was discovered in the Black Sea in 1947 (Drapkin, 1953), and has subsequently spread throughout the Sea of Azov, the Adriatic Sea (since 1973) (Ghisotti, 1974), the Aegean Sea (since 1990) (Koutsoubas & Voultsiadou-Koukoura, 1999) and recently in America (since 1998) (Mann & Harding, 2000; Pastorino et al., 2000). The first record of *R. venosa* in America was made in the Chesapeake Bay on the East Coast of the United States in 1998 (Harding & Mann, 1999). In 1999 was found in the Río de la Plata Estuary at the Samborombón Bay (Argentina) (Pastorino et al., 2000). Furthermore, in the Uruguayan coastal zone was detected in Maldonado Department (intertidal zone) (Scarabino et al., 1999), in the subtidal Río de la Plata area (Rodríguez-Capítulo et al., 2003), and recently colonising successfully the outer estuarine Río de la Plata area (Carranza et al., in press) (Figure 18). This gastropod is a successful invader of marine coastal/brackish ecosystems, being tolerant to wide variations in temperature, salinity and oxygen concentration (Chung et al., 1993;

Mann & Harding, 2003). Their high fertility (Chung et al., 1993), dispersion assisted by a planktonic larvae (Mann & Harding, 2003), and fast growth (Harding & Mann, 1999) make *R. venosa* a potentially successful invader worldwide (Savini et al., 2004). This species is a broad predator of subtidal Mollusca, it usually feeds on bivalves of economic interest like oysters, mussels and clams (Harding & Mann, 1999; Savini et al., 2002; Savini & Occhipinti-Ambriogi, 2006), and has been identified as the main reason for the collapse of several banks of mussels and oysters in the Black Sea (ICES, 2004).

This species colonised the Río de la Plata Estuary (Pastorino et al., 2000) and is successfully breeding in this area (Carranza et al., in press). There, it was reported in a salinity range of 14.93 and 28.24, between temperatures from 18.12 to 22.04 °C and in a depth range of 4-12 m (Carranza et al., in press). Furthermore, recent unpublished data suggests the potential predation of rapa whelk on autochthonous species of the Río de la Plata such as *Mactra isabelleana* and *Ostrea puelchana* based on their coexistence with *Rapana venosa* in this area.

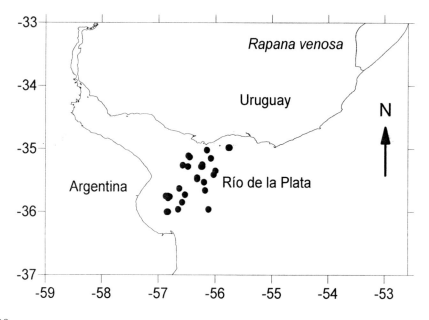

Figure 18.

Conclusion

Although the available information about marine biodiversity and environmental perturbation in the Uruguayan coastal zone was improved during the last decades, it is still restricted to isolated areas and to some aspects of aquatic ecosystems only. The increase in the number of studies concerning benthic processes and faunal community patterns (including macro, meio and microfauna), principally in subtidal coastal zones, the continental shelf, slope and abyssal plains, compartments were they are almost non existent, is of primary importance. Thus, the implementation and development of integrative baseline studies on these topics are highly relevant, in order to contribute to the conservation of benthic biodiversity in the coastal zone of Uruguay.

As invader species are often main factors in ecological degradation, the establishment of national and international research programmes to minimise the impacts of exotic species, and to develop models that predict the introduction of new invaders is crucial. It is clear that once established in a new environment, the eradication of exotic species is very difficult. With the knowledge of the distribution, ecology, life history and impacts of alien species, both in a national and regional scale, it would be possible to improve management and control. Since shipping traffic appears to be the most important introduction vector, we emphasise the need for effective controls on ballast water discharges in the harbours of the region. The prevention of such a problem is always easier, less expensive and more effective than its eradication.

ACKNOWLEDGMENTS

These studies were partially funded by the Administración Nacional de Puertos (ANP) Uruguay, Comisión Sectorial de Investigación Científica (CSIC) and Intendencia Municipal de Montevideo (IMM) of uruguay. Also thanks to IAI-SACC and INCOFISH projects. The Grupo de Buceo de la Armada Nacional (GRUBU) helped with the sampling. Special thanks are also due to our numerous field collaborators from the Facultad de Ciencias, Montevideo, the Dirección Nacional de Tecnología Nuclear (DINATEN) and the Dirección Nacional de Recursos Acuáticos (DINARA). The authors are grateful for the kindly invitation of Dr. Columbus to participate as collaborators in the book.

REFERENCES

Agard, JBR; Gobin, J; Warwick RM. Analysis of marine macrobenthic community structure in relation to pollution, natural oil seepage and seasonal disturbance in a tropical environment (Trinidad, West Indies). *Marine Ecology Progress Series,* 1993 92: 233-243.

Ahumada, R; Contreras S. Contenido de metales (Ba, Cd, Co, Cr, Cu, Ni, Pb, Sr, V y Zn) en sedimentos de los fiordos y canales adyacentes a Campos de Hielo Sur. *Ciencia y Tecnología Marina,* 1999 22: 47-58.

Allen, FE. Distribution of marine invertebrates by ships. *Australian Journal of Marine and Freshwater Research,* 1953 4, 307-316.

Amaral, ACZ; Morgado, EH; Salvador, LB. Poliquetas indicadores de poluição orgânica em praias paulistas. *Revista Brasileira de Biologia,* 1998 58: 307-316.

Amestoy, F; Spinetti, M; Fabiano, G. Aquatic species introduced in Uruguay. *Verhandlungen International Vere in Limnology,* 1988 26: 2170-2173.

Baisch, PRN; Nienchéski, LFH; Lacerda, LD. Trace metal distribution in sediments of the Patos Lagoon estuary, Brazil. In: *Metals in Coastal Environments of Latin America,* Seeliger, U.; Lacerda, L.D.; Patchineelam, S.R. (eds.). Springer Verlag, Berlin, 1988.

Balech, E. División zoogeográfica del litoral sudamericano. Revista Biología Marina, 1954 4: 184-195.

Balech, E. Caracteres biogeográficos de la Argentina y Uruguay. *Boletín del Instituto Biología Marina* 1964 7: 107-112.

Bastidas, C; Boné, D; García, EM. Sedimentation rates and metal content of sediments in a Venezuelan Coral Reef. *Marine Pollution Bulletin* 1999 38: 16-24.

Baptista Neto, JA; Smith, BJ; McAllister, JJ. Heavy metal concentrations in surface sediments in a nearshore environment, Jurujuba Sound, Southeast Brazil. *Environmental Pollution*, 2000 109: 1-9.

Benvenuti, CE. Benthic invertebrates. In: *Subtropical convergence marine ecosystem. The coast and the sea in the warm temperate southwestern Atlantic*, Seeliger, U; C. Odebrecht and J.P. Castello (eds.), pp. 43-46. New York: Springer Verlag, Heidelberg, 1997.

Blackburn, TM; Gaston, KJ. Spatial patterns in the species richness of birds in the *New World. Ecography,* 1996 19: 369-376.

Boltovskoy, E. Masas de agua (características, distribución, movimiento) en la superficie del Atlántico sudoeste según indicadores biológico-foraminíferos. Servicio Hidrografía naval, Buenos Aires, 1970 (H643), 99p.

Boltovskoy, E; Wright, R. *Recent Foraminífera.* Junk, The Hague, 1976.

Boltovskoy, E; Lena, H. Foraminíferos del Río de la Plata: Argentina, Servicio Hidrografía Naval, Argentina, 1974 660:1- 30.

Boltovskoy, D; Gibbons, MJ; Hutchings, L; Binet, D. General biological features of South Atlantic. South Atlantic. In: *Zooplankton.* Boltovskoy, D. (ed.), 1-42 p. Backhyus Publishers, Leiden, The Netherlands, 1999.

Borja, A; Franco, J; Pérez, V. A marine biotic index to establish the ecological quality of soft-bottom benthos within European estuarine and coastal environments. *Marine Pollution Bulletin,* 2000 40: 1100-1114.

Borja, A; Muxika, I; Franco, F. The application of a marine Biotic Index to different impact sources affecting soft-bottom benthic communities along European coasts. *Marine Pollution Bulletin*, 2003 46: 835-845.

Borja, A; Franco, J; Muxika, I. The Biotic Indices and the Water Framework Directive: the required consensus in the new benthic monitoring tools. *Marine Pollution Bulletin,* 2004 48: 405-408.

Borja, A; Muxica, I; Franco, J. Long-term recovery of soft-bottom benthos following urban and industrial sewage treatment in the Nervión estuary (southern Bay of Biscay). *Marine Ecology Progress Series,* 2006 313: 43-55.

Borthagaray, A; Clemente, JM; Boccardi, L; Brugnoli, E; Muniz P. Impacto potencial de invasión de *Ficopomatus enigmaticus* (Fauvel) (Polychaeta: Serpulidae) en la Laguna de Rocha, Uruguay. *Pan-American Journal of Aquatic Sciences,* 2006 1: 57-65.

Boschi, EE. Nuevos aportes al conocimiento de la distribución geográfica de los crustáceos decápodos del Mar Argentino. *Physis, Buenos Aires, 1976* 35A: 59-68.

Boschi, EE. El ecosistema estuarial del Río de la Plata (Argentina y Uruguay). Anuario del Instituto de Ciencias del Mar y Limnología Universidad Nacional Autónoma de México, 1988 15: 159-182.

Boschi, EE. Species of decapod crustaceans and their distribution in the American marine zoogeographic provinces. Revista de Investigación y Desarrollo Pesquero (Chile), 2000 13: 7-136.

Brown, JH; Lomolino, MV. Biogeography, 2nd edn,. Sinauer Associates, Sunderland, MA, 1998, p. 691.

Brown, CJ; Cooper, KM; Meadows, WJ; Limpenny, DS; Rees, EIS. Small-scale mapping of sea-bed assemblages in the eastern English Channel using sidescan sonar and remote sensing sampling techniques. Estuarine, *Coastal and Shelf Science,* 2002 54: 263-274.

Brugnoli, E; Clemente, J; Boccardi, L; Borthagaray, A; Scarabino, F. Update and prediction of golden mussel (*Limnoperna fortunei*): distribution in the principal hydrographic basin of Uruguay. *Anais da Academia Brasileira de Ciências,* 2005 77: 235-244.

Brugnoli, E; Clemente, J; Riestra, G; Boccardi, L; Borthagaray, A. Especies acuáticas exóticas en Uruguay: situación, problemática y gestión. In: Menafra, R; Rodríguez, L; Scarabino, F; Conde, D (eds.). Bases para la conservación y manejo de la costa uruguaya. Montevideo. (In press).

Burone, L; Muniz, P, Venturini, N; Sprechmann, P. Estúdio de la fauna de foraminíferos bentônicos como herramienta en la interpretación de los disturbios antropogénicos de la zona costera del departamento de Montevideo – Uruguay. In: II Simpósio Brasileiro de Oceanografia, 2004. São Paulo, Brasil.

Burone, L; Venturini, N; Sprechmann, P; Valente, P; Muniz, P. Foraminiferal responses to polluted sediments in the Montevideo coastal zone, Uruguay. *Marine Pollution Bulletin,* 2006 52: 61-73.

Burone, L; Pires-Vanin, AMS. Foraminiferal assemblages in the Ubatuba Bay, Southeastern Brazilian coast. *Scientia Marina,* 2006 70: 203-217.

Calliari, D; Cervetto, G; Gómez, M. Short-term variability in abundance and vertical distribution of the opossum shrimp *Neomysis americana* in the Solis Grande river estuary, Uruguay. Atlantica, Rio Grande, 2001 23: 117-125.

Calliari, D; Defeo, O; Cervetto, G; Gómez, M; Giménez, L; Scarabino, F; Brazeiro, A; Norbis, W. Marine life of Uruguay: critical update and priorities for future research. *Gayana,* 2003 67: 341-370.

Cardellino, R; Ferrando L. Carta geológica del Uruguay, segundo segmento (Montevideo). Sector XCVII. Universidad de la República Depto. de Publicaciones Col Ciencias Serie Especial, Montevideo, Uruguay, 1969 82pp.

Carlton, JT. Man´s role in changing the face of the ocean: biological invasions and implications for conservation of nearshore environments. *Conservation Biology*, 1989 3: 265-273.

Carlton JT. Pattern, process, and prediction in marine invasion ecology. *Biological Conservation,* 1996 78: 97-106.

Carlton, JT. Molluscan invasions in marine and estuarine communities. *Malacologia,* 1999 41: 439-454.

CARP (Comisión Administradora del Río de la Plata). Estudio para la Evaluación de la Contaminación en el Río de la Plata. Informe de Avance, Montevideo, 1990, 422 pp.

Carranza, A; Scarabino, F; Ortega L. Distribution of Large Benthic Gastropods in the Uruguayan Continental Shelf and Río de la Plata Estuary. *Journal of Coastal Research* (in press).

Carreón-Martínez, LB; Huerta-Díaz, MA; Nava-López, C; Siqueriros-Valencias A. Mercury and silver concentrations in sediments from the Port of Ensenada, Baja California, Mexico. *Marine Pollution Bulletin,* 2001 42: 415-418.

Cataldo, DH; Boltovskoy, D. Yearly reproductive activity of *Limnoperna fortunei* (Bivalvia) as inferred from the occurrence of its larvae in the plankton of the lower Paraná river and the Río de la Plata estuary (Argentina). *Aquatic Ecology,* 2000 34: 307-317.

Chung, EY; Kim, SY; Kim, YG. Reproductive ecology of the purple shell *Rapana venosa* (Gastropoda: Muricidae), with special reference to the reproductive cycle, depositions of egg capsules and hatchings of larvae. *Korean Journal of Malacology,* 1993 9: 1-15.

Clemente, J; Brugnoli, E. First record of *Limnoperna fortunei* (Dunker 1857) (Bivalvia: Mytilidae) in continental waters of Uruguay. *Boletín Sociedad Zoológica,* 2002 29-33.

Conde, D; Bonilla, S; Aubriot, L; de León, R; Pintos, W. Comparison of the areal amount of chlorophyll a of planktonic and attached microalgae in a shallow coastal lagoon. *Hydrobiologia,* 1999 408/409: 285-291.

Constanza, R; D'arge, R; Degroot, R; Farber, S; Grasso, M; Hannon, B; Limburg, K; Aeem, SN; O'neill, RV; Paruelo, J; Raskin, RG; Sutton, P; Van Den Belt, M. The value of the world's ecosystem services and natural capital, *Nature,* 1997 387: 253-260.

Cranston, RE; Masello, AM; Kurucz, AP. Comparing anthropogenic metal distribution in three urban regions: Montevideo (Uruguay), Halifax and Vancouver (Canada). *EcoPlata II Scientific Papers,* 2002 1:1-8.

Currie, DJ. Energy and large scale patterns of animal and plant species richness. *American Naturalist,* 1991 137: 27-49.

Dabezies, M; Gómez-Erache, M; Sans K; Brugnoli, E. Distribución de larvas planctónicas de la especie invasora *Limnoperna fortunei* en el estuario del Río de la Plata. Proceedings of the 8[vas] Jornadas de Zoología de Uruguay, II Encuentro de Ecología de Uruguay, Montevideo, 2005.

Dana, JD. On an isothermal oceanic chart, illustrating the geographical distribution of marine animal. American Journal Sciences. *Articles series 2,* 1853 16: 153-167.

Danulat; E; Muniz, P; García-Alonso, J; Yannicelli B. First assessment of the highly contaminated harbour of Montevideo, Uruguay. *Marine Pollution Bulletin,* 2002 44: 551-576.

Darrigran, G. Macroinvertebrates associated with *Limnoperna fortunei* (Dunker, 1857) (Bivalvia, Mytilidae) in Río de la Plata, *Argentina. Hydrobiologia,* 1998 367: 223–230.

Darrigran, G. Potential impact of filter-feeding invaders on temperate inland freshwater environments. *Biological Invasions,* 2002 4: 145-156.

Darrigran, G; Pastorino, G. The recent introduction of asiatic bivalve, *Limnoperna fortunei* (Mytilidae) in to South America. *The Veliger,* 1995 38: 183-187.

Darrigran, G; Ezcurra de Drago, I. Invasion of the exotic freshwater mussel *Limnoperna fortunei* (Dunker 1857) (Bivalvia: Mytilidae) in South America. *The Nautilus,* 2000 114: 69-73.

Dauer, DM. Biological criteria, environmental health and estuarine macrobenthic community structure. *Marine Pollution Bulletin,* 1993 26: 249-257.

Dauer, DM; Alden, RV. Long-term in the macrobenthos and water quality of the Lower Chesapeake bay. *Marine Pollution Bulletin,* 1995 30: 840-850.

Dauer, DM; Conner, WG. Effects of moderate sewage input on benthic polychaete populations. *Estuarine and Marine Sciences,* 1980 10: 335-346.

Day, JW; Hall, CAS;Kemp, JM; Yañez-Arancibia, A. Estuarine ecology. Wiley & Sons publications, 1989, 558 pp.

de Poorter, M. Borrador de Guías para la prevención de pérdidasde diversidad biológica ocasionadas por invasion biológica. Cuarta Reunión del Órgano Subsidiario de Asesoramiento Científico, Técnico y Tecnológico. Documento de Base. Unión Internacional para laConservación de la Naturaleza (UICN). 1999, 15 pp.

Debeney, J. P; Duleba, W; Bonetti, C; Melo e Souza, SH; Eichler, B. Pararotalia cananeiaensis N. SP.: indicator of marine influence and water circulation in Brazilian coastal and paralaic environments. *Journal of Foraminiferal Research,* 2001 31: 152-163.

Defeo, O; de Alava, A. Effects of human activities on long-term trends in sandy beach populations: the wedge clam *Donax hanleyanus* in Uruguay. *Marine Ecology Progress Series,* 1995 123: 73-82.

Defeo, O; de Alava, A. Impacto humano en la biodiversidad de playas arenosas. Un estudio de caso. In: Manejo de zonas costeras y su impacto en la biodiversidad: estudios de caso. Moreno-Casasola P., M.L. Martínez & O. Defeo (eds.). Red Iberoamericana de Biodiversidad de Ecosistemas Costeros CYTED-Instituto de Ecología, Xalapa, Veracruz, 2002: pp. 169-179.

Defeo, O; McLachlan, A. Patterns, processes and regulatory mechanisms in sandy beach macrofauna: a multi-scale analysis. *Marine Ecology Progress Series*, 2005 295: 1-20.

Defeo, O; Brazeiro, A; Riestra, G. Impacto de la descarga de un canal artificial en la biodiversidad de gasterópodos en una playa de arena de la costa atlántica uruguaya. Comunicaciones de la Sociedad Malacológica del Uruguay, 1996 8: 13-18.

D'Orbigny, AD. 1835-43. Voyage dans l'Ameriwue Méridionale execute pendant les annés 1826, 1827, 1828, 1829, 1830, 1831, 1832 et 1833. Tome V. partie 3. Mollusques. Paris et Strasbourg, 1845-55. Mollusques viviant et fossils. Paris. 1853. histoire physique, politique et naturalle de l' Ile de Cuba par M. Ramón de la Sagra. Mollusques para A. D. d'Orbigny. Paris.

Drake, P; Baldó, F; Sáenz, V; Arias, AM. Macrobenthic Community Structure in Estuarine Pollution Assessment on the Gulf of Cádiz (SW Spain): is the Phylum-level Meta-analysis Approach Applicable?. *Marine Pollution Bulletin,* 1999 38: 1038-1047.

Drapkin, E. Novii mollusc V. Cemom more. Prirada, 1953 8: 92-95.

Engle, VD; Summers, JK; Gaston, GR Benthic indices of environmental condition of Gulf of Mexico estuaries. *Estuarios*, 1994 17: 372-389.

Escofet, A; Gianuca, N; Maytía, S; Scaraino, V. Playas arenosas del Atlántico Sudoccidental entre los 298 y 438 LS.: consideraciones generales y esquema biocenológico. Memorias del Seminario sobre Ecología Bentónica y Sedimentación de la Plataforma Continental del Atlántico Sur, 1979 1, 245–258.

Fischer, AG. Latitudinal variations in organic diversity. *Evolution,* 1959 14: 64-81.

Gaines, SD; Lubchenco, J. A united approach to marine plant herbivore interactions. II. Biogeography. *Annual Review in Ecology and Systematics,* 1982. 13: 111-138.

Ghisotti, F. *Rapana venosa* (Valenciennes) nuova ospite adiratica? Conchiglie Milano, 1974 10: 125-126.

Giberto, DA; Bremec, CS; Acha, EM; Mianzan, H. Large-scale spatial patterns of benthic assemblages in the SW Atlantic: the Río de la Plata estuary and adjacent shelf waters. Estuarine, *Coastal and Shelf Science,* 2004 61: 1-13.

Giménez, L; Borthagaray, A; Rodríguez, M; Brazeiro, A; Dimitriadis, K. Scale-dependent patterns of macrofaunal distribution in soft-sediment intertidal habitats along a large-scale estuarine gradient. *Helgoland Marine Research,* 2005 59: 224-236.

Giménez, L; Dimitriadis, C; Carranza, A; Borthagaray, AI; Rodríguez, M. Unravelling the complex structure of a benthic community: A multiscale-multianalytical approach to an estuarine sandflat. Estuarine, *Coastal and Shelf Science*, 2006 68: 462-472.

Grall, J; Glémarec, M. Using biotic indices to estimate macrobenthic community perturbations in the bay of Brest. Estuarine, *Coastal and Shelf Science,* 1997 44: 43-53.

Grasshoff, K. Determination of oxygen. In: *Methods of Seawater Analysis.* Grasshoff E & Kremling (eds.), 1983, pp 61-72. 2nd ed. Verlag Chemie, Weinheim.

Gray, JS. Pollution-induced changes in populations. Phil Transactions of the Royal Society of London B, 1979 286: 545-561.

Guerrero, RA; Acha, EM; Framiñan, MB; Lasta, CA. Physical oceanography of the Rio de la Plata estuary, Argentina. *Continental Shelf Research,* 1997 17: 727-742.

Hall, JA; Raffaelli, DG,;Thrush, SF. Patchiness and disturbance in shallow water benthic assemblages. In: Giller P Hildrew A raffaelli D (eds.*), Aquatic ecology: Scale, pattern and Process.* Blackwell Science, 1994, Oxford.

Harding, JM; Mann, R. Observations on the biology of the veneid rapa whelk, *Rapana venosa* (Valenciennes, 1846) in the Chesapeake Bay. *Journal of Shelfish Research,* 1999 18:9-17.

Heip, C. Benthic studies: summary and conclusions. *Marine Ecology Progress Series,* 1992.91: 265-269.

Heip, C. Eutrophication and zoobenthos dynamics. *Ophelia,* 1995 41: 113-136.

Heip, C; Basford, DJ; Craeymeersch, JA; Dewarumez, JM; Dorjes, J; de Wilde, PAJ; Duineveld, GCA; Eleftheriou, A; Herman, PMJ; Niemann, U; Kunitzer, A; Rachor, E; Rumohr, H; Soetaert, K; Soltwedel, T. Trends in biomass, density and diversity of North Sea macrofauna. *ICES Journal of Marine Science,* 1992 49: 13-22.

Hillebrand, H. Strenght, slope and variability of marine latitudinal gradients. *Marine Ecology Progress Series,* 2004 273: 251-276.

Hilly, C; Le Bris, H; Glémarec, M. Impacts biologiques des emissaries urbains sur les écosystèmes benthiques. *Oceanis,* 1986 12: 419-426.

Howarth, RW; Boyer, EW; Pabich, WJ; Galloway, JN. Nitrogen use in the United States from 1961-2000 and Potencial Future Trends. *A Journal of the Human Environment,* 2000 31: 88-96.

ICES. Alien Species Alert: *Rapana venosa* (veined whelk). Edited by Roger Mann, Anna Occhipinti, and Juliana M. Harding. ICES, 2004, Cooperative Research Report N° 264. 14 pp.

Ieno, EN; Bastida, RO. Spatial and temporal patterns in coastal macrobenthos of Samboronbon Bay, Argentina: a case study of very low diversity. *Estuaries,* 1998 21: 690-699.

Jørgensen, B. Material flux in the sediment. In: Jørgensen B and Richardson K. (eds.). Coastal and estuarine studies. *American Geophysical Union.* 1996, p: 115-135.

Koustsoubas, D; Voultsiadou-Koukoura, E. The occurrence of *Rapana venosa* (Valeciennes, 1846)(Gastropoda, Thaididae) in the Aegean Sea. *Bolletino Malacologico,* Milano, 1999 26: 201-204.

Lacerda, LD; Huertas, R; Moresco, HF; Carrasco, G; Viana, F; Lucas, R; Pessí, M. Trace metal concentrations and geochemical partitioning in arroyo Carrasco wetlands, Montevideo, Uruguay. *Geochimica Brasiliensis,* 1998 12: 63-74.

Lacerda, LD; Pfeiffer, WC; Fiszman, M. Níveis de metais pesados em sedimentos marinhos da Baía da Ribeira, Angra dos Reis. *Ciência e Cultura,* 1982 34: 921-924.

Langone, JA. Notas sobre el mejillón dorado *Limnoperna fortunei* (Dunker 1857)(Bivalvia, Mytilidae) en Uruguay. Publicación extra Museo Nacional de Historia Natural y Antropología, Montevideo, 2005 1:1-18.

Lalli, CM; Parsons, TR. Biological Oceanography. An introduction. Butterworth Heinemann, Oxford, 1997, 314 pp.

Lercari, D; Defeo, O. Effects of freshwater discharge in sandy beach populations: the mole crab *Emerita brasiliensis* in Uruguay. *Estuarine, Coastal and Shelf Science*, 1999 49: 457-468.

Lercari, D; Defeo, O. Variation of a sandy beach macrobenthic community along a human-induced environmental gradient. *Estuarine, Coastal and Shelf Science*, 2003 58: 17-24.

Lercari, D; Defeo, O. Large-scale diversity and abundance trends in sandy each macrofauna along full gradients of salinity and morphodynamics. Estuarine, Coastal and Shelf Science, 2006 68: 27-35.

Lercari, D; Defeo, O; Celentano, E. Consequences of a freshwater canal discharge on the benthic community and its habitat on an exposed sandy beach. *Marine Pollution Bulletin,* 2002 44: 1392-1399.

Loeblich, A R; Tappan, H. *Foraminiferal genera and their classification.* Van Nostrand Reinold, New York, 1988 vol. 1 e 2, 270p. + 212p, 847 pls.

López-Armengol, MF; Casciotta, J. First record of the predation of the introduced freshwater bivalve *Limnoperna fortunei* (Mytilidae) by the native fish Micropogonias (Scianidae) in the Río de la Plata estuary, South America. *Iberus,* 1998, 105-108.

López-Jamar, E. Distribución espacial del poliqueto *Spiochaetopterus costarum* en las Rías Bajas de Galicia y su posible utilización como indicador de contaminación orgánica en el sedimento. *Boletín del Instituto Español de Oceanografía,* 1985 2: 68-76.

López-Laborde, E; Nagy, G. Hydrography and sediment transport characteristics in the Río de la Plata. In: Perillo Piccolo & Pino-Quivira (eds.). Estuaries of the South America: their geomorphology and dynamics. Springer-Verlag, Berlin, 1999.

López-Laborde, J; Perdomo, A; Gómez-Erache, M. Diagnóstico Ambiental y Socio-Demográfico de la Zona costera Uruguaya del Río de la Plata. Compendio de los principales resultados. ECOPLATA. Montevideo, 2000, 177 pp.

MacDonald, DD; Carr, SR; Calder, FD; Long, ER; Ingersol, CG. Development and evaluation of sediment quality guidelines for Florida coastal waters. *Ecotoxicology,* 1996 5: 253-278.

Machado, W; Silva-Filho, EV; Oliveira, RR; Lacerda, LD. Trace metal retention in mangrove ecosystems in Guanabara Bay, SE Brazil. *Marine Pollution Bulletin*, 2002 44: 1277-1280.

Mann, R; Harding, JM. Salinity tolerance of larval *Rapana venosa*: implications for dispersal and establishment of an invading predatory gastropod on the North American Atlantic coast. *Biological Bulletin,* 2003 204: 96-103.

Mann, R; Harding, JM. Invasion of the North American Atlantic coast by a large predatory Asian mollusk. *Biological Invasions,* 2000 2: 7-22.

Mansur, MC; Valer, RM; Aires, N. *Limnoperna fortunei* (Dunker 1857) molusco bivalve invasor na bacia do Guaíba, Río Grande do Sul, Brasil. Biociencias, Porto Alegre, 1999 7: 147-149.

Mansur, MC; Santos, CP; Darrigran, G; Heydrich, I; Callil, CT; Cardoso, FR. Primeiros dados quali-quantitativos do "mexilhao dourado", Limnoperna fortunei (Dunker, 1857), no lago Guaíba, Bacia da laguna dos Patos, Rio Grande do Sul, Brasil e alguns aspectos de sua invasao no novo ambiente. Revista Brasileira de Zoologia, *Curitiba*, 2003 22: 75-84.

Masello, A; Menafra R. Macrobenthic community of the uruguayan coast zone and adjacent areas. In: EcoPlata Team (Eds.) *The Río de la Plata. An environmental overview.* Dalhousie University, Halifax, Nova Scotia, 1996 pp:113- 162.

Maytia, S; Scarabino, V. Las comunidades del litoral rocoso del Uruguay: zonación, distribución local y consideraciones biogeográficas. Memorias del Seminario sobre Ecología Bentónica y Sedimentación de la Plataforma Continental del Atlántico Sur. UNESCO, 1979, pp: 149-160.

Méndez, SM; Anciaux, F. Efectos en las características del agua costera provocados por la descarga del Canal Andreoni en la playa de La Coronilla (Rocha, Uruguay). Frente. Marítimo, 1995 8:101-107.

Méndez, N; Flos, J; Romero, J. Littoral soft-bottom polychaete communities in a pollution gradient in front of Barcelona (Western Mediterranean, Spain). *Bulletin of Marine Science*, 1998 63: 167-178.

Mianzán, H; Lasta, C; Achas, M; GUERRERO, R; Macchi, G; Bremec, C. The Río de la Plata estuary, Argentina-Uruguay: In: Seeliger, U. & B. Kjerfve (eds.) Coastal Marine ecosystems of Latin America, 2001. Springer-Verlag, Berlin. *Ecological Studies Series* 144: 185-204.

Mirlaen, N; Baraj, B; Niencheski, LF; Baisch, P; Robinson, D. Effect of accidental sulphuric acid on metal distributions in estuarine sediment of Patos Lagoon. *Marine Pollution Bulletin*, 2001 42: 1114-1117.

Montalto, L; Oliveros, O; Ezcurra de Drago, I; Demonte, l. Peces del río Paraná medio, predadores de una especie invasora *Limnoperna fortunei* (Bivalvia: Mytilidae). Revista FABICIB, 1999 3: 85-101.

Monro, CCA. On a small collection of Polychaeta from Uruguay. *Annals and Magazine of Nature History,* 1938 2: 311-314.

Moresco, H; Dol, I. Metales en sedimentos de la Bahía de Montevideo. *Revista de la Asociación de Ciencias Naturales del Litoral,* 1996 27: 1-5.

Morton, B. Freshwater fouling bivalves. Proceed. First International *Corbicula* Sympossium. Texas Univ, 1997 p: 1-14

Moyano, M; Moresco, H; Blanco, J; Rosadilla, M; Caballero, A. Baseline studies of coastal pollution by heavy metals, oil and PAHs in Montevideo. *Marine Pollution Bulletin,* 1993 26: 461-464.

Mucha, AP; Vasconcelos, MTDS; Bordalo, AA. Macrobenthic community in Douro estuary: relations with heavy metals and natural sediment characteristics. *Environmental Pollution,* 2002 121: 169-180.

Muniz, P; Pires, AMS. Polychaete associations in a subtropical environment (São Sebastião Channel, Brazil): a structural analysis. *Marine Ecology* PSZN, 2000 21: 145-160.

Muniz, P; Venturini, N. Spatial distribution of the macrozoobenthos in the Solís Grande Stream estuary (Canelones-Maldonado, Uruguay). *Brazilian Journal of Biology,* 2001 61: 409-420.

Muniz, P; Venturini, N; Rodríguez, M; Martínez, A; Lacerot, G; Gómez, M. Benthic communities in a highly polluted urban bay. In: Milón Delgado Paredes Paredes & Benavides (eds.). Ecología y Desarrollo Sostenible: Reto de América Latina para el Tercer Milenio. Arequipa, Perú, 2000 274pp.

Muniz, P; Venturini, N; Martínez, A. Physico-chemical characteristics and pollutants of the benthic environment in the Montevideo coastal zone, Uruguay. *Marine Pollution Bulletin*, 2002 44: 962- 968.

Muniz, P; Danulat, E; Yannicelli, B; Garcia-Alonso, J; Medina, G; Bícego, MC. Assessment of contamination by heavy metals and petroleum hydrocarbons in sediments of Montevideo Harbour (Uruguay). *Environment International*, 2004a 29: 1019-1028.

Muniz, P; Venturini, N; Gómez-Erache, M. Spatial distribution of chromium and lead in the benthic environment of coastal areas of the Río de la Plata estuary (Montevideo, Uruguay). *Brazilian Journal of Biology*, 2004b 64: 103-116.

Muniz, P ; Venturini, N ; Pires-Vanin, A M S ; Tommasi, L R ; Borja, A. Testing the applicability of a Marine Biotic Index (AMBI) to assessing the ecological quality of soft-bottom benthic communities, in the South America Atlantic region. *Marine Pollution Bulletin*, 2005a 50: 624-637.

Muniz, P; Clemente, J; Brugnoli, E. Benthic invasive pests in Uruguay: a new problem or an old one recently perceived? *Marine Pollution Bulletin*, 2005b 50: 1014-1018.

Murray, JW. Ecology and Paleoecology of Benthic Foraminifera. Longman Scientific and Technical/Wiley U K/ New York, 1991 397pp.

Muxika, I; Borja, A; Bonne, W. The suitability of the marine biotic index (AMBI) to new impact sources along European coasts. *Ecological Indicators*, 2005 5:19–31.

Nagy, G; Martínez, C; Caffera, MR; Pedrosa, G; Forbes, EA; Perdomo, A. The hydrological and climatic setting of the Río de la Plata. In: Wells & Daborn (eds.). The Río de la Plata, an environmental overview. An EcoPlata project background report, pp: 17-70. Dalhousie University, Halifax, Nova Scotia, 1997.

Nagy, GJ; Gómez-Erache, M; López, CH; Perdomo, AC. Distribution patterns of nutrients and symptoms of eutrophication in the Rio de la Plata estuary. *Hydrobiologia*, 2002 475/476: 125-139.

Naylor, RL; Williams, SL; Strong, DR. Aquaculture—a gateway for exotic species. *Science*, 2001 294:1655–1656.

Nión, H. Zonación del macrobentos en un sistema lagunar litoral oceánico. Seminario sobre Ecología Bentónica y Sedimentación de la Plataforma Continental del Atlántico Sur. (Montevideo, Uruguay), Memória UNESCO, 1979 1: 225-235.

Obenat, SM; Pezzani, EE. Life cycle and population structure of the polychaete *Ficopomatus enigmaticus* (Serpulidae) in Mar Chiquita coastal lagoon, Argentina. *Estuaries*, 1994 17: 263-270.

Obenat, S. Biología del anélido introducido *Ficopomatus enigmaticus* (Polychaeta, Serpulidae). In: Iribarne, O (ed). Reserva de Biósfera Mar Chiquita: Características físicas, biológicas y ecológicas. Editorial Martin. Mar del Plata, 2001.

Orensanz, JM; Schwindt, E; Pastorino, G; Bortolus, A; Casas, G; Darrigran, G; Elías, R; López-Gappa, JJ; Obenat, S; Pascual, S; Penchaszadeh, P; Piriz, ML; Scarabino, F; Spivak, ED; Vallarino, E. No longer the pristine confines of the world ocean: a survey of exotic marine species in the southwestern Atlantic. *Biological Invasions*, 2002 4: 115-143.

Oug, E; Kristoffer, N; Brage, R. Relationship between soft bottom macrofauna and polycyclic aromatic hydrocarbons (PAH) from smelter discharge in Norwegian fjords and coastal waters. *Marine Ecology Progress Series*, 1998 173: 39-52.

Palacio, FJ. Revisión zoogeográfica marina del sur de Brasil. Boletim do Instituto Oceanográfico, São Paulo, 1982 31: 69-92.

Pastorino, G; Darrigran, G; Martin, S; Lunaschi, G. *Limnoperna fortunei* (Dunker 1857) (Mytilidae), nuevo bivalvo invasor en aguas del Río de la Plata. *Neotrópica*, 1993 39: 34.

Pastorino, G; Penchaszadeh, PE; Schejter, L; Bremec, C. *Rapana venosa* (Valenciennes, 1846) (Mollusca: Muricidae): a new gastropod in south Atlantic waters. *Journal of Shelfish Research*, 2000 19: 897-899.

Pearson, TH; Rosenberg, R. Macrobenthic succession in relation to organic enrichment and pollution of the marine environment. *Oceanography and Marine Biology Annual Review*, 1978 16: 29-331.

Penchaszadeh, PE; Darrigran, G; Angulo, C; Averbuj, A; Brignoccoli, N; Brogger, M; Dogliotti, A; Pirez, N. Predation on the invasive freshwater mussel *Limnoperna fortunei* (Dunker 1857) (Mytilidae) by the fish *Leporinus obtusidens* Valenciennes 1846 (Anostomidae) in the Rio de la Plata, Argentina. *Journal of Shellfish Research*, 2000 19: 229-231.

Pianka, ER. Latitudinal gradients in species diversity: a review of concepts. *American Naturalist*, 1966 100: 33-46.

Raffaelli, D; Hawkins, S. Intertidal ecology. Chapman & Hall. London, 1997, 356pp.

Ricciardi, A; Atkinson SK. Distinctiveness magnifies the impact of biological invaders in aquatic ecosystems. *Ecology Letters*, 2004 7: 781-784.

Rakocinski, CF; Brown, SS; Gaston, GR; Heard, RW; Walker, WW; Summers, JK. Species-abundance-biomass responses by estuarine macrobenthos to sediment chemical contamination. *Journal of Aquatic Ecosystem Stress and Recovery*, 2000 7: 201-214.

Rees, HL; Pendle, MA; Waldock, O; Limpenny, DS; Boyd, SE. A comparison of benthic biodiversity in the North Sea, English Channel and Celtic Seas. *ICES Journal of Marine Science*, 1999 56: 228-246.

Rex, MA; Stuart, CT; Hessler, RR; Allen, JA; Sanders, HL; WILSON, GDF. Global scale latitudinal patterns of species diversity in the deep sea benthos. *Nature*, 1994 365: 636-639.

Riestra, G; Giménez, LJ; Scarabino, V. Análisis de la comunidad macrobentónica infralitoral de fondo rocoso en Isla Gorriti e Isla de Lobos (Maldonado, Uruguay). Frente Marítimo (Uruguay), 1992 11: 123-127.

Rohde, K. Latitudinal gradients in species diversity: the search for the primary cause. *Oikos*, 1992 65, 514-527.

Ritter, C; Montagna, PA. Seasonal hypoxia and models of benthic response in a Texas Bay. *Estuaries*, 1999 22: 7-20.

Rodriguez-Capitulo, A; Cortelezzi, A; Paggi, AC; Tangorra, M. Phytoplankton and Benthos of the environmental survey of the Río de la Plata. N° 2. Benthos. Technical report United Nations Development Programme-Global Environmental Facilities. PNUD Project/Gef RLA/99/G31, 2003. 48 pp. (In Spanish). http://www.freplata.org/documentos/tecnico.asp.

Rodríguez-Gallego, L; Conde, D; Rodríguez-Graña, L. Las Lagunas Costeras de Uruguay: estado actual del conocimiento. Simposio CARP-COFREMAR (V Jornadas de Ciencias del Mar). Mar del Plata, Argentina, 2003.

Ruiz, GM; Fofonoff, PW; Carlton, JT; Wonham, MJ; Hines, AH. Invasion of coastal marine communities in North America: apparent patterns, processes, and biases. *Annual Review of Ecology and Systematics,* 2000 31: 481-531.

Saiz-Salinas, JI. Evaluation of adverse biological effects induced by pollution in the Bilbao Estuary (Spain). *Environmental Pollution,* 1997 96: 351-359.

Sánchez-Mata, A; Glémarec, M; Mora, J. Physico-chemical structure of the benthic environment of a Galician ría (Ría de Ares-Betanzos, north-west Spain). *Journal of the Marine Biological Association of the United Kingdom,* 1999 79: 1-21.

Sanvicente-Anorve, L; Lepretre, A; Davoult, D. Diversity of benthic macrofauna in the eastern English Channel: comparison among and within communities. *Biodiversity and Conservation,* 2002 11: 265-282.

Savini, D; Harding, JM; Mann, R. Rapa whelk *Rapana venosa* (Valencienne, 1846) predation rates on hard clams *Mercenaria mercenaria* (Linnaeus, 1758). *Journa of Shellfishes Research,* 2002 21: 777-779.

Savini, D; Occhipinti-Ambriogi, A. Spreading potential of an invader: *Rapana venosa* in the Northern Adriatic sea. *Rapp Comm Int Mer Médit.,* 2004 37: 548.

Savini, D; Occhipinti-Ambriogi, A. Comsumption rates and prey preference of the invasive gastropod *Rapana venosa* in the Northern Adriatic Sea. *Helgoland Marine Research,* 2006 60: 153-159.

Savage, C; Field, JG; Warwick, RM. Comparative meta-analysis of the impact of offshore marine mining on macrobenthic communities versus organic pollution studies. *Marine Ecology Progress Series,* 2001 221: 265-275.

Scarabino, V; Maytía, S; Caches, M. Carta binómica litoral del departamento de Montevideo. I. Niveles superiores del sistema litoral. Comunicaciones de la Sociedad Malacológica del Uruguay, 1975 4: 117-129.

Scarabino, F; Menafra, R; Etchegaray, P. Presencia de *Rapana venosa* (Valenciennes, 1846) (Gastropoda: Muricidae) en el Río de la Plata. Boletín de la Sociedad Zoológica del Uruguay (Segunda Epoca) 11 (Actas de las V Jornadas de Zoología del Uruguay), Montevideo, 1999 11: 40.

Scarabino, F; Verde, M. *Limnoperna fortunei* (Dunker 1857) en la costa uruguaya del Río de la Plata (Bivalvia: Mytilidae). Comunicaciones de la Sociedad Malacológica del Uruguay, 1995 7: 374-375.

Scarabino, F. Conservación de la malacofauna uruguaya. Comunicaciones de la Sociedad *Malacológica de Uruguay,* 2004 8: 267-273

Schwindt, E. Impacto de un poliqueto exótico y formador de arrecifes. In: Iribarne, O. (Ed), Reserva de la Biosfera mar Chiquita: Características físicas y ecológicas. Ed. Martin, Mar del Plata, 2001: 109-113.

Schwindt, E; Iribarne, O. Settlement sites, survival and effects on benthos of an introduce reef-building polychaete in a SW Atlantic coastal lagoon. *Bulletin of Marine Science,* 2000 67: 73-82.

Schwindt, E; Iribarne, O; Isla, F. Physical effects of an invading reef-building polychaete on an Argentina estuarine environment. *Estuarine Coastal and Shelf Science,* 2004a 59: 109-120.

Schwindt, E; De Francesco, C; Iribarne, O. Individual and reef growth of the invasive reef-building polychaete *Ficopomatus enigmaticus* in a south-western Atlantic coastal lagoon. *Journal of the Marine Association of the United Kingdom,* 2004b 84: 987-993.

Sealey, K S; Bustamante, G. Setting geographic priorities for marine conservation in Latin America and the Caribbean. *The Nature conservancy,* Arlington, Virginia, 1999, 125 pp.

Semmens, BX; Buhle, ER; Salomon, AK; Pattengill-Semmens, CV. A hotspot of non-native marine fishes: evidence for the aquarium trade as invasion pathway. *Marine Ecology Progress Series,* 2004 266: 239-244.

Silva, JSV da; Souza, RCCL de. Água de lastro e bioinvasão. Editora Interciência, Rio de Janeiro, Brazil, 2004.

Silva, J SV da; Costa, F; Correa, RC; Sampaio, KT; Danelon, OM. Agua de lastro e bioinvaşao. In: Silva, JVS da; Souza, RCCL (eds.). Agua de lastro e bioinvaşao. pp 1-10. Interciencia, Río de Janeiro, 2004.

Tam, J; Carrasco, FD. Macrobenthic sublittoral species assemblages from Central Chile along a global scale of perturbation. *Gayana Oceanología,* 1997 5: 107-113.

ten Hove, HA; Weerdenburg, CA. A generic revision of the brackish-water serpulid *Ficopomatus* Southern 1921 (Polychaeta: Serpulidae), including *Mercierella* Fauvel 1923, *Sphaeropomatus* Treadwell 1934, *Mercierellopsis* Rioja 1954, and *Neopomatus* Pillai 1960. *Biological Bulletin,* 1978 154: 96-120.

Tenore, KR. Macrobenthos of the Pamlico River Estuary, North Carolina. *Ecological Monographs,* 1972 42: 51-69.

Thouzeau, G; Robert, G; Ugarte, R. Faunal assemblages of benthic megainvertebrates inhabiting sea scallop grounds from eastern Georges Bank in relation to environmental factors. *Marine Ecology Progress Series,* 1991 74: 61-82.

Thrush, SF; Hewitt, JE; Norkko, A; Nicholls, PE; Funnell, GA; Ellis, JI. Habitat change in estuaries: predicting broad-scale responses of intertidal macrofauna to sediment mud content. *Marine Ecology Progress Series,* 2003 263: 101-112.

Tundisi, JG; Matsumura-Tundisi, T ; Rocha, O. Ecossistemas de águas interiores. p. 153-194. In: Águas doces no Brasil. Capital ecológico, uso e conservação. Rebouças, A; Braga, B; Tundisi, JG (eds.), São Paulo. Escritura Editora, 1999.

Turner, SJ; Trush, SF; Pridmore, RD; Hewitt, JE; Cummings, VJ; Maskery, M. Are soft-sediment communities stable? An example from a windy harbour. *Marine Ecology Progress Series,* 1995 120: 219-230.

Van Dolah, RF; Hyland, JL; Holland, AF; Rosen, JS; Snoots, TR. A benthic index of biological integrity for assessing habitat quality in estuaries of the southeastern USA. *Marine Environmental Research,* 1999 48: 269-283.

Venturini, N; Muniz, P.Status of Sediment Pollution and Macrobenthic Subtidal Communities of Montevideo Coastal Zone, Uruguay. In: IX CONGRESO LATINOAMERICANO SOBRE CIENCIAS DEL MAR, 2001, San Andrés. Anales del IX Congreso Latinoamericano sobre Ciencias del Mar, 2001.

Venturini, N; Muniz, P; Rodríguez, M. Macrobenthic subtidal communities in relation to sediment pollution: the phylum-level meta-analysis approach in a south-eastern coastal region of South America. *Marine Biology,* 2004 144: 119-126.

Viana, F; Huertas, R; Danulta, E. Heavy Metal Levels in Fish from Coastal Waters of Uruguay. *Archives of Environmental Contamination and Toxicology,* 2005 48: 530-537.

Villa, N. Spatial distribution of heavy metals in seawater and sediments from the coastal areas of the southern Buenos Aires Province, Argentina. In: Seeliger, U.; Lacerda. L.D.; Patchineelam, S.R. (eds.) Metals in coastal environments of Latin America. Springer Verlag, Berlin, 1988.

Walton, W R. Techniques for recognition of living foraminifera. *Contr. Cushman. Fnd. form. Res* 1952 3:56- 60.

Warwick, RM. A new method for detecting pollution effects on marine macrobenthic communities. *Marine Biology,* 1986 92: 557-562.

Warwick, RM; Clarke, KR. Comparing the severity of disturbance: a meta-analysis of marine macrobenthic community data. *Marine Ecology Progress Series*, 1993 92: 221-231.

Weisberg, SB; Ranasinghe, JA; Dauer, DM; Schaffner, LC; Diaz, RJ; Frithsen, JB. An estuarine benthic index of biotic integrity (B-IBI) for Chesapeake bay. *Estuaries*, 1997 20: 149-158.

Willig, MR; Selcer, KW. Bat species density gradients in the New World: a statistical assessment. *Journal of Biogeography,* 1989 16: 189:195.

Wilson, JG. The role of bioindicators in estuarine management. *Estuaries*, 1994 17, 1A: 94-101.

In: Progress in Environmental Research
Editor: Irma C. Willis, pp. 127-154

ISBN 978-1-60021-618-3
© 2007 Nova Science Publishers, Inc.

Chapter 3

INFLUENCE OF GRASS COVER ON THE LEACHING OF HERBICIDES IN BURGUNDY VINEYARDS

Sylvie Dousset[1], David Landry[2], Astrid Jacobson[3], Philippe Baveye[3] and Francis Andreux[1]*

[1]UMR 1229 INRA/Université de Bourgogne, Géosol - Centre des Sciences de la Terre,
Université de Bourgogne, 6 bd Gabriel 21000 Dijon, France
[2]UMR INRA - INA-PG - Unité Environnement et Grandes Cultures,
BP 01, 78850 Thiverval-Grignon, France
[3]Department of Crop and Soil Sciences, Cornell University, 1002 Bradfield Hall,
Ithaca NY 14853, USA

ABSTRACT

Field studies monitoring herbicide pollution in the vineyards of Burgundy (France) have revealed that the drinking water reservoirs are contaminated with several herbicides. The purpose of this work is to assess the effectiveness of alternative soil management practices, such as grass cover, for reducing the leaching of oryzalin, diuron, glyphosate and some of their metabolites in soils, and subsequently in preserving groundwater quality. The mobility of these herbicides and their metabolites was studied in structured soil columns (15 x 20 cm) under outdoor conditions. In the first experiment the leaching of diuron, oryzalin and glyphosate was monitoring from May 2001 to May 2002. The soil was a calcaric cambisol under two vineyard soil management practices: a bare soil chemically-treated, and a vegetated soil in which grass was planted between the vine rows. A second experiment was conducted with diuron applied in early May 2002 on three calcaric cambisols and monitored until November 2002. Grass was planted on three of the six columns. In the second experiment, higher amounts of diuron residues were found in the effluents from the grass-covered columns (from 0.10% to 0.48%) than from

---

* Please contact author at: UMR MGS 1229 INRA - Université de Bourgogne, Centre des Sciences de la Terre, 6 boulevard Gabriel - 21 000 DIJON – France, Tel: 03 80 39 68 88 - Fax: 03 80 39 63 87; Email: Sylvie.Dousset@u-bourgogne.fr

the bare-soil columns (0.09% - 0.16%). At the end of the monitoring period, less of the diuron residues were recovered in the vegetated soil profiles, from 4.3% to 10.6%, than in the bare soils, from 10.8% to 34.9%. In the first experiment, greater amounts of herbicide residues were measured in the percolates of the bare soil, 0.96%, 0.10% and 0.21% for diuron, oryzalin and glyphosate respectively than in the percolates of the vegetated soil, 0.16%, 0.10% and 0.05%, respectively. At the end of the monitoring period, more residues from diuron, oryzalin and glyphosate were recovered in the bare soil profiles: respectively 7.6%, 2.4% and 0.007% than in the vegetated soil, 12.5%, 4.4% and 0.01%, respectively. The dissipation of the three herbicide residues seems faster in the bare soil than in the vegetated soil, contrary to the diuron dissipation in the experiment 2.

This study showed that diuron, oryzalin and to a lesser extent, glyphosate, AMPA, DCPMU and DCPU, leach through the soils; thus, these molecules may be potential contaminants of groundwater. The alternative soil management practice of planting a grass cover under the vine rows could reduce groundwater contamination by pesticides by reducing the infiltrating amounts of diuron, oryzalin, glyphosate and some of their metabolites. Nevertheless, after one year, persistence of herbicide residues was greater in the grass cover soil than in the bare soil. The difference in the results for diuron may be explained by the difference in the duration of the grass cover, 4 years in the experiment 1 versus 4 weeks in the experiment 2.

Keywords: Glyphosate; Diuron, Oryzalin, Transport; Lysimeter; Grass cover

1. INTRODUCTION

As a result of numerous sources of pollution, including the use of agricultural pesticides, drinking waters resources are becoming increasingly scarce and a crucial issue for developed countries. It is critical that solutions be proposed to better protect water quality, in particular that of surface waters, which are generally the most contaminated and also the most sensitive to contamination (IFEN, 2004). Viticulture is an important agricultural sector in France, and a great consumer of pesticides to control disease, insect damage and weed competition in the vineyards. Consequently, many recent studies have reported the presence of herbicide residues in surface- or ground-waters near several vineyards; for example in Languedoc-Roussillon (Lennartz et al., 1997; Louchart et al., 2004) and in Champagne (Garmouma et al., 1998). In Burgundy, water analyses of the drinking water reservoir in Vosne-Romanée have reported the presence of herbicides at concentrations higher than the European regulatory limit of 0.1 µg L^{-1} (DIREN et al., 2004; ECC, 1998). Consequently, agricultural institutions advise wine producers to use alternative practices to chemical weeding, such as grass cover.

Numerous works have shown that the grass cover reduces erosion and run-off due to sediment deposition and increases water infiltration within the vegetated zone (Dillaha et al. 1989; VanDijk et al., 1996). More recent studies have been concerned with the use of these buffer zones to limit surface water contamination by pesticides. A number of authors have reported that the amounts of pesticides in the run-off from vegetated buffer zones are lower than the amounts entering the zone (Patty et al., 1997; Schmitt *et al.*, 1999; Watanabe *et al.*, 2001). The effectiveness of the vegetated buffer zones at reducing the amounts of pesticide in the run-off may be explained by the processes of retention and/or infiltration within the zone

(Kloppel *et al.*, 1997 ; Mersie *et al.*, 1999 ; Seybold *et al.*, 2001 ; Delphin and Chapot, 2001 ; Souiller *et al.*, 2002 ; Benoit *et al.*, 2003 ; Mersie *et al.*, 2003). The respective proportions of the two processes in reduction of pesticide run-off would depend on the affinity of the molecules for the soil (sorption) (Arora *et al.*, 2003; Boyd *et al.*, 2003). For example, a reduction in the run-off of relatively water soluble herbicides such as atrazine and metolachlor, from a vegetated zone, could be explained by increased infiltration, whereas reduction in chlopyrifos run-off would be due to sorption onto the sediments retained by the buffer zone vegetation (Arora *et al.*, 2003). Similar conclusions were reached by Boyd et al. (2003) who showed that chlorpyrifos was retained on sediments deposited on vegetated buffer zones, whereas, at the same slope, atrazine and acetochlor infiltrated the soil and were detected in the drains.

Reducing the quantity of pesticides found in surface waters does not mean that groundwater contamination must increase due to increased infiltration. Various studies have shown that the sorption and degradation of pesticides is higher in vegetated zones than in cultivated soils (Benoit *et al.*, 1999; Madrigal *et al.*, 2002; Krutz *et al.*, 2003; Krutz *et al.*, 2004) or fallow soils (Staddon *et al.*, 2001). Benoit *et al.* (2000) also show that the mineralization of isoproturon and the formation of bound residues increase in vegetated soils compared to cultivated soils. Of course, these were laboratory batch studies and therefore somewhat removed from the reality of field conditions. If the rate of infiltration through a vegetative soil increases so that equilibrium is rarely achieved, potential sorption and degradation of the compounds will decrease relative to the results of batch studies in which steady-state conditions are attained. Very few studies have been conducted on the amounts of pesticides leached from vegetated soils and the results of these studies are contradictory. In some cases, the use of grass covers on soil reduces the amounts of pesticides leached compared with the amounts leached from cultivated or fallow soils (Liaghat and Prasher, 1996 ; Benoit *et al.*, 2000), in other cases, there was no difference (Belden and Coats, 2004).

The objective of this work was to assess the role of the grass cover on the leaching of several herbicides used in vineyards: oryzalin, diuron, glyphosate and some of their metabolites. The mobility of these herbicides was studied using structured soil columns (15 x 20 cm) from Meuilley and Vosne-Romanée (Burgundy) under outdoor conditions. In a first experiment the leaching of diuron, oryzalin and glyphosate was monitoring from May 2001 to May 2002. The soil was sampled in a grass-covered plot and in an adjacent, chemically-weeded plot. A second experiment was conducted with diuron applied in May 2002 on three calcareous soils and monitored until November 2002. Grass was planted on three of the six columns. Concentrations of the parent molecules and their metabolites were monitored in the column leachates during the monitoring periods, and throughout the soil profile at the end of the experiments. The results permitted an assessment of the potential benefits of grass cover on the leaching of herbicides through soil and subsequently on the preservation of groundwater quality.

2. MATERIALS AND METHODS

2.1. Chemicals

Oryzalin [4-(dipropylamino)-3,5-dinitrobenzenesulphonamide], diuron [3-(3,4-dichlorophenyl)-1,1-dimethylurea] with its main metabolites, DCPMU [N'-3,4-dichlorophenyl-N-methylurea], DCPU [3,4-dichlorophenylurea], DCA [3,4-dichloroaniline], and glyphosate [N-(phosphono-methyl-glycine] and its metabolite, AMPA [amino-methyl-phosphonic acid] were obtained from Cluzeau (Sainte-Foy-La-Grande, France) with > 98% certified purity. The water solubility of glyphosate is much higher than that of diuron and oryzalin, (11.6 g L^{-1} vs 36.4 mg L^{-1} and 2.6 mg L^{-1}, respectively) (Tomlin, 1997). Its half-life measured under laboratory conditions was shorter, from 7.9 to 14.4 d (Eberbach, 1999; Accinelli et al., 2004) than that of diuron and oryzalin, 4-8 months and 10 days to 1.2 months, respectively (Tomlin, 1997). Glyphosate shows a higher adsorption coefficient, K_{oc} from 8.5 to 10231 L kg^{-1} (Gerritse et al, 1996; Cheah et al, 1997; De Jonge and De Jonge, 1999) compared to K_{oc} of diuron and oryzalin, 400 L kg^{-1} and 700-1100 L kg^{-1}, respectively (Tomlin, 1997). All the other chemicals used in the experiments were of analytical reagent grade or higher and were used without further purification. The solvents were HPCL grade except the methanol used to extract the herbicides from the soil, which was technical grade. All solutions were prepared in milliQ^{+}-water (Millipore).

2.2. Soil Sampling

In the first experiment (Exp. I), a calcaric cambisol (FAO, 1998) from a Vosne-Romanée vineyard (named after cambisol 1) was chosen from the bottom of a topolithosequence (Landry et al., 2005, 2006). The soil was sampled from the 0-20 cm layer of two adjacent plots: a grass-covered cambisol with rye grass between the vine rows, and a chemically-weeded, bare cambisol. Rye grass was planted between the vine rows of the vegetated cambisol in 1997, and induced an increase in the organic carbon content of the 0-2.5 cm surface layer of soil (3.5%) relative to the soil sampled under the row (1.8%, data not shown). The chemically-weeded cambisol contained an organic layer composed of grape seeds and large partly-decomposed plant fragments in the 3-6 cm depth, explaining the high organic carbon content and high content of coarse materials in this layer (Table 1). The dissolved organic carbon (DOC) from these two organic matter-rich soil layers was extracted using the method proposed by Zsolnay (1996). Oryzalin (trade name: Surflan) and glyphosate (Roundup and Buggy) had been applied to both vineyard plots every year since 1997, at 3.5 kg ha^{-1} for oryzalin and 5.7 kg ha^{-1} for glyphosate. Diuron had not been applied to these soils since, at least, 1997. In order to measure possible herbicide residues from previous field applications, samples of each soil were collected from the 0-20 cm layer with a shovel, at the same time and from the same location that the soil columns were isolated. Each soil sample was sieved to < 2 mm and analyzed for oryzalin, diuron and its metabolites, and glyphosate and AMPA. Triplicate analyses were performed. Neither herbicide was detected in the soils.

In the second experiment (Exp. II), three calcaric cambisol were sampled: one from the 0-20 cm soil layer of a Meuilley vineyard (cambisol 2) and of two from Vosne-Romanée vineyards (cambisol 3 and cambisol 4). A commercially available grass blend (25% perennial rye grass *Barcredo*, 25% perennial rye grass *Capri*, and 50% tall fescue *Barbizon*) was planted and cultivated for 4 weeks in three of the six sampled soil columns to represent the effects of alternative field management practices.

Table 1. Main characteristics of the soils in the columns from 0-20 cm

Soil	Soil Texture (%)					OC	CaCO$_3$	Total bulk	Porosity
	> 2mm	sand	silt	clay	class	%	%	density	%
								g. cm^{-3}	
Soil columns treated with oryzalin (exp. I)									
Bare cambisol 1	34	31	38	33	Clay-loam	2.2	33	1.49	44
Veg. cambisol 1	22	27	41	33	Clay-loam	2.1	28	1.55	42
Soil columns treated with diuron (exp.I)									
Bare cambisol 1	38	27	37	37	Clay-loam	2.8	26	1.56	41
Veg. cambisol 1	26	23	43	35	Clay-loam	2.0	22	1.55	42
Soil columns treated with glyphosate (exp. I)									
Bare cambisol 1	38	29	38	33	Clay-loam	3.1	29	1.51	43
Veg. cambisol 1	35	31	38	31	Clay-loam	2.0	32	1.30	51
Soil columns treated with diuron (exp. II)									
Bare cambisol 2	30	13	45	43	silty clay	3.4	12	1.14	57
Veg. cambisol 2	27	12	45	43	silty clay	3.7	8	0.95	64
Bare cambisol 3	27	22	36	42	clay	2.1	21	1.22	54
Veg. cambisol 3	26	23	34	44	clay	2.2	21	1.14	57
Bare cambisol 4	33	27	31	43	clay	1.8	22	1.15	57
Veg. cambisol 4	30	38	26	35	clay	1.4	30	1.08	59

2.3. Column Set-up

Column extraction of the soils occurred in March 2001 (Exp. I) and in March 2002 (Exp. II), before herbicide treatment of the fields, and was facilitated by the use of a shovel to carefully excavate the surrounding soil. Final carving of the soil was carefully performed by hand resulting in 15 cm diameter cylinders of structured soil. A 25-cm long polyvinyl chloride (PVC) pipe with an internal diameter of 20 cm was placed around each soil cylinder. The space between the pipe and soil was filled with expandable foam to prevent water from moving preferentially down the side of the pipe rather than through the soil. Then minimal-expansion foam was allowed to cure overnight. The columns were then removed from the field by digging under the PVC pipe, and placing nylon mesh (105 μm openings) at the bottom of each column base to retain the soil. Three undisturbed soil columns were extracted from each field plot for Exp. I and two undisturbed soil columns were extracted from each field plot for Exp. II.

The columns were brought to the experimental site at INRA-Dijon (15 km from Vosne-Romanée) for installation in an outdoor, in-ground lysimeter collection system, consisting of a perforated PVC support connecting the columns to funnels and PFTE-lined collection tubing leading to high-density polyethylene bottles in an underground pit. The volume around the columns was backfilled with sand to mimic field conditions and buffer temperature changes. The soil columns were subjected to natural weather conditions for 1 year, from May 2001 to May 2002 (Exp. I), or for 6 months, from May to November 2002 (Exp. II). During the monitoring periods, weekly precipitation collected by an on-site rain gauge was recorded at the INRA site and the average weekly temperatures were provided by Dijon-Longvic meteorological station (5 km from the experimental site).

2.4. Herbicide Application

Before herbicide treatment of the vegetated soil in Exp. I, the aerial part of the rye grass was cut to reduce herbicide sorption by the grass so that the results between the two soils columns could be better compared. During the monitoring period, the grass was periodically cut as is typically done by viticulturists with the natural grass cover in the field. On 27 April 2001, oryzalin, diuron or glyphosate was applied to each column by pipetting 30 mL of a methanolic solution (or aqueous solution for glyphosate) containing 290 mg L^{-1} of oryzalin, 137 mg L^{-1} of diuron or 340 mg L^{-1} of glyphosate onto the surface of the soils to simulate application rates of 3.5 kg ha^{-1} active ingredient (a.i.), 1.8 kg ha^{-1} a.i. or 5.7 kg ha^{-1} a.i., respectively (i.e., the equivalent of the dose applied on vineyard). To avoid possible interaction between the herbicides, only one was applied to each column.

In Exp. II the grass was kept trimmed level with the top of the PVC pipe and grass clippings were returned to the soil surface prior to the application of the herbicide. On 7 May 2002, diuron was applied to the top of each column by pipetting 20 mL of a methanolic solution containing 225 mg L^{-1} of diuron to simulate an application rate of 2.0 kg ha^{-1} (a.i.).

In both experiments, herbicides were applied to the columns at least 24 hrs after the last rain and 48 hrs before the next precipitation event.

2.5. Leachate Collection and Analyses

Leachates were collected weekly or after large precipitation events from May 2001 to May 2002 (Exp. I), and from May 2002 to November 2002 (Exp. II). Leachate volumes were determined gravimetrically and then stored at -18°C prior to analysis.

For the analysis of oryzalin, diuron and its three metabolites, DCPMU, DCPU, and DCA, the leachate samples were concentrated by solid-phase extraction with LC-18 bonded silica cartridges (Supelclean, Supelco) according to the procedure outlined by Landry et al. (2004). Oryzalin, diuron, DCPMU, DCPU, and DCA were analyzed using a Waters 600 HPLC equipped with a Diode Array Detector and a 25 cm x 4.6 mm C18-column packed with Kromasil 5 μm. The mobile phase was a 70:30 (v/v) acetonitrile:water solution eluted at 0.8 mL min^{-1}. UV absorbance occurred at 254 nm for oryzalin, at 251 nm for diuron and

DCPMU, 247 nm for DCPU, and 245.5 nm for DCA. Recovery mean rates were 95.1% for oryzalin, 90.8% for diuron, 91.7% for DCPMU, 90.6% for DCPU, and 5.5% for DCA. Due to its poor recovery from the leachates, DCA data were not used. All the sample concentrations were corrected for these recovery values. The detection limits were 0.05 μg L^{-1} for diuron and DCPMU, and 0.1 μg L^{-1} for oryzalin, DCPU and DCA. Glyphosate and AMPA were analyzed by the Institut Pasteur (Lille, France), using the procedure by Vreeken et al. (1998). Glyphosate and AMPA were first derivatised by adding 9-fluorenyl methoxycarbonyl chloride (FMOC-Cl) and borate buffer to the sample, and reacting overnight at 37°C. The reaction was stopped by adding phosphoric acid, i.e., by lowering the pH. Glyphosate and AMPA were analyzed by LC-ESI-MS-MS (Agilent Technologies, Palo Alto, CA, USA). Minimum detectable levels were 0.1 μg L^{-1} for glyphosate and AMPA in leachates.

2.6. Soil Column Sectioning and Soil Analyses

The columns were removed from the outdoor lysimeter-support system in May 2002 (Exp. I), and in early November 2002 (Exp. II). At the end of the monitoring period, the soil columns were divided into 5 sections (0-2.5 cm; 2.5-5 cm, 5-10 cm; 10-15 cm; 15-20 cm), air-dried, weighed, and sieved to 2 mm. The >2 mm fractions were weighed as the coarse fraction. The <2 mm fraction were characterized by determinations of texture (NFX 31-107), pH (NF ISO 10390), carbonates (NF ISO 10693), and total organic C (NF ISO 10694) at INRA-Arras (Table 1).

After homogenization, 50-g sub-samples were taken from each section, put in 500-mL polypropylene bottles, and extracted three times successively with 100 ml of a 80/20 (v/v) methanol/water solution by agitation on a rotary shaker for 10 h in order to extract oryzalin or diuron residues from soil. In Exp. II, a preliminary soil extraction with water was performed (Landry et al., 2004). Each suspension was then centrifuged for 20 min at 6700 g. The three successive extracts were combined and evaporated to dryness in a rotary evaporator at 30°C. The residues were then dissolved in 2 ml of acetonitrile and stored at -18°C prior to analysis. To extract glyphosate residues from the soil, 500-mL polypropylene bottles containing 50 g soil and 100 mL distilled water were agitated on a rotary shaker for 10 hours. The suspensions were then centrifuged for 20 min at 6700 g. The extracts were stored at -18°C prior to analysis. Oryzalin, diuron, DCPMU, DCPU, DCA, glyphosate and AMPA were analyzed as described above. Recovery rates were from 91% to 93% for oryzalin, from 72% to 96% for diuron, 75% to 89% for DCPMU, 68% to 78% for DCPU, 3% to 11% for DCA, and 95% for glyphosate and AMPA, depending on the type of soil. All the herbicide concentrations for the samples have been corrected based on the appropriate recovery value. Detection limits expressed as a function of soil dry weight were 1.5 μg kg^{-1} for diuron and DCPMU, 3.0 μg kg^{-1} for oryzalin, DCPU and DCA, and 0.2 μg kg^{-1} for glyphosate and AMPA.

3. RESULTS AND DISCUSSION

3.1. Soil Characterization

The soils are shallow with depths of 20 cm to 60 cm, and have developed on karstic terrain thereby increasing the risk of groundwater contamination by pesticides applied to the soils. The four calcareous soils had similar characteristics typified by high clay contents (31% to 44%), pH values of 8.2, high concentrations of total carbonate (8% to 33%), and low concentrations of total organic carbon (1.4% to 2.8%) (Table 1). The soil column porosities were estimated from the bulk and the particle densities. The bulk densities varied from 0.9 to 1.7 g cm^3 depending of the soil and a particle density of 2.65 was assumed. The porosity was, therefore, from 41% to 64%. The cambisol 2 (Meuilley) was relatively rich in organic matter (3.4% to 3.7%) because it had been forested just 3 years before. Its high organic carbon content and the relatively recent crossing of tractors enhancing the soil compaction may explain why this soil had the lowest bulk density of the soils studied (Table 1).

3.2. Cumulative Rainfall and Water Eluted

In Exp. 1, cumulative rainfall for the period between 27 April 2001 and 1 May 2002 was 673 mm, below the 30-yr average of 740 mm. The deficit occurred throughout the year except during the months of July and October 2001 and January 2002, which were wetter than the corresponding months of the 30-yr norm. The average total cumulative leachate volume collected for the columns was 8.7 L ± 1.2 L. Water movement was very poor in two lysimeters: the bare cambisol 1 treated with oryzalin and the vegetated cambisol 1 treated with glyphosate for which the leachates were below average, 2.6 L and 4.7 L, respectively. Indeed, for the bare cambisol 1 treated with oryzalin and the vegetated cambisol 1 treated with glyphosate, very little leachate was recovered from the beginning of the experiment.

In Exp. II, which was conducted from 7 May 2002 to 5 Nov. 2002, 376 mm of rain fell at the experimental site at INRA-Dijon, which is slightly less precipitation than the 30-yr average (401 mm) for the same time period measured at the meteorological station Dijon-Longvic. The deficit occurred during the period from May to July, which was slightly drier than the 30-yr average. By comparison, the period from May 2001 to November 2001 for Exp. 1 was 16% wetter (438 mm rain) than the equivalent period for Exp. 2 (376 mm). Total cumulative leachate volumes recovered for all columns averaged 6.1 L ± 0.2 L. Percolation through these columns was quite similar and the effect of precipitation events is clearly visible in the stepped-response of the leachate collected. In the bare cambisol 4, only 3.3 L of leachate was collected. This lower leachate recovery was associated with slower percolation rates and visible surface ponding after storms. In both experiments 1 and 2, the amounts of leachate recovered in the columns were not correlated with the presence or absence of grass cover (Table 1).

3.3. Herbicide Residue Concentrations in Leachates

The weekly rainfall and the leachate concentrations of oryzalin, diuron, glyphosate and their metabolites for Exp. I are presented in Figure 1. The first rainfall (3.4 mm) occurred the day after the herbicide treatments and the first drainage events (from 2.5 mm to 26 mm depending of the soil) occurred 12 to 19 days after herbicide treatments. During the monitoring period, diuron and oryzalin were detected in 60% and 22% of the leachates from the bare cambisol 1, but only 18% and 16% of the leachates from the vegetated cambisol 1. However, the percentage of detection of oryzalin detected in the bare cambisol 1 leachates was lower (14%) than of the amount detected in the vegetated cambisol 1 leachates (24%). This may be due to the very poor water circulation in the bare cambisol 1 treated with oryzalin. The herbicides were first detected in the percolates less than 4 weeks after application, after less than 3 cm net drainage from the bare cambisol 1 and less than 1.5 cm net drainage from the vegetated cambisol 1. If the total pore volume of the soil column had been involved in the migration of these herbicides, approximately 8 cm of drainage would have been required before any surface-applied chemical reached the percolates. The early herbicide detection is consistent with preferential-flow concepts and suggests that a small fraction of the pore volume was active in rapid transport. Such presumptive evidence for macropore flow has also been seen by Isensee et al. (1990) and Dousset et al. (2004). Additional evidence for the movement of the herbicides by preferential flow is that herbicides reached the drains at the same time despite differences in their sorption coefficients. More sorptive compounds are generally expected to move more slowly and at attenuated concentrations. The same phenomenon has been reported by Kladivko et al. (1991), who showed that pesticides with different sorption coefficients such as alachlor, carbofuran, cyanazine, and atrazine drained from a silt loam soil at the same time. Diuron, glyphosate and oryzalin appeared later in the leachates of the bare cambisol 1 (1 week, 2 weeks and > 1 month, respectively) than in the leachates of the vegetated cambisol 1 (1 week for the three herbicides). Roy et al. (2001) and Raturi et al. (2003), who reported similar results, found that grass cover improved soil structure. Thus, water infiltration might be faster in the vegetated soil due to the extensive root systems present in the grass-covered soils. Herbicide residues were detected in leachates of the bare cambisol 1 longer (32 to 37 weeks after treatment) than in the leachates of the vegetated cambisol 1, where they were detected up to 15 to 21 weeks after treatment (Figure 1). In addition, the concentrations of herbicide residues in column leachates were greater in the leachates of the bare soil, from 0.3 µg L^{-1} to 24.3 µg L^{-1} for oryzalin, from 0.8 µg L^{-1} to 16.1 µg L^{-1} for diuron, and from 0.1 µg L^{-1} to 17 µg L^{-1} for glyphosate, compared to those of the vegetated soil, from 0.4 µg L^{-1} to 4.2 µg L^{-1}, from 0.2 µg L^{-1} to 5.2 µg L^{-1}, and from 0.1 µg L^{-1} to 2.7 µg L^{-1}, respectively. The highest concentrations of oryzalin and glyphosate measured in the leachates were 24.3 µg L^{-1} and 17 µg L^{-1} in June 2001 from the bare soil and 4.2 µg L^{-1} in 2.7 µg L^{-1} in July 2001 from the vegetated soil, as a consequence of storms. This may be explained by greater sorption and faster degradation of the parent molecules in the vegetated cambisol 1 than in the bare cambisol 1, similar to the results reported by Benoit et al. (1999), Staddon et al. (2001) and Krutz et al. (2003, 2004).

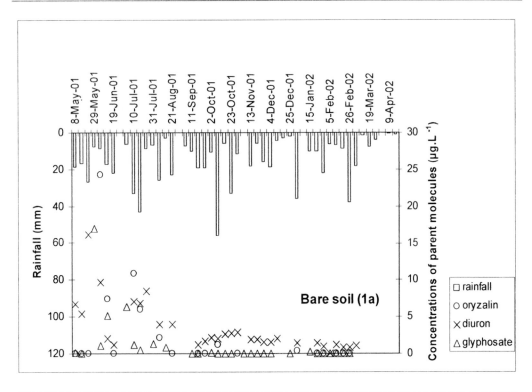

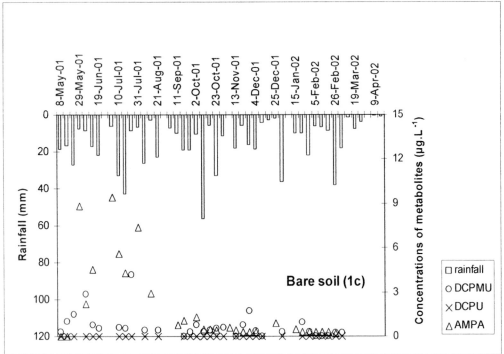

Figure 1. Rain events (□) and concentrations of oryzalin (O), diuron (X), glyphosate (Δ) (1a, 1b) and their metabolites (DCPMU (O), DCPU (X) and AMPA (Δ)) (1c, 1d) measured in the leachates of the bare and vegetated cambisol 1 from May 2001 to May 2002 (exp. I).

Of the metabolites, DCPMU was detected in the leachates 1 week after treatment and AMPA from 1 to 5 weeks after treatment, suggesting a rapid degradation of diuron and glyphosate and the presence of preferential flow in the cambisol 1. These results confirmed those obtained under laboratory conditions in which DCPMU appeared in the leachates 4 to 7 days after diuron application (Landry et al., 2004). Similarly, Dousset et al. (2004) showed that AMPA appeared from 3 to 12 days after glyphosate treatment in leachates of loamy sand and sandy loam, soil columns under laboratory conditions. DCPU was detected much later, 7 weeks after diuron application, and less frequently (from 0% to 8% of leachates) than DCPMU (from 24% to 50% of leachates) in agreement with the results of Tixier et al. (2002). AMPA, DCPMU and DCPU were detected in 54%, 50% and 0% of the leachates from the bare soil, but only 24%, 24% and 8% of the leachates from the vegetated soil. The concentrations of the metabolite residues, which were as high as 9.4 μg L^{-1} were noteworthy relative to the concentrations of parent molecule (up to 24.3 μg L^{-1}). The metabolites appeared at the same time or later in the leachates of the bare cambisol 1 than in the leachates of the vegetated cambisol 1, invalidating the assumption that the degradation was faster in the vegetated soil than in the bare soil. This result might also have been due to the greater leaching of metabolites through the bare soil than through the vegetated soil. The metabolite concentrations varied from 0.2 μg L^{-1} to 4.2 μg L^{-1} for DCPMU, no DCPU detection, and from 0.2 μg L^{-1} to 9.4 μg L^{-1} for AMPA in the leachates of the bare soil, and from 0.1 μg L^{-1} to 4.8 μg L^{-1} for DCPMU, from 0.2 μg L^{-1} to 4.4 μg L^{-1} for DCPU, and from 0.1 μg L^{-1} to 3.5 μg L^{-1} for AMPA in the leachates of the vegetated soil. Similar results were found by Goody et al. (2002) who measured concentrations of diuron, DCPMU and DCPU of 17 μg L^{-1}, 0.6 μg L^{-1} and 0.2 μg L^{-1} respectively, at a 54 cm depth in calcareous soil solution, 8 days after the application of diuron in outdoor conditions. Several studies conducted under outdoor conditions have reported quite short glyphosate half-lives of 10 to 60 days (Roy et al., 1989; Feng and Thompson, 1990). AMPA was measured in the leachates of both soil columns at concentrations similar to those of glyphosate (Figure 1), and in agreement with the results of Veiga et al. (2001) who measured mean concentrations of glyphosate at 0.14 mg L^{-1} and of AMPA at 0.11 mg L^{-1} in 30-cm depth soil solution using ceramic cup lysimeters. Similar to the results for glyphosate, AMPA was measured at higher concentrations in the leachates of the bare soil than in the leachates of the vegetated soil.

The weekly rainfall and the diuron and its metabolites concentrations in the leachates of Exp. II are presented in Figure 2. The first rain event occurred within 2 days (1.2 mm) and the first drainage event (from 5.0 mm to 36.6 mm depending on the soil) occurred from 14 to 21 days after diuron treatment. During the monitoring period, the detection frequency of diuron was lower in the leachates of the bare soils, from 8% and 23% than in the leachates of the vegetated soils, from 8% to 38%, except for the bare cambisol 4. Diuron appeared earlier and at lower concentrations in the leachates of the bare soils, (from 0.3 μg L^{-1} to 3.6 μg L^{-1}, 2 to 3 weeks after treatment) than in the leachates of the vegetated soils (from 0.1 μg L^{-1} to 20.2 μg L^{-1}, 3 to 6 weeks after treatment). Diuron was detected later in the soil leachates of Exp. II than those of Exp. I, probably because the soils were drier in Exp. II than in Exp. I at the beginning of the leaching experiment. In addition, diuron residues were detected in the leachates of the bare soils for shorter periods (on average 15 weeks after treatment) than in the leachates of the vegetated soils (on average 22 weeks after treatment) (Figure 2). These results for diuron were contradictory to those of Exp. I.

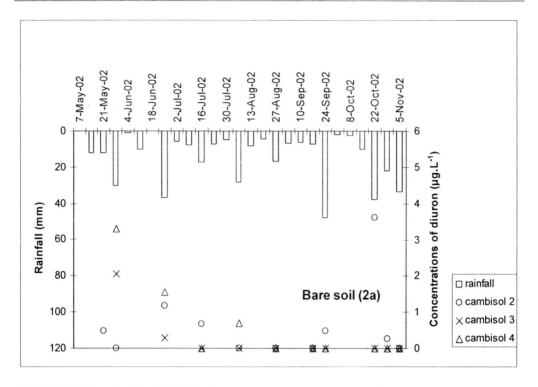

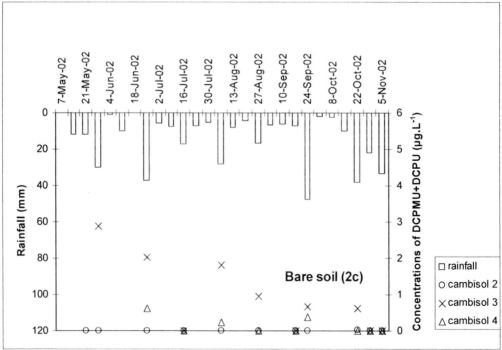

Figure 2. Rain events (□) and concentrations of diuron (2a, 2b) and its metabolites (DCPMU + DCPU) (2c, 2d) measured in the leachates of the bare and vegetated cambisol 2 (O), 3 (X) and 4 (Δ) from May 2002 to Nov. 2002 (exp. II).

In Exp. II, DCPMU appeared later and less frequently in the leachates from the bare soils, from 0% to 4% than in the leachates from the vegetated soils, from 0% to 23%, except for the bare cambisol 4. The metabolite concentrations varied from 0.0 µg L^{-1} to 0.1 µg L^{-1} for DCPMU and from 0.6 µg L^{-1} to 6 µg L^{-1} DCPU in the leachate of the bare soils, and from 0.0 µg L^{-1} to 3.6 µg L^{-1} for DCPMU and from 0.0 µg L^{-1} to 2.2 µg L^{-1} for DCPU in the leachates of the vegetated soils. Contradictory to Exp. 1, these results suggest that the degradation of diuron was faster in the vegetated soils than in the bare soils, in relation with a microbial activity enhanced in the grass rhizosphere or with the migration of DCPMU faster through the vegetated soil.

3.4. Percolated Amounts of Herbicide Residues

The cumulative percentages of oryzalin, diuron and glyphosate eluted in Exp. I, expressed as functions of percolated water height, are shown in Figure 3. At the end of the monitoring period, similar or greater herbicide residues were recovered in the leachates of the bare soil, 0.10% for oryzalin, 0.84% for diuron, and 0.06% for glyphosate, than in the leachates of the vegetated soil, 0.10%, 0.10% and 0.02%, respectively (Figure 3). These values run counter to those of Fishel and Coat (1993) who reported that, under natural conditions, oryzalin did not leach below 7.5 cm in a sandy loam soil with 1.8% OC, even after 241 mm of rainfall. However, the results are in good agreement with those obtained by Gooddy et al. (2002) who detected diuron residues in a calcareous soil solution to a depth of 54 cm. Concerning glyphosate, our results are in agreement with several studies performed under laboratory conditions. Cheah et al. (1997) recovered 0.04% to 0.07% applied glyphosate in the leachates of soil columns following 200 mm of simulated rainfall. Similarly, Dousset et al. (2004) found that, after 160 mm of simulated rainfall, 0.001% and 0.01% of the glyphosate applied to sandy soils was recovered as glyphosate in the leachate and 0.001% and 0.03% was recovered as AMPA. The values of increasing mobility between diuron, oryzalin and glyphosate are not surprising given the sorption coefficients of the herbicides, which vary from 0.5 to 40 L kg^{-1} for oryzalin (Jacques and Harvey, 1979), from 0.15 to 15 L kg^{-1} for diuron (Madhun et al., 1986; Rae et al., 1998), and from 0.25 to 5000 for glyphosate (Gerritse et al., 1996; Mamy and Barriuso, 2005). The amounts of herbicide residues recovered were greater in the leachates of the bare cambisol 1 than the vegetated cambisol 1. A coarse material-rich layer, containing from 32% to 46% grape seeds in the bare cambisol, but only 16% to 34% coarse material in the vegetated cambisol, could have improved drainage through the bare soil (Landry et al. 2005; 2006). However, this hypothesis was not confirmed by the amount of water that percolated through the column. Another possibility is that the mobility of the herbicide residues was enhanced by the formation of complexes with DOC. Indeed, the DOC content was higher in the grape seed layer (3.7 mg DOC per g of soil) than in the vegetated layer (2.6 mg DOC per g of soil). Gonzalès-Pradas et al. (1998) showed that diuron mobility could be promoted by the incorporation of peat in the soil surface layer, which resulted in DOC enriched leachates. On the other hand, grass cover increased the OC content of the 0-2.5 cm soil surface layers (from 2.9% to 4.2%), relative to that of the bare soils (from 1.9% to 3.6%) (Landry et al. 2005; 2006), which could enhance the retention of herbicide residues in the upper layers. Thus, the lower amounts of herbicide residues that leached through the vegetated cambisol might be due to greater diuron sorption in the root-rich soil surface layer, and its accelerated degradation. Benoit et al. (2000) showed that grass cover caused a decrease of about 50% in the amount of isoproturon, another phenylurea herbicide that percolated from undisturbed soil columns due to increased isoproturon adsorption and degradation.

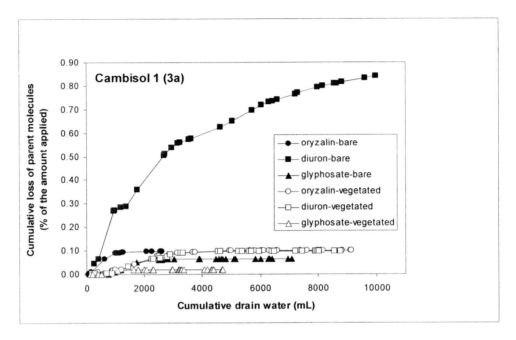

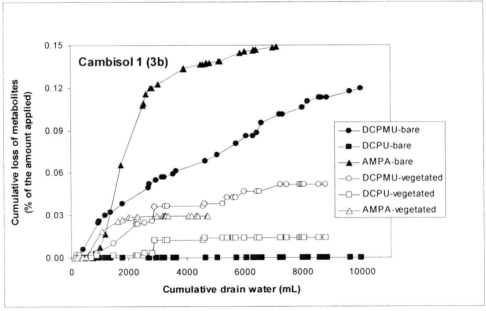

Figure 3. Cumulative amount of leached oryzalin (O), diuron (□), glyphosate (Δ) (3a) and their metabolites (DCPMU (O), DCPU (□) and AMPA(Δ)) (3b) as a function of cumulative leachate collected from the columns of the bare (black symbols) and vegetated (white symbols) cambisol 1 (exp. I).

The amounts of DCPMU (relative to the amount of diuron applied) were greater in the leachates of the bare cambisol 1, 0.12% than in those of the vegetated cambisol 1, 0.05% (Figure 3). The amounts of DCPU were 0.01% and 0% in the leachates of the vegetated cambisol and bare cambisol, respectively. The amount of DCPU recovered in the leachates was at least 3 times lower than the DCPMU, which is the first metabolite formed by microbial degradation of diuron. Under field conditions, Gooddy et al. (2002) found values of the same order of magnitude with a DCPMU/ DCPU ratio of about 11. On the other hand, lower extractable amounts of DCPU and the absence of DCA could be explained by its incorporation into humic acid due to the presence of the NH_2 group (Andreux et al., 1992). Like DCPMU, AMPA was recovered in greater amounts in the leachates of the bare soil, 0.15%, than in the leachates of the vegetated soil, 0.03% (Figure 3). Similarly, Dousset et al. (2004) found that, after 160 mm of simulated rainfall, 0.001% and 0.03% of the glyphosate applied to sandy soils was recovered as AMPA. Thus, in the vegetated cambisol 1, herbicide degradation might be enhanced by increased microbial activity in the rhizosphere. Benoit et al. (2000) also found that isoproturon (phenylurea herbicide) degraded more quickly in a grass-covered soil than in a cropped soil.

In Exp. II, after 376 mm of rainfall, less than 0.5% of diuron was recovered in percolates, as diuron or as its metabolites. More diuron was found in the leachate of the soil columns that were initially covered with grass, from 0.1% to 0.6% than in the bare soil columns, from 0.02% to 0.2% (Figure 4). As in Exp. I, DCA could not be determined in the soil leachates due to poor analytical recovery. Like their parent molecule, DCPMU and DCPU were recovered in greater amounts in the leachates of the grass covered soils, from 0.02% to 0.35% and from 0.05% to 0.28%, respectively, than in the leachates of the bare soils, 0% and from 0.04% to 0.14%, respectively. The elevated amounts of metabolites recovered in the grass-covered columns suggest a higher degradation rate in these columns as observed by Benoit et al. (1999) in their study on the sorption and degradation of a phenyl-urea, isoproturon, in a grassed buffer strip (Jacobson et al., 2005). The increased herbicide residues recovered in the grass-covered soil columns were also correlated with an increased total dissolved organic carbon content in the percolates, and thus could be due to facilitated transport of the herbicide. Another explanation could have been that there was more water infiltration through the vegetated soils than the bare soils; however that was not the case except for the bare cambisol 4, where diuron leaching was similar through the bare and vegetated soils.

Although the order of magnitude for the cumulative losses of diuron and its metabolites was the same for both experiments 1 and 2, the results are contradictory. Indeed, when considering equivalent durations for the two experiments (from May to November 2001), less diuron and less of its metabolites were measured in the Exp. I leachates of the vegetated soil (0.01% to 0.1%) than in the leachates of the bare soil (0% to 0.7%). The difference could be due to the length of time that grass had been grown on the soil. In Exp. I the soil had a field-established grass-cover for 4 years, whereas, in Exp. II, grass-cover was established in the soil columns 4 weeks prior to herbicide application. In Exp. I, grass cover would enhance the microbial degradation of diuron and its sorption to the soil, and therefore, would reduce the amounts of diuron leached. In Exp. II, roots of the young grass cover could exude high amounts of organic compounds in soil which could complex diuron and increase its facilitated transport relative to that in the bare soil.

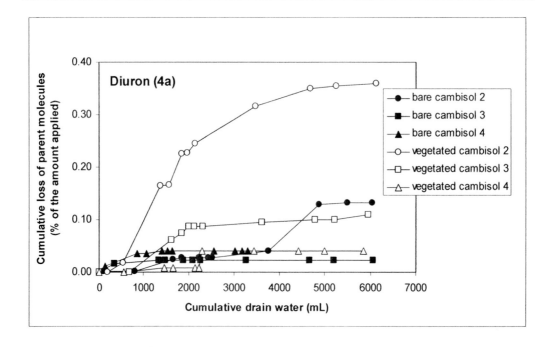

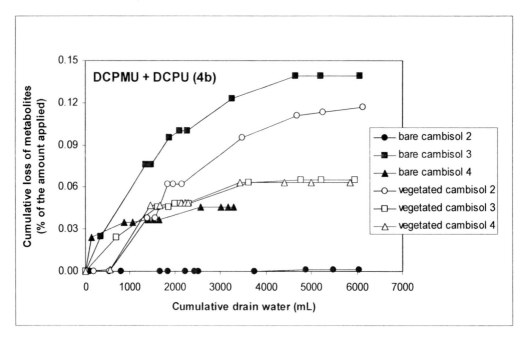

Figure 4. Cumulative amount of leached diuron (4a) and its metabolites (DCPMU + DCPU) (4b) as a function of cumulative leachate collected from the columns of the bare (black symbols) and vegetated (white symbols) cambisol 2 (O), cambisol 3 (☐) and cambisol 4 (Δ) (exp. II).

3.5. Distribution of Herbicide Residues in Soil Profiles

The distribution of herbicide residues in the 20-cm long soil columns after 673 mm of rainfall in Exp. I is shown in Figure 5. No extractable glyphosate was detected in the bare or vegetated soil (Figure 5). These results are in agreement with those of Newton et al. (1994) who did not detect any glyphosate down to 30 cm in a forest soil 398 days after application. They also agree with findings by Müller et al. (1981) and Feng and Thompson (1990) that glyphosate degrades rapidly and persists less than one year under outdoor conditions. Oryzalin reached 10 to 15 cm depths and diuron was detected at 20 cm depth in the vegetated and the bare cambisol 1, respectively. Similarly, Golab et al. (1975) detected oryzalin residues in the 0-15 cm layer of a sandy loam soil, under outdoor conditions, one year after its application (1.5 kg ha^{-1}), and numerous studies have shown diuron migration to depths greater than 30 cm one year after application under natural conditions (Pätzold and Brümmer, 1996; Field et al., 2003). The results of Exp. I showing greater movement of diuron than oryzalin through the soils are consistent with those obtained by Futch and Singh (1999) who reported that diuron reached a deeper layer (16 cm) than oryzalin (4 cm) in a sandy soil after 130 mm of simulated rain. The greater adsorption coefficient of oryzalin than diuron explains its higher retention by soil surfaces and thus, its lower leaching potential. The deeper herbicide migration in the vegetated soil was inconsistent with the results of several studies that reported decreases in herbicide mobility through grass covered soils (Benoit et al., 2000; Roy et al., 2001). However, grass cover could improve soil structure which would improve the infiltration of water and herbicides in the root zone as shown by Raturi et al. (2003). In addition, when comparing the amounts of oryzalin and diuron in the soils layer by layer, the results from Exp. I showed higher levels in the vegetated soils (1.06% to 2.23%, and from 0.78% to 4.23%) than in the bare soils (0.06% to 0.90%, and 0.27% to 2.57%) (Figure 5). In total, 4.4% of oryzalin and 9% of diuron were recovered in the vegetated soil profiles, but only 2.4% of oryzalin and 6.5% of diuron in the bare soil profiles. The difference between molecules' recoveries correlates well with the difference in their half-lives, from 5 days for oryzalin to the much longer 70 days for diuron, reported by Giry and Ayele (1998) for a sandy loam soil with a 2.4% OC content. Comparing the vegetated and bare soils, the results of Exp. I showed that there were higher amounts of herbicides in the leachates of the bare soil. In large part, the herbicides were confined to the upper 0-2.5 cm layer of the vegetated soil where 51% of the oryzalin and 47% of the diuron were recovered compared with 38% and 40%, respectively, in the bare soil. This finding was consistent with the higher OC content in the upper layer of the vegetated cambisol, induced by grass cover. It was also in agreement with the findings of Benoît et al. (2000) who noted that grass buffer strips enhance the retention of ^{14}C-isoproturon, especially in the form of bound residues.

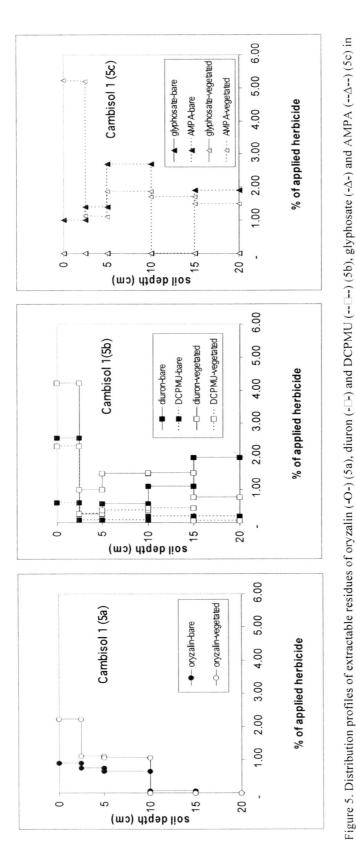

Figure 5. Distribution profiles of extractable residues of oryzalin (-O-) (5a), diuron (-□-) and DCPMU (--□--) (5b), glyphosate (-Δ-) and AMPA (--Δ--) (5c) in the bare (black symbols) and vegetated (white symbols) cambisol 1, one year after treatment (exp. I).

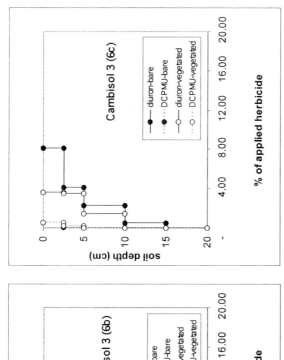
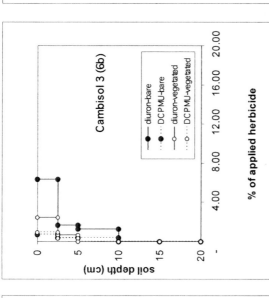
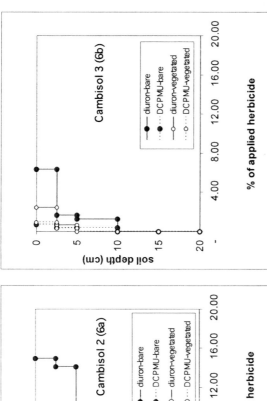

Figure 6. Distribution profiles of extractable residues of diuron (–O–) and DCPMU (--O--) in the bare (black symbols) and vegetated (white symbols) cambisol 2 (6a), cambisol 3 (6b) and cambisol 4 (6c), 6 months after treatment (exp. 1).

The only diuron metabolite detected in the soil profiles was DCPMU, in accord with the findings of Field et al. (2003). Other studies have shown that DCA and DCPU become rapidly soil bound (Kearney and Plimmer, 1972) perhaps due to the binding of humic acids and the NH_2 group (Andreux et al., 1992). DCPMU and AMPA were distributed throughout the column, i.e. reached the 20 cm depth (Figure 5). As with the parent molecules, the percentages of DCPMU and AMPA recovered were higher in the vegetated soil profile, 3.5% and 0.011%, than in the bare soil profile, 1.2% and 0.007% of the herbicide initially applied (Figure 5). This would explain the higher amounts leached of DCPMU and AMPA measured in the leachates of the bare soil, 0.12% and 0.15% compared to 0.05% and 0.03% for the vegetated cambisol. These results are in good agreement with those obtained by Gomez de Barreda et al. (1993) who report that DCPMU reached the 12-18 cm layer of a sandy soil under laboratory conditions. Like the parent molecules, the 0-2.5 cm surface layer of the vegetated cambisol contained the greatest amount of DCPMU, 2.3%, and AMPA, 0.005% than in the bare soil surface, 0.6% and 0.001%, respectively, for similar or higher organic carbon contents in the vegetated soil (Table 1). The grass cover of the vegetated soil seemed to enhance the adsorption of metabolites in the soil surface. In addition, the grass cover of this soil could stimulate microbial activity as shown by several authors (Benoit et al., 2000; Roy et al., 2001; Landry et al., 2005).

Diuron migrated through the soil columns in Exp. II to a lesser extent than it did in Exp. I, which might be due to the shorter experimental period duration of the second experiment (6 months) compared with the first (12 months). Unlike in Exp. I, diuron reached deeper layers (15 cm) in the bare soil profiles of Exp. II than in the vegetated soil profiles (10 cm). The only exception was the cambisol 2 through which diuron migration was similar in both the vegetated and bare soil (Figure 6). Smaller amounts of herbicide residues were recovered in the vegetated soils (4.3% to 10.6%) than in the bare soils (10.8% to 34.9%). These results agree with the greater amounts of diuron measured in the vegetated soil leachates than in the bare soil leachates. In addition, smaller amounts of diuron were recovered from the surface (0-2.5 cm) of the vegetated soils (3.1% to 4.4%) than the corresponding bare soil columns (8.2% to 19.3%) (Figure 6). The longer persistence of diuron in the bare soil than in the vegetated soil could be explained by its faster degradation in the rhizosphere of vegetated soil, as has been demonstrated in other studies (Benoit et al., 1999; Staddon et al., 2001 Jacobson et al., 2005). As mentioned previously, differences between the results of Exp. I and II may be explained by the more recent introduction of grass cover on the Exp. II soil columns (1 month) than in Exp. I (3 years).

As in Exp. I, the only diuron metabolite detected in the soil profiles, was DCPMU. Like diuron, DCPMU was recovered in greater amounts and at greater depths in the bare soils (0.7% to 5.5% at 15 cm) than in the vegetated soils (0.9% to 4.7% at 10 cm) (Figure 6). This would explain the higher amounts leached of DCPMU measured in the leachates of the grass covered soils, (0.02% to 0.35%) compared to the bare soil (0.0%). Also like diuron, the 0-2.5 cm surface layer of the vegetated cambisols contained less DCPMU (0.7% to 1.6%) than the bare soil surfaces, (0.7% to 2.4%) (Table 1), which contradicted our hypothesis that diuron would be more rapidly degraded in the rhizosphere of the vegetated soil.

3.6. Mass Balance

The amounts of oryzalin, diuron and glyphosate residues measured in the five soil-layer extracts and in the column leachates were summed for each column in Exp. I (Figure 7). The total amounts of oryzalin and diuron residues recovered were greater for the vegetated soil, 4.5% and 12.7% than for the bare soil, 2.5% and 8.6%, respectively. In this case, the dissipation of oryzalin and diuron was faster in the bare cambisol than the vegetated cambisol. It was the opposite for glyphosate. More residues were measured in the bare soil (0.28%) than in the vegetated soil (0.06%) one year after the glyphosate was applied. Nevertheless the glyphosate recoveries were very low compared to the oryzalin and diuron recoveries. In agreement with our results, Luchini et al. (1993) recovered 11%-13% of the applied ^{14}C-diuron in undisturbed sandy clay and clay soil columns, after one year under field conditions. Our oryzalin recovery values are slightly lower than those of Golab et al. (1975) who recovered only 8% of the ^{14}C-oryzalin applied to a sandy soil after one year under natural conditions. We did not find any information in the literature about the recovery rate of glyphosate under outdoor or field conditions. The total amounts of the herbicides applied to the soil columns in Exp. I and II were not recovered. Losses were huge, 95.6% and 97.6% for oryzalin, 87.5% and 92.4% for diuron and 99.9% and 99.8% for glyphosate in the bare and vegetated soils, respectively. These losses could have occurred during handling or could be due to volatilization, photodecomposition, mineralization, and/or formation of non-extractable residues. Volatilization of the herbicide molecules was probably negligible due to their low vapor pressures: $<1.3.10^{-3}$ mPa for oryzalin, 1.1×10^{-3} mPa for diuron and 9.3×10^{-3} mPa for glyphosate (Tomlin, 1997). The data from the literature about the UV decomposition of these three herbicides are not relevant to our studies because they were determined under conditions extremely different from those in the field. The formation of oryzalin metabolites (not identified due to the unavailability of analytical standards) could explain most of the losses observed for that molecule. Similarly, the degradation of diuron into metabolites other than DCPMU and DCPU could also explain its observed losses. Few data regarding oryzalin mineralization are available in the literature. Krieger et al. (1998) measured 10% ^{14}C-oryzalin mineralization in 6 months. However, based on its half-life values, which ranged from 17 to 77 days under outdoor conditions (Golab et al., 1975; Stoller and Wax, 1977), oryzalin appears to degrade quickly. Reported diuron mineralization ranges from 20% in 100 days (Walker and Robert, 1978) to 29-36% in 60 days (Luchini et al., 1993). Glyphosate mineralization, although measured under laboratory conditions very different from those in the field, is high, ranging from 43.5% to 46.5% in 42 days of incubation, in the bare and vegetated soils (Landry et al. 2005). Other authors working under laboratory conditions have also found high amounts of ^{14}C-glyphosate mineralized: 90% in 60 days of incubation (Cheah et al., 1998) and 69% to 75% in 90 days of incubation (Smith and Aubin, 1993). Thus a fast rate of glyphosate mineralization could explain the huge glyphosate losses measured in this study. The rapid mineralization of glyphosate could be due to 5 years of repeated glyphosate treatments to the bare and the vegetated cambisols, resulting in its enhanced degradation by the acclimated microbial biomass (Kilbride and Paveglio, 2001). In addition, non-extractable residues may have formed. There is no data available in the literature regarding non-extractable residue formation under outdoor conditions. However, under natural conditions, 55% of the total ^{14}C-oryzalin (Golab et al., 1975) and 21%- 28% of the total ^{14}C-diuron (Luchini et al., 1993) applied to a soil were in the form of bound residues within 1 year.

Working with similar soils, Eberbach (1999) showed that within 24 hours, 50% of the total glyphosate applied was non-extractable.

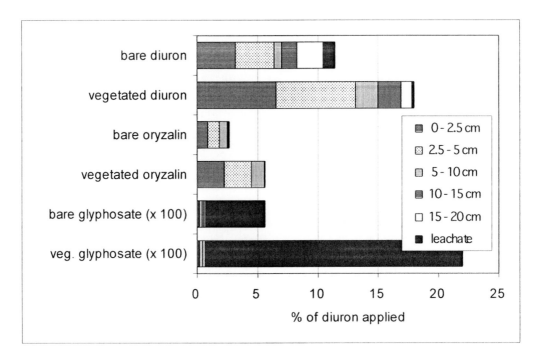

Figure 7. Mass balance of residues of oryzalin, diuron and glyphosate in the cambisol 1 (% herbicide applied).

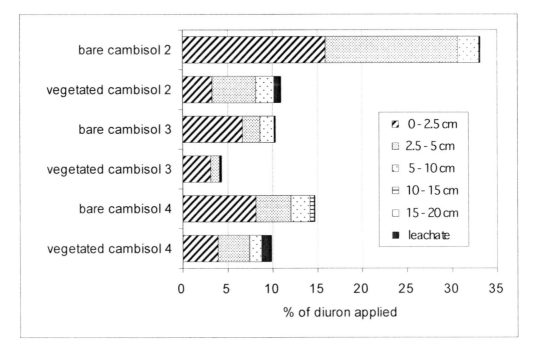

Figure 8. Mass balance of residues of diuron in the cambisol 2, 3 and 4 (% herbicide applied).

In Exp. II, the mass balance performed on diuron residues recovered in the soil columns and leachates was lower for the vegetated soils, from 10.9% to 35% than for the bare soils, from 4.5% to 11.8% (Figure 8). This result was contrary to what was seen in Exp. I. The difference is probably due to the accelerated degradation of diuron in the Exp. II vegetated soils. In addition to more rapid diuron mineralization of diuron, the formation of non-extractable residues might have been higher in the vegetated soil than in the bare soil, as was observed by Benoit et al. (2000) in their studies with isoproturon.

3. CONCLUSION

One year after treatment, similar amounts or more herbicide residues were recovered in the leachate from the bare cambisol 1, 0.10% for oryzalin, 0.96% for diuron + its metabolites, and 0.21% for glyphosate + AMPA than in the leachates of the vegetated cambisol 1, 0.10%, 0.16% and 0.05%, respectively. A second experiment, monitored for 6 months, showed some contradictory results with the previous study. For example, more diuron residues were found in the leachate of the soil columns that were initially covered with grass, from 0.10% to 0.48% than in the bare soil columns, from 0.09% to 0.16%. This difference in the results between the two experiments may be explained by the different duration of grass cover, 4 weeks for Exp. II versus 3-4 years for the Exp. I. Nevertheless, in general the soil grass cover seems to reduce the leaching of the three herbicides studied, and would decrease subsequently reduce the risk of groundwater contamination by pesticides.

In Exp. I, 4.4% of oryzalin, 12.5% of diuron + its metabolites, 0.01% of glyphosate + AMPA were recovered in the vegetated soil profile, and only 2.4%, 7.6% and 0.007% in the bare soil profile, respectively. Contrary to the first study, a second study found that greater amounts of diuron residues were recovered in the bare soils, from 10.7% to 34.9% than in the corresponding vegetated soils, from 4.3% to 10.6%. Finally, the dissipation of oryzalin and diuron seems faster in the bare soils than in the vegetated soil, whereas, glyphosate disappeared faster in the vegetated soils.

This work performed under outdoor conditions has demonstrated that the three herbicides studied and some of their metabolites may be transported through 20 cm of surface soil. This suggests that when applied to the shallow soils on the karstic terrain of Vosne-Romanée, the herbicides could reach groundwater. The results of the experiments suggest that the presence of well-established grass cover would reduce herbicide transport through soil to groundwater. With the proviso that these results be confirmed for other pesticides, the use of alternative soil management practice, such as grass cover under the vine rows should be encouraged to preserve groundwater quality under the vineyards of Burgundy.

ACKNOWLEDGEMENTS

This work was made possible thanks to a Chateaubriand Fellowship (2001-2002) awarded to Astrid Jacobson, a PhD grant from the Ministère de l'Education Nationale, de la Recherche et de la Technologie awarded to David Landry, and funding from the Region of

Burgundy. We also thank Météo France (Longvic, 21) for providing the meteorological data and the viticulturists who so graciously granted us access to their fields.

REFERENCES

Accinelli, C., Screpanti, C., Vicari A., & Catizone, P. 2004. Influence of insecticidal toxins from *Bacillus thruringiensis* subsp. *Kurstaki* on the degradation of glyphosate and glufosinate-ammonium in soil samples. *Agric. Ecosyst. Environ.* 103, 497-507.

Andreux, F., Portal, J.M., Schiavon, M., & Bertin, G. 1992. The binding of atrazine and its dealkylated derivatives to humic-like polymers derived from catechol. *Sci. Total Environ.* 117-118, 207-217.

Arora, K., Mickelson, SK., & Baker, J.L. 2003. Effectiveness of vegetated buffer strips in reducing pesticide transport in simulated runoff. *Trans. ASAE* 46, 635-644.

Belden, J.B., & Coats, J.R. 2004. Effect of grasses on herbicide fate in the soil column : infiltration of runoff, movement, and degradation. *Environ. Toxicol. Chem.* 23, 2251-2258.

Benoit, P., Souiller, C., Madrigal, I., Pot, V., Réal, B., Coquet, Y., Margoum, C., Laillet, B., Dutertre, A., Gril, J.J., & Barriuso, E. 2003. Fonctions environnementales des dispositifs enherbés en vue de la gestion et de la maîtrise des impacts d'origine agricole. *Etude et Gestion des Sols* 10, 299-312.

Benoit, P., Barriuso, E., Vidon, Ph., & Réal, B. 1999. Isoproturon sorption and degradation in a soil from grassed buffer strip. *J. Environ. Qual.* 28, 121-129.

Benoit, P., Barriuso, E., Vidon, Ph., & Réal, B. 2000. Isoproturon movement and dissipation in undisturbed soil cores from a grassed buffer strip. *Agronomie* 20, 297-307.

Boyd, P.M., Baker, J.L., Mickelson, S.K. & Ahmed, S.I. 2003. Pesticide transport with surface runoff and subsurface drainage through a vegetative filter strip. *Trans. ASAE* 46(3), 675-684.

Cheah, U., Kirkwood, R.C., & Lum, K. 1997. Adsorption, desorption and mobility of four commonly used pesticides in Malaysian agricultural soils. *Pestic. Sci.* 50, 53-63.

Cheah, U., Kirkwood, R.C., & Lum, K. 1998. Degradation of four commonly used pesticides in Malaysian agricultural soils. *J. Agric. Food Chem.* 46, 1217-1223.

De Jonge, H., & De Jonge, L.W. 1999. Influence of pH and solution composition on the sorption of glyphosate and prochloraz to a sandy loam soil. *Chemosphere* 39, 753-763.

Delphin, J.E., & Chapot, J.Y. 2001. Leaching of atrazine and deethylatrazine under a vegetative filter strip. *Agronomie* 21, 461-470.

Dillaha, T.A., Reneau, R.B., Mostaghimi, S. & Lee, D. 1989. Vegetative filter strips for agricultural nonpoint source pollution control. *Trans. ASAE* 32, 513-519.

DIREN, DRAF, DRASS, & FREDON Bourgogne 2004. Réseau de suivi des pesticides dans les eaux en région Bourgogne. Rapport 2003 de présentation des résultats de l'année hydrologique d'août 2002 à juillet 2003, 105p.

Dousset, S., Chauvin, C., Durlet, P., & Thévenot, M. 2004. Transfer of hexazinone and glyphosate through undisturbed soil columns in soils under Christmas tree cultivation. *Chemosphere* 57, 265-272.

Eberbach, P.L. 1999. Influence of incubation temperature on the behavior of triethylamine-extractable glyphosate (N-Phosphomethylglycine) in four soils. *J. Agric. Food Chem.* 47, 2459-2467.

European Community Council. 1998. Directive concerning the quality of water intended for human consumption. Official Journal of the European Communities, 98/83 EEC L330, European Union's publisher, *Luxembourg*, L, 32-54.

Feng, J.C., & Thompson, D.G. 1990. Fate of glyphosate in a Canadian forest watershed : 2. Persistence in foliage and soils. *J. Agric. Food Chem.* 38, 1118-1125.

Field, J.A., Reed, R.L., Sawyer, T.E., Griffith, S.M., & Wigington, P.J. 2003. Diuron occurrence and distribution in soil and surface and ground water associated with grass seed production. *J. Environ. Qual.* 32, 171-179.

Fishel, F.M., & Coats, G.E. 1993. Effect of commonly used turfgrass on bermudagrass (*Codon dactylon*) root growth. *Weed Sci.* 41, 641-647.

Food and Agriculture Organization of the United Nations, 1998. World reference base for soil resources. ISSS-ISRIC-FAO, FAO, Rome, Italy.

Futch, S.H., & Singh, M., 1999. Herbicide mobility using soil leaching columns. *Bull. Environ. Contam. Toxicol.* 62, 520-529.

Garmouma, M., Teil, M. J., Blanchard, M., & Chevreuil, M. 1998. Spatial and temporal variations of herbicide (triazine and phenylureas) concentrations in the catchment basin in Marne river (France). *Sci. Total Environ.* 224, 93-107.

Gerritse, R.G., Beltran, J., & Hernandez, F. 1996. Adsorption of atrazine, simazine, and glyphosate in soils of the Gnangara Mound, Western Australia. *Aust. J Soil Res.* 34, 599-607.

Giry, G. & Ayele, J. 1998. Détermination des paramètres de sorption sur le sol de trois herbicides: oryzalin, isoxaben et norflurazon. *Eur. J. Water Qual.* 29, 17-32.

Golab, T., Bishop, C.E., Donoho, A.L., Manthey, J.A., & Zornes, L.L. 1975. Behaviour of ^{14}C oryzalin in soil and plants. *Pestic. Biochem. Physiol.* 5, 196-204.

Gomez de Barreda, D., Gamon, M., Lorenzo, E. & Saez, A. 1993. Residual herbicide movement in soil columns. *Sci. Total Environ.* 132, 155-165.

Gonzáles-Pradas, E., Villafranca-Sánchez, M., Fernández-Pérez, M., Socías-Vicinia, M., & Ureña-Amate, M.D. 1998. Sorption and leaching of diuron on natural and peat-amended calcareous soil from spain. *Wat. Res.* 32, 2814-2820.

Gooddy, D.C., Chilton, P.J., & Harrison, I., 2002. A field study to assess the degradation and transport of diuron and its metabolites in a calcareous soil. *Sci. Total Environ.* 297(1-3), 67-83.

IFEN. 2004. Les pesticides dans les eaux. Sixième bilan annuel. Données 2002. Etudes et travaux n°42, 33p.

Isensee, A.R., Nash, R.G., & Helling, C.S. 1990. Effect of conventional vs. no-tillage on pesticide leaching to shallow groundwater. *J. Environ. Qual.* 19, 434-440.

Jacques, G.L., & Harvey, R.G. 1979. Adsorption and diffusion of dinitroaniline herbicides in soils. *Weed Sci.* 27, 450-455.

Jabobson, A.R., Dousset, S., Guichard, N., Baveye, P., & Andreux, F. 2005. Diuron mobility through vineyard soils contaminated with copper. *Environ. Pollut.* 138, 250-259.

Kearney, P.C., & Plimmer, J.R. 1972. Metabolism of 3,4-dichloroaniline in soils. *J. Agric. Food Chem.* 20, 584-585.

Kilbride, K.M. & Paveglio, F.L. 2001. Long-term fate of glyphosate associated with repeated Rodeo applications to control smooth cordgrass (Spartina alterniflora) in Willapa Bay, Washington. *Arch. Environ. Con. Tox.* 40, 179-183.

Kladivko, E.J., Van Scoyoc, G.E., Monke, E.J., Oates, K.M. & Pask, W. 1991. Pesticide and nutrient movement into subsurface tile drains on a silt loam soil in Indiana. *J. Environ. Qual.* 20: 264-270.

Kloppel, H., Kördel, W., & Stein, B. 1997. Herbicide transport by surface runoff and herbicide retention in a filter strip - rainfall and runoff simulation studies. *Chemosphere* 35, 129-141.

Krieger, M., Merritt, D.A., Wolt, J.D., Patterson, V.L. 1998. Concurrent patterns of sorption-degradation for oryzalin and degradates. *J. Agric. Food Chem.* 46, 3292-3299.

Krutz, L.J., Senseman, S.A., McInnes, K.J., Hoffmann, D.W. & Tierney, D.P. 2004. Adsorption and desorption of metolachlor and metolachlor metabolites in vegetated filter strip and cultivated soil. *J. Environ. Qual.* 33, 939-945.

Krutz, L.J., Senseman, S.A., McInnes, K.J., Zuberer, D.A., & Tierney, D.P. 2003. Adsorption and desorption of atrazine, desethylatrazine, deisopropylatrazine, and hydroxyatrazine in vegetated filter strip and cultivated soil. *J. Agric. Food Chem.* 51, 7379-7384.

Landry, D., Dousset, S., & Andreux, F. 2004. Laboratory leaching studies of oryzalin and diuron through three undisturbed vineyard soil columns. *Chemosphere* 54(6), 735-742.

Landry, D., Dousset, S., Fournier, J.C., & Andreux, F. 2005. Leaching of Glyphosate and AMPA under two soil management practices in Burgundy vineyards (Vosne-Romanée, 21-France). *Environ. Pollut.* 138, 191-200.

Landry, D., Dousset, S. & Andreux, F. 2006. Leaching of oryzalin and diuron through undisturbed vineyard soil columns under outdoor conditions. *Chemosphere* 62, 1736-1747.

Lennartz, B., Louchart, P., Andrieux, P., & Voltz, M. 1997. Diuron and simazine losses to runoff water in Mediterranean vineyards. *J. Environ. Qual.* 26, 1493-1502.

Liaghat, A., & Prasher, S.O. 1996. A lysimeter study of grass cover and water table depth effects on pesticides residues in drainage water. *Trans. ASAE* 39, 1731-1738.

Louchart, X., Voltz, M., Coulouma, G., & Andrieux, P. 2004. Oryzalin fate and transport in runoff water in Mediterranean vineyards. *Chemosphere* 57, 921-930.

Luchini, L.C., Costa, M.C., Ostiz, S.B., Musumeci, M.R., Nakagawa, L.E., De Andrea, M.M., & Matallo, M. 1993. Behaviour of diuron in a sandy clay and clay soils from Sao Paulo state, Brazil. In: Biagini-Lucca, G. (Ed), IX Symposium Pesticide Chemistry, Mobility and degradation of xenobiotics, *Piacenza*, pp. 127-133.

Madhun, Y.A., Freed, V.H., Young, J.L., & Fang, S.C. 1986. Sorption of bromacil, chlortoluron and diuron by soil. *Soil Sci. Soc. Am. J.* 50, 1467-1471.

Madrigal, I., Benoit, P., Barriuso, E., Etiévant, V., Souiller, C., Réal, B., & Dutertre, A. 2002. Capacités de stockage et d'épuration des sols de dispositifs enherbés vis-à-vis des produits phytosanitaires. Deuxième partie propriétés de rétention de deux herbicides, l'isoproturon et le diflufénicanil dans différents sols de bandes enherbées. *Etude et Gestion des Sols* 9, 287-302.

Mamy, L., & Barriuso, E. 2005. Glyphosate adsorption in soils compared to herbicides replaced with the introduction of glyphosate resistant crops. *Chemosphere* 61, 844-855.

Mersie, W., Seybold, C.A., McNamee, C. & Huang, J. 1999. Effectiveness of switchgrass filter strips in removing dissolved atrazine and metolachlor from runoff. *J. Environ. Qual.* 28, 816-821.

Mersie, W., Seybold, C.A., McNamee, C., & Lawson, M.A. 2003. Abating endosulfan from runoff using vegetative filter strips : the importance of plant species and flow rate. *Agric. Ecosyst. Environ.* 97, 215-223.

Müller, M.M., Rosenberg, C., Siltanen, H., & Wartiovaara, T. 1981. Fate of glyphosate and its influence on nitrogen cycling in two Finnish agricultural soils. *B. Environ. Contam. Tox.* 27, 724-730.

Newton, M., Horner, L.M., Cowell J.E., White, D.E., & Cole, E.C. 1994. Dissipation of glyphosate and aminomethylphosphonic acid in north American forests. *J. Agric. Food Chem.* 42, 1795-1802.

Patty, L., Réal, B., & Gril, J.J. 1997. The use of grassed buffer strips to remove pesticides, nitrate and soluble phosphorus compounds from runoff water. *Pestic. Sci.* 49, 243-251.

Pätzold, S., & Brümmer, G.W. 1996. Abbau-, sorptions- und verlagerungsverhalaten des herbizides diuron in einer obstbaulich genutzten parabraunerde aus Löb. Z. Pflanzenernähr. *Bodenk.* 160, 165-170.

Rae, J.E., Cooper, C.S., Parker, A., & Peters, A. 1998. Pesticide sorption onto aquifer sediments. *J. Geochem. Explor.* 64, 263-276.

Raturi, S., Carroll, M. J., & Hill, R. L. 2003. Turfgrass thatch effects on pesticide leaching: a laboratory and modeling study. *J. Environ. Qual.* 32, 215-223.

Roy, D.N., Konar, S.K., Banerjee, S., Charles, D.A., 1989. Persistence, movement and degradation of glyphosate in selected Canadian boreal forest soils. *J. Agric. Food Chem.* 37, 437-440.

Roy, J. W., Hall, J. C., Parkin, G. W., Wagner-Riddle, C. & Clegg, B.S. 2001. Seasonal leaching and biodegradation of dicamba in turfgrass. *J. Environ. Qual.* 30, 1360-1370.

Schmitt, T.J., Dosskey, M.G., & Hoagland, K.D. 1999. Filter strip performance and processes for different vegetation, widths, and contaminants. *J. Environ. Qual.* 28, 1479-1489.

Seybold, C., Mersie, W., & Delorem, D. 2001. Removal and degradation of atrazine and metolachlor by vegetative filter strips on clay loam soil. Commun. *Soil Sci. Plant Anal.* 32, 723-737.

Smith, A.E., & Aubin, A.J. 1993. Degradation of ^{14}C-glyphosate in Saskatchewan soils. *B. Environ. Contam. Tox.* 50, 499-505.

Souiller, C., Coquet, Y., Pot, V., Benoit, P., Réal, B., Margoum, C., Laillet, B., Labat, C., Vachier, P., & Dutertre, A. 2002. Capacités de stockage et d'épuration des sols de dispositifs enherbés vis-à-vis des produits phytosanitaires. Première partie : dissipation des produits phytosanitaires à travers un dispositif enherbé; mise en évidence des processus mis en jeu par simulation de ruissellement et infiltrométrie. *Etude et Gestion des Sols*, 9, 269-285.

Staddon, W.J., Locke, M.A., Zablotowicz, R.M. 2001. Microbiological characteristics of a vegetative buffer strip soil and degradation and sorption of metolachlor. *Soil Sci. Soc. Am. J.* 65, 1136-1142.

Stoller, E.W., & Wax, L.M., 1977. Persistence and activity of dinitroaniline herbicides in soil. *J. Environ. Qual.* 6: 124-127.

Tixier, C., Sancelme, M., Aït-Aïssa, S., Widehem, P., Bonnemoy, F., Cuer, A., Truffaut, N., & Veschambre, H. 2002. Biotransformation of phenylurea herbicides by a soil bacterial strain, Arthrobacter sp. N2: structure, ecotoxicity and fate of diuron metabolite with soil fungi. *Chemosphere* 46, 519-526.

Tomlin, C.D.S. 1997. *The Pesticide Manual,* 11th ed. British Crop Protection Council and The Royal Society of Chemistry, UK.

VanDijk, P.M., Kwadd, F.J.P.M., & Klapwijk, M. 1996. Retention of water and sediment by grass strips. *Hydrol. Process.* 10, 1069-1080.

Veiga, F., Zapata, J.M., Fernandez Marcos, M.L., & Alvarez, E. 2001. Dynamics of glyphosate and aminomethylphosphonic acid in a forest soil in Galicia, north-west Spain. *Sci. Total Environ.* 271, 135-144.

Vreeken, R. J., Speksnijder, P., Bobeldijk-Pastorova, I., & Noij, Th. H. M. 1998. Selective analysis of the herbicides glyphosate and aminomethylphosphonic acid in water by on-line solid-phase extraction-high-performance liquid chromatography-electrospray ionization mass spectrometry. *J. Chromatogr.* A 794, 187-199.

Walker, A., & Robert, M.G. 1978. The degradation of methazole in soil. II. Studies with methazole, methazole degradation products and diuron. Pestic. Sci. 9, 333-341.

Watanabe, H., & Grismer, ME. 2001. Diazinon transport through inter-row vegetative filter strips : micro-ecosystem modeling. *J. Hydrol.* 247, 183-199.

Zsolnay, A. 1996. Dissolved humus in soil waters. In Piccolo, A. (Ed.), *Humic substances in terrestrial ecosystems.* Elsevier, Amsterdam, pp. 171-224.

In: Progress in Environmental Research
Editor: Irma C. Willis, pp. 155-173

ISBN 978-1-60021-618-3
© 2007 Nova Science Publishers, Inc.

Chapter 4

BUREAU OF LAND MANAGEMENT (BLM) LANDS AND NATIONAL FORESTS[*]

Ross W. Gorte and Carol Hardy Vincent

SUMMARY

The 109th Congress is considering issues related to the public lands managed by the Bureau of Land Management (BLM) and the national forests managed by the Forest Service (FS). The Administration is addressing issues through budgetary, regulatory, and other actions. Several key issues of congressional and administrative interest are covered here.

Energy Resources

The Energy Policy Act of 2005 has been enacted into law and affects energy development on federal lands in a variety of ways, including through changes to the federal oil, gas, and coal leasing programs and the application of environmental laws to certain energy-related agency actions. Significant changes at the administrative level may be forthcoming in response to the legislation's enactment. New legislation has also been introduced to address evolving energy issues related to public lands.

Wild Horses and Burros

Controversial changes to the Wild Free-Roaming Horses and Burros Act of 1971 gave the agencies authority to sell certain old and unadoptable animals and removed the ban on selling wild horses and burros and their remains for commercial products. BLM has resumed animal sales with provisions to prevent their slaughter. Bills have been

---

[*] Excerpted from CRS Report IB10076, dated October 26, 2005.

introduced to overturn the changes (H.R. 297/S. 576) and to foster adoptions and sales (H.R. 2993/S. 1273).

Wilderness

Many wilderness recommendations for federal lands are pending. Questions persist about wilderness review and managing wilderness study areas (WSAs). Bills to designate areas have been introduced, and the 109th Congress may address wilderness review and WSA protection.

Wildfire Protection

The Healthy Forests Restoration Act of 2003 (P.L. 108-148), President Bush's Healthy Forests Initiative, and other provisions may help protect communities from wildfires by expediting fuel reduction. Wildfire protection also has been addressed through changes in regulations. The 2005 fire season is on a pace to be the worst on record. The 109th Congress is conducting oversight on fire protection and implementation of the law and regulations. Hurricane Katrina damaged southern forests, exacerbating fuel problems.

Southern Nevada Land Sales

The Southern Nevada Public Land Management Act allows the Secretary of the Interior to sell land near Las Vegas, with the proceeds permanently appropriated for certain purposes. The President has proposed altering the distribution of receipts, with 70% going to the Treasury rather than to a special account. No related legislation has been introduced.

R.S. 2477 Rights of Way

Revised Statute (R.S.) 2477 granted rights of way to construct highways across unreserved federal lands, but the extent of valid rights of way is unclear in some states. Bush Administration regulations on "disclaimers of interest" may be used to clear title to R.S. 2477 highway easements; this may "pertain to" R.S. 2477 which has been prohibited by Congress.

Other Issues

The Administration and Congress have addressed other issues, as well, including competitive sourcing, grazing management, national forest planning, FS NEPA categorical exclusions, national forest roadless areas, hardrock mining, and national monuments.

MOST RECENT DEVELOPMENTS

P.L. 109-54, the FY2006 Interior appropriations act, capped funds for DOI and FS competitive sourcing studies; barred funds for energy leasing activities in national monuments; and provided funds for wildfire protection for management of wild horses and burros.

As of October 14, wildfires in 2005 have burned 8.2 million acres, more than any of the past five years, nearly 80% above the 10-year average, and on a pace to exceed the post-1960 record of 8.4 million acres (in 2000).

As of September 20, 2005, 1,445 wild horses and burros have been sold under a new authority enacted in December of 2004.

On August 9, 2005, BLM announced an intent to prepare a supplement to its final environmental impact statement on proposed changes to regulations on livestock grazing.

On July 2, 2005, a U.S. District Court ruled that certain FS regulations related to categorical exclusions from NEPA for certain decisions (and thus also exempt from certain public challenges) violated the law. The FS has responded by suspending more than 1,500 permits, projects, and contracts.

BACKGROUND AND ANALYSIS

The Bureau of Land Management (BLM) in the Department of the Interior (DOI) and the Forest Service (FS) in the Department of Agriculture (USDA) manage 454 million acres of land, two-thirds of the land owned by the federal government and one-fifth of the total U.S. land area. The BLM manages 261.5 million acres of land, predominantly in the West. The FS administers 192.5 million acres of federal land, also concentrated in the West.

The BLM and FS have similar management responsibilities for their lands, and many key issues affect both agencies' lands. However, each agency also has unique emphases and functions. For instance, most BLM lands are rangelands, and the BLM administers mineral development on all federal lands. Most federal forests are managed by the FS, and only the FS has a cooperative program to assist nonfederal landowners. Moreover, development of the two agencies has differed, and historically they have focused on different issues.

History of the Bureau of Land Management

For the BLM, many of the issues traditionally center on the agency's responsibilities for land disposal, range management (particularly grazing), and minerals development. These three key functions were assumed by the BLM when it was created in 1946, by the merger of the General Land Office (itself created in 1812) and the U.S. Grazing Service (created in 1934). The General Land Office had helped convey land to settlers and issued leases and administered mining claims on the public lands, among other functions. The U.S. Grazing Service had been established to manage the public lands best suited for livestock grazing under the Taylor Grazing Act of 1934 (TGA, 43 U.S.C. §§315, et seq.).

Congress frequently has debated how to manage federal lands, and whether to retain or dispose of the remaining public lands. In 1976, Congress enacted the Federal Land Policy and Management Act of 1976 (FLPMA, 43 U.S.C. §§1701, et seq.), sometimes called BLM's Organic Act because it consolidated and articulated the agency's responsibilities. Among other provisions, the law establishes a general national policy that the BLM-managed public lands be retained in federal ownership, establishes management of the public lands based on the principles of multiple use and sustained yield, and generally requires that the federal government receive fair market value for the use of public lands and resources. Today BLM public land management encompasses diverse uses, resources, and values, such as energy and mineral development, timber harvesting, livestock grazing, recreation, wild horses and burros, fish and wildlife habitat, and preservation of natural and cultural resources.

History of the Forest Service

The FS was created in 1905, when forest lands reserved by the President (beginning in 1891) were transferred from DOI into the existing USDA Bureau of Forestry (initially an agency for private forestry assistance and forestry research). Management direction for the national forests, first enacted in 1897 and expanded in 1960, identifies the purposes for which the lands controlled by the Forest Service are to be managed and directs "harmonious and coordinated management" to provide sustained yields of the many resources found in the national forests — including timber, grazing, recreation, wildlife and fish, and water.

Many issues concerning national forest management and use have focused on the appropriate level and location of timber harvesting. Major conflicts over clearcutting began in the 1960s, and litigation in the early 1970s successfully challenged FS clearcutting in West Virginia and elsewhere. In part to address these issues, Congress enacted the National Forest Management Act of 1976 (NFMA; P.L. 94-588) to revise timber sale authorities and to elaborate on considerations and requirements in land and resource management plans. This NFMA planning has been widely criticized as expensive, time-consuming, and ineffective for making decisions and informing the public. (See "Other Issues," below.)

Wilderness protection also is a continuing issue for the FS. The Multiple-Use Sustained-Yield Act of 1960 (16 U.S.C. §528-531) included wilderness as an appropriate use of national forest lands, and possible national forest wilderness areas have been reviewed under the 1964 Wilderness Act as well as in the national forest planning process. Pressure to protect pending wilderness recommendations and other areas contributed to the Clinton Administration's decision to protect "roadless areas" not designated as wilderness. (See "Other Issues," below.)

Scope of Issue Brief

Many issues affecting BLM and FS lands are similar, and the missions of the agencies are nearly identical. By law, the BLM and FS lands are to be administered for multiple uses, although slightly different uses are specified for each agency. In practice, the land uses considered by the agencies include recreation, range, timber, minerals, watershed, wildlife and fish, and conservation. BLM and FS lands also are required to be managed for sustained yield — a high level of resource outputs in perpetuity, without impairing the productivity of the lands. Further, many issues, programs, and policies affect both agencies. For these reasons, BLM and FS lands often are discussed together, as in this issue brief.

This brief focuses on several issues affecting BLM and FS lands that are of interest to the 109th Congress. While in some cases the issues discussed here are relevant to other federal lands and agencies, this brief does not comprehensively cover issues primarily affecting other federal lands, such as the National Park System (managed by the National Park Service, DOI) or the National Wildlife Refuge System (managed by the Fish and Wildlife Service, DOI). For background on federal land management generally, see CRS Report RL32393, *Federal Land Management Agencies: Background on Land and Resources Management*. For brief, general information on natural resource issues, see CRS Report RL32699, *Natural Resources: Selected Issues for the 109th Congress*. Information on FY2006 appropriations for the BLM and FS (and other agencies and programs funded by Interior and Related Agencies appropriations bills) is included in CRS Report RL32893, *Interior, Environment, and Related Agencies: FY2006 Appropriations*. For information on park and recreation issues, see CRS Issue Brief IB10145, *National Park Management*, and CRS Issue Brief IB10141, *Recreation on Federal Lands*. For information on oil and gas leasing in the Arctic National Wildlife Refuge (ANWR), see CRS Issue Brief IB10136, *Arctic National Wildlife Refuge (ANWR): Controversies for the 109th Congress*. For information on local compensation for the tax-exempt status of federal lands, see CRS Report RL31392, *PILT (Payments in Lieu of Taxes): Somewhat Simplified*, and CRS Report RS22004, *The Secure Rural Schools and Community Self-Determination Act of 2000: Forest Service Payments to Counties*. For information on other related issues, see the CRS web page at [http://www.crs.gov/].

Energy Resources (by Aaron M. Flynn)

Background

BLM administers the Mineral Leasing Act of 1920, which governs the leasing of onshore oil and gas, coal, and other minerals on many federal lands, including lands managed by the BLM and the FS. Leasing on BLM lands goes through a multi-step approval process. If the minerals are located on FS lands, the FS must perform a leasing analysis and approve leasing decisions for specific lands before BLM may lease minerals.

A controversial issue is whether and how to increase access to federal lands for energy and mineral development. A BLM study (Dec. 1, 2000) determined that, of the roughly 700 million acres of federal minerals, (1) about 165 million acres (24%) have been withdrawn from mineral entry, leasing, and sale, subject to valid existing rights, and (2) mineral development on another 182 million acres (26%) is subject to the approval of the surface management agency and must not be in conflict with land designations and plans. In January 2003, several federal agencies issued a similar assessment, *Scientific Inventory of Onshore Federal Lands' Oil and Gas Resources and Reserves and the Extent and Nature of Restrictions or Impediments to Their Development*. Some assert that these reports show that more federal lands currently are available for energy development than generally had been realized, while others focus on the amount of lands withdrawn.

The oil and gas industry contends that entry into areas off-limits to development, particularly in the Rocky Mountain region, is necessary to ensure future domestic oil and gas supplies. Opponents maintain that the restricted lands are environmentally sensitive or unique, and the United States could realize equivalent energy gains through conservation and increased exploration elsewhere. (For more information, see CRS Report RL33014, *Leasing and Permitting for Oil and Gas Development on Federal Public Domain Lands*.)

Administrative Actions

Executive Order 13212 (May 18, 2001) established a policy of encouraging increased energy production on federal lands, and a series of administrative actions have followed to implement this policy. Recent FS land management planning regulations (70 *Fed. Reg.* 1023, Jan. 5, 2005) have been promulgated to increase management flexibility and streamline energy project permitting, among other things. (See "Other Issues," below.) The BLM and FS have also proposed significant changes to regulations governing the approval of oil and gas leases (70 *Fed. Reg.* 43349, July 27, 2005). The changes would include new requirements for development on split estates, a new approval process for multiple wells based on a single environmental review and a Master Development Plan, and additional bonding requirements. The proposal would also encourage the use of various best management practices aimed at reducing surface, visual, and wildlife impacts.

Additionally, BLM has issued a final rule (70 *Fed. Reg.* 58854, Oct. 7, 2005) governing the fees the agency will charge for processing documents associated with mineral development. In accordance with the Energy Policy Act of 2005, certain fee changes associated with oil and gas applications for permits to drill and geothermal exploration and drilling permits have been deferred until the completion of a permitting pilot program.

Legislative Activity

The Energy Policy Act of 2005 (P.L. 109-58) has been enacted into law. The new law affects federal lands in a variety of ways. It significantly amends the Geothermal Steam Act of 1970, providing new guidelines for geothermal development of BLM and FS lands, imposing new royalty rates, and deeming geothermal leasing consistent with existing land management plans. The law also amends the Mineral Leasing Act to modify statutory requirements governing federal coal leases, ending the 160-acre limit on coal lease modifications and allowing certain mining operations to continue beyond the current 40-year limitation. In addition, the law makes several changes in the regulation of onshore federal oil and gas. The Secretary of the Interior must evaluate and streamline the existing oil and gas leasing and permitting processes and establish a Federal Permit Streamlining Pilot Project. Oil and gas lease acreage limitations are also relaxed and a five-year reclamation program for abandoned well sites on federal lands is authorized. Additional provisions include requirements that DOI and USDA establish utility corridors on federal lands and that the Secretary of Interior take additional steps to move forward with oil shale leasing.

The House has recently passed new energy policy legislation, and it too would affect the administration of BLM and FS lands, if enacted. The Gasoline for America's Security Act, H.R. 3893, would, among other things, require the President to designate certain federal lands, which might include lands under the jurisdiction of BLM or FS, as suitable for refinery construction or expansion. Upon such designation, an expedited permitting process would be available for a refinery sited in the designated area.

Whether to open the Arctic National Wildlife Refuge (ANWR) to oil and gas development continues to be one of the most contentious issues in the energy debate. This issue may be addressed through the budget reconciliation process.

Finally, legislation affecting energy development on federal lands has been introduced in response to Hurricane Katrina. The Hurricane Katrina Energy Emergency Relief Act, H.R. 3710, would require the suspension of any royalty relief program applicable to oil or natural gas production from federal lands, so long as specified commodity prices were maintained. The resulting royalty payments would then be available at the discretion of the President for authorized disaster relief and for the Low-Income Home Energy Assistance Program. Additional bills have been introduced in the 109th Congress addressing specific energy and other mineral leasing issues, such as geothermal energy access, potash or soda ash royalties, and coal leasing procedures. However, the numerous bills on specific energy and other mineral leasing issues are not listed in the "Legislation" section of this article.

Wild Horses and Burros (by Carol Hardy Vincent)

Background

The Wild Free-Roaming Horses and Burros Act of 1971 (16 U.S.C. §§1331, et seq.) sought to protect wild horses and burros on federal land and placed them under the jurisdiction of BLM and the FS. For years, management of wild horses and burros has generated controversy and lawsuits. Controversies have involved the method of determining the "appropriate management levels" (AMLs) for herd sizes, as the statute requires; whether and how to remove animals from the range to achieve AMLs; alternatives to adoption for reducing wild horses and burros on the range, particularly fertility control and holding animals in long-term facilities; and whether appropriations for managing wild horses and burros are adequate. There was particular concern that adopted horses were slaughtered, despite prohibitions on that practice. (For background, see CRS Report RS21423, *Wild Horse and Burro Issues.*)

The 108th Congress enacted changes to wild horse and burro management on federal lands (§142, P.L. 108-447). These changes have intensified controversies. One change gave the agencies new authority to sell, "without limitation," excess animals (or their remains) that essentially are deemed too old (more than 10 years old) or otherwise unable to be adopted (tried unsuccessfully at least three times). Proceeds are to be used for the BLM wild horse and burro adoption program. A second change removed the ban on wild horses and burros and their remains being sold for processing into commercial products. A third change removed criminal penalties for processing into commercial products the remains of a wild horse or burro, if it is sold under the new authority. Also, the law did not expressly prohibit BLM from slaughtering healthy wild horses and burros, as annual appropriations bills apparently had since FY1988. These changes have been supported as providing a cost-effective way to help the agencies achieve AML, to improve the health of the animals, to protect range resources, and to restore a natural ecological balance on federal lands. The changes have been opposed as potentially leading to the slaughter of healthy animals.

Administrative Actions

On April 25, 2005, BLM temporarily suspended sale and delivery of wild horses and burros, due to concerns about the slaughter of some animals sold under the new authority. According to BLM, 41 animals that were sold under the new authority were subsequently resold or traded, and then sent to slaughterhouses by the new owners. Another 52 animals also had been sold to slaughterhouses, but Ford Motor Co. committed to purchasing them. On May 19, 2005, the agency resumed sales after revising its bill of sale and pre-sale negotiation procedures to protect sold animals from slaughter. Purchasers formerly gave written affirmation of an intent to provide humane care, and now also must agree not to knowingly sell or transfer ownership of animals to persons or organizations that intend to resell, trade, or give away animals for processing into commercial products. Sales contracts also now incorporate criminal penalties for anyone who knowingly or willfully falsifies or conceals information. Some horse advocates have questioned whether the penalties would withstand legal challenge because the law provides for the sale of animals without limitation. Also, according to BLM, purchased animals are classified as private property free of federal protection. BLM also pursued agreements with the three U.S. horse processing plants to not purchase horses sold under the new law. While there are no written agreements, the plants apparently have taken steps to preclude accepting these animals.

According to the BLM, 8,400 wild horses and burros initially were affected by the new law. There are about 7,000 animals available for sale currently, with 1,445 having been sold and delivered as of September 20, 2005. BLM has been negotiating sales of groups of excess animals, for instance with ranchers, tribes, and horse, humane, and other organizations, with the price determined on a case-by-case basis. Ranchers, horse advocates, and other prospective purchasers are considering or promoting several related ideas. They include outsourcing the sale of wild horses and burros; creating private sanctuaries as tourist attractions; raising funds for wild horses and burros by selling horse sponsorships; and allowing proceeds of land disposals to be used for wild horse and burro management.

As of February 2005, there were about 32,000 wild horses and burros on the range, with the national maximum AML set at 28,000, according to BLM. BLM has been pursuing a multi-year effort to achieve AML. Some critics assert that the current AMLs are set low in favor of livestock. BLM manages another 24,500 animals in holding facilities, as of October 2005. For management of wild horses and burros during FY2006, BLM requested $36.9 million, a reduction of $2.1 million (5%) from the FY2005 level of $39.0 million. The agency asserted that the reduction can be accomplished through program efficiencies, such as a reduction in the cost of the adoption program; an increase in animals adopted; and an expected reduction during FY2005 of 5,000 animals in long-term holding facilities. The cost per animal per year in these facilities is $465-$500, according to varying BLM estimates.

Legislative Activity

P.L. 109-54, the FY2006 Interior appropriations law, provided $36.9 million for BLM management of wild horses and burros (excluding a rescission of 0.476%), and an additional $1.2 million in fees collected from adoptions. It did not prohibit funds for the sale or slaughter of wild horses and burros, as originally passed by the House. In addition, bills have been introduced (H.R. 297 and S. 576) to overturn the changes to wild horse and burro management enacted during the 108th Congress. (See "Background," above.) Other bills (H.R. 2993 and S. 1273) seek to foster the sale and adoption of wild horses and burros while establishing further protections. Changes include eliminating the limit of four animals per adopter per year; reducing the minimum adoption fee from $125 to $25 per animal; removing the provision that excess, unadoptable animals be destroyed in a humane and cost-effective manner and making them available for sale; imposing a one-year wait period before buyers obtain title to sold animals, and removing the provision for sale of animals "without limitation." Some opponents fear that additional sales or adoptions could increase the risk of slaughter.

Wilderness (by Ross W. Gorte and Pamela Baldwin)

Background

The Wilderness Act established the National Wilderness Preservation System in 1964 and directed that only Congress could designate federal lands as wilderness. Designations are often controversial because commercial activities, motorized access, and roads, structures, and facilities generally are restricted in wilderness areas. (See CRS Report RS22025, *Wilderness Laws: Permitted and Prohibited Uses.*) Similarly, agency wilderness studies are controversial because many uses also are restricted in the study areas to preserve wilderness characteristics while Congress considers possible designations.

Some observers believe that the Clinton rule protecting national forest roadless areas (discussed below) was prompted by a belief that Congress had lagged in designating areas which "should" be wilderness. Others assert that the Bush Administration — in addressing R.S. 2477 rights-of-way (discussed below), promulgating new guidance to end additional, formal BLM wilderness study areas, and eliminating the nationwide national forest roadless area protections of the Clinton Administration — is attempting to open areas with wilderness attributes to roads, energy and mineral exploration, and development, thereby making them ineligible to be added to the Wilderness System.

Administrative Actions

The Wilderness Act directed the Secretary of Agriculture to review the wilderness potential of administratively designated national forest primitive areas and the Secretary of the Interior to review the wilderness potential of National Park System and National Wildlife Refuge System lands. The Forest Service expanded its review and sent recommendations to the President and Congress in 1979. *Release language*, in statutes designating national forest wilderness areas, and the new FS planning regulations (36 C.F.R. §219.7(a)(5)(ii)) provide for periodic review of potential national forest wilderness areas in the agency's planning process.

The Secretary of the Interior was directed to review the wilderness potential of BLM lands in §603 of FLPMA, and to maintain the wilderness character of wilderness study areas (WSAs) "until Congress has determined otherwise." In 1996, following debate over additional wilderness areas proposed in legislation for Utah, then-Secretary Bruce Babbitt used the BLM authority to inventory its lands and resources (§201 of FLPMA; 43 U.S.C. §1711) to identify an additional 2.6 million acres in Utah as having wilderness qualities. The state of Utah filed suit alleging that the inventory was illegal. On September 29, 2003, Interior Secretary Gale Norton settled the case and issued new wilderness guidance (Instruction Memoranda Nos. 2003-274 and 2003-275) prohibiting further wilderness reviews and limiting the *nonimpairment* standard of management to the BLM's previously designated WSAs. (See CRS Report RS21917, *Bureau of Land Management (BLM) Wilderness Review Issues.*)

Legislative Activity

Many wilderness recommendations remain pending, including some FS areas and many BLM and Park System areas. Nearly 20 such bills for wilderness areas in more than a dozen states have been introduced this Congress. The "Legislation" section of this article does *not* identify these bills; it identifies bills to substantively amend the Wilderness Act or alter wilderness or WSA management.

Bills were introduced in the 106th, 107th, and 108th Congresses to prohibit future BLM wilderness reviews and to place time limits on WSA status, generally terminating WSAs 10 years after the bills' enactment or after Congress establishes new WSAs. The House Committee on Resources reported bills in the 106th and 107th Congresses, but there was no floor consideration. No action occurred in the 108th Congress. To date, no wilderness review or WSA legislation had been introduced in the 109th Congress.

Wildfire Protection (by Ross W. Gorte)

Background

Recent fire seasons have killed firefighters, burned homes, threatened communities, and destroyed trees. The 2005 fire season is on a pace to be the worst since record-keeping began in 1960, with 8.2 million acres burned through October 14, nearly 80% above the 10-year average. Many assert that the threat of severe wildfires has grown, because many forests have unnaturally high fuel loads (e.g., dense undergrowth and dead trees) and increasing numbers of structures are in and near the forests (the *wildland-urban interface*). Reducing fuels on federal lands has been urged as a way to reduce the threats from fire. Proponents of fuel reduction contend that needed treatments often are delayed by environmental studies, administrative appeals, and litigation. Opponents of accelerated review processes argue that *streamlining* fuel projects could increase logging on federal lands, that such projects might not receive proper environmental review, and that reducing fire risk in the interface requires reducing fuels and modifying structures on private lands. The National Fire Plan is the program of wildfire protection activities and funding for the FS and BLM.

Administrative Actions

In August 2002 (107th Congress), President Bush proposed a Healthy Forests Initiative to improve wildfire protection by expediting projects to reduce hazardous fuels on federal lands. Congress enacted the Healthy Forests Restoration Act of 2003 (P.L. 108-148) with many of the proposals in the President's initiative and other provisions (described below under "Legislative Activity").

The Administration has made several regulatory changes to facilitate fire protection activities, which are unaffected by P.L. 108-148. First, two new categories of actions can be excluded from NEPA analysis and documentation: fuel reduction and post-fire rehabilitation activities (68 *Fed. Reg.* 33814, June 5, 2003). These categorical exclusions are limited in scale, and cannot be used in certain areas or under certain circumstances, but may be used for timber sales if fuel reduction is the primary purpose. Second, the administrative review processes were revised (68 *Fed. Reg.* 33582, June 4, 2003, for the FS; 68 *Fed. Reg.* 33794, June 5, 2003, for the BLM). The revisions sought (1) to clarify that some emergency actions may be implemented immediately and others after complying with publication requirements; and (2) to expand emergencies to include those "that would result in substantial loss of economic value to the Government if implementation of the proposed action were delayed." A U.S. District Court found these and other regulations violate the legal requirements for public review of FS decisions. (See "Other Issues," below.)

The Administration has made other regulatory changes that could affect fuel reduction, public involvement, and environmental impacts. For example, new categorical exclusions for small timber harvesting projects (68 *Fed. Reg.* 44598, July 29, 2003) and new regulations for FS planning (70 *Fed. Reg.* 1023, Jan. 5, 2005; see "Other Issues," below) have been completed. The total impact of the regulatory changes is generally greater discretion for FS action without environmental studies and with fewer opportunities for the public to comment on, or to request administrative review of, those actions.

Legislative Activity

H.R. 1904, the Healthy Forests Restoration Act of 2003, was signed into law (P.L. 108-148) on December 3, 2003. (See CRS Report RS22024, *Wildfire Protection in the 108th Congress.*) Title I authorized a new, alternative process for reducing fuels on FS or BLM lands in many areas. The act contained five other titles that indirectly relate to fire protection.

The 109th Congress is overseeing the implementation of this law. On February 17, 2005, a House Resources subcommittee hearing focused on a Government Accountability Office (GAO) review of progress in wildfire protection. The GAO report (GAO-05-147) and testimony (GAO-05-353T) found some progress, but noted that the agencies lack a cohesive long-term strategy for addressing excess fuels and other wildfire threats. On April 26, 2005, the Senate Energy and Natural Resources Committee held a hearing on wildfire preparedness. GAO's report (GAO-05-380) and testimony (GAO-05-627T) noted that the knowledge and technology for protecting structures and improving communications exist, but their adoption by landowners and governmental agencies has been slow. Also, questions were raised about the airworthiness of firefighting airtankers, in the wake of an airtanker crash in California on April 20, 2005. On July 14, 2005, witnesses testified before a House Appropriations subcommittee on progress in implementing the National Fire Plan; GAO testified on the need for identifying long-term options and funding needs (GAO-05-923T). On August 31, 2005, the House Resources Subcommittee on Forests and Forest Health held a field hearing on the health of the Black Hills (SD) National Forest.

Congress also has addressed wildfire protection through appropriations. The FY2006 Interior appropriations act (P.L. 109-54) included $2.59 billion for the National Fire Plan, $76.7 million (3%) more than the Administration requested and $413.2 million (14%) less than the FY2005 appropriations (including $524.1 million of emergency and supplemental funds for FY2005).[1] The law also included provisions requiring a report on the Biscuit fire (OR) rehabilitation and consideration of the effects of competitive sourcing (see below) on wildfire protection. (See CRS Report RL32893, *Interior, Environment, and Related Agencies: FY2006 Appropriations.*) In addition, bills have been introduced to alter firefighter and fire organization compensation and safety practices, and a provision was included in the Energy Policy Act of 2005 (§210) authorizing grants for producing energy from biomass fuels removed from forests to reduce wildfire risks.

Southern Nevada Public Land Management Act (by Carol Hardy Vincent)

Background

Historically, proceeds from the sale of BLM lands under various laws were deposited in the general fund of the Treasury. Certain recent laws have provided for land sales and established separate Treasury accounts available to the Secretary for subsequent land acquisition and other purposes. A proposal in the President's FY2006 budget seeks to change one such law — the Southern Nevada Public Land Management Act (SNPLMA, P.L. 105-263) — to send most proceeds to the Treasury.

SNPLMA allows the Secretary of the Interior to sell or exchange certain lands around Las Vegas, NV. The Secretary and the relevant local government unit jointly choose the lands offered for sale or exchange. In practice, these responsibilities of the Secretary are performed by the BLM. State and local governments get priority to acquire lands for local purposes under the Recreation and Public Purposes Act (43 U.S.C. §869). Proceeds are distributed in different ways, depending on which lands are sold. In general, 85% is deposited into a special account, which is permanently appropriated for certain purposes, including (1) federal acquisition of environmentally sensitive lands in Nevada; (2) development of a multi-species habitat conservation plan in Clark County, NV; (3) conservation initiatives on federal land in Clark County; (4) capital improvements at certain federal areas; and (5) development of parks, trails, and natural areas in Clark County. The other 15% of the revenues are provided to the state of Nevada and certain local entities for state and local purposes, such as the Nevada general education program.

The law was enacted in part to promote sale of federal land for development near fast-growing Las Vegas, to acquire environmentally sensitive land, and to foster competition in land disposals in response to criticisms that the government did not consistently receive a fair price for land it sold. Collections from SNPLMA land sales in FY2005 are estimated at $1.2 billion, vastly exceeding expectations at the time the law was enacted ($70 million annually) and more than double the amount collected in FY2004 ($530.5 million).

Administrative Actions

The President's FY2006 budget request supported amending SNPLMA to change the allocation of revenue to the special account. The Administration recommended that 15% of the receipts go to the special account and 70% go to the Treasury, with the remaining 15% to the state of Nevada and local entities as under current law. The Administration stated that because SNPLMA land sales have produced receipts far beyond expectations, there is significantly more revenue than is needed for land acquisition in Nevada. Consequently, proceeds of land sales increasingly are being used for local projects which are not overseen by Congress, thus reducing accountability, and do not reflect the highest needs of the nation, according to the Administration. Further, the change would still provide for far more money for Nevada than anticipated when the law was enacted, according to the Administration. The SNPLMA proposal could be opposed as impeding development in the Las Vegas area, federal acquisition of land with valuable resources, and conservation and recreation initiatives in Clark County. It is one of many changes advocated by the Administration that affect receipts or spending levels in FY2006 or subsequent years.

Legislative Activity

Administration budget documents for FY2006 stated that the President intends to submit a legislative proposal to accomplish his desired change regarding SNPLMA receipts. The Administration has not done so to date, according to the BLM, and no related legislation has been introduced to amend SNPLMA. P.L. 109-54, the FY2006 Interior appropriations act, did not include a House-passed provision to require the Secretary of the Interior to report on expenditures under SNPLMA during FY2003 and FY2004.

R.S. 2477: Rights of Way Across Public Lands (by Pamela Baldwin)

Background

In 1866, in an act that became Revised Statute (R.S.) 2477, Congress granted rights of way across unreserved public lands "for the construction of highways." This grant was repealed in 1976, but existing rights were protected. What constitutes construction of highways and whether a qualifying right of way existed by the time of repeal in 1976 can be contentious. These issues are important because possible rights of way may affect the management of federal lands, perhaps degrading their wilderness suitability while increasing access for recreation and other uses. Section 108 of the FY1997 Interior appropriations act (P.L. 104-208) states that final regulations "pertaining to" R.S. 2477 rights of way cannot take effect unless expressly authorized by an act of Congress.

Administrative Actions

On January 6, 2003 (68 *Fed. Reg.* 494), the BLM finalized changes to its regulations for issuing "disclaimers of interest," a procedure to help clear title to property or interests in property with respect to possible interests of the United States. This procedure is to be used to acknowledge R.S. 2477 rights of way. Interior Secretary Norton and the state of Utah executed a Memorandum of Understanding on April 9, 2003, under which the DOI will acknowledge the existence of R.S. 2477 rights of way in Utah, by disclaiming any federal interest. Other states also have requested MOUs. The MOU does not fully clarify what criteria will be used to validate right of way claims. Critics assert that the disclaimer regulations "pertain to" R.S. 2477 rights of way and are unlawful under §108 of P.L. 104-208. GAO has concluded that the Utah MOU itself is an unlawful regulation pertaining to R.S. 2477 (GAO Opinion B-300912, *Recognition of R.S. 2477 Rights-of-Way Under the Department of the Interior's FLPMA Disclaimer Rules and Its Memorandum of Understanding with the State of Utah*, Feb. 6, 2004). The first notice of an application for a disclaimer (filed in regard to a Utah road) was published on February 9, 2004 (69 *Fed. Reg.* 6000); Utah withdrew the application on September 16, 2004. Two new Utah applications have been filed (70 *Fed. Reg.* 19500, April 13, 2005). A recent case concluded that state law plays a significant role in determining the validity of R.S. 2477 highways, but also cast doubt on the use of administrative disclaimers to validate such rights of way. (See *SUWA v. BLM*, 2005 U.S. App. LEXIS 19381 (10th Cir. 2005).)

Legislative Activity

The 108th Congress considered, but did not enact, legislation to establish a process for resolving R.S. 2477 claims and define certain terms critical to evaluating the validity of such claims. Also in the 108th Congress, the House approved an amendment to FY2004 Interior appropriations legislation to prohibit implementation of the 2003 changes to the disclaimer regulations in certain federal conservation areas, but this language was eliminated in conference. H.R. 3447 in the 109th Congress would establish an administrative process and criteria for resolving R.S. 2477 claims.

Other Issues

Several other federal lands topics could be addressed through legislation or oversight. These include agency competitive sourcing initiatives, grazing management, hardrock mining, national forest planning, national monuments and the Antiquities Act, and roadless areas of the National Forest System.

Competitive Sourcing. (by Carol Hardy Vincent)

The Bush Administration's Competitive Sourcing Initiative would subject federal agency activities judged to be commercial in nature to public-private competition. This government-wide effort could affect diverse activities in agencies including BLM and the FS. The goal is to save money through competition, particularly in areas where private business might provide better services (e.g., administration and maintenance). The plan is controversial, with concerns as to whether it would save the government money, the private sector could provide the same quality of service, or it is being used to accomplish policy objectives. Through December 2004, BLM had studied 415 full-time equivalents (FTEs) to determine whether they should be subject to competitive bidding. That is 12% of the agency's 3,340.5 FTEs identified as commercial. While 176 FTEs were subjected to competitive bidding, none were contracted out. For the FS, similar information is not readily available.

P.L. 109-54, the FY2006 Interior appropriations law, capped DOI competitive sourcing studies during FY2006 at $3.45 million. It did not specify the portion for BLM. BLM had sought $562,000, for planning and competitive sourcing studies during FY2006 on up to 150 FTEs. The law limited FS spending on competitive sourcing during FY2006 to no more than $3.0 million. The Administration had urged removing the funding limitations. Further, the law directed the Secretary of Agriculture to assess the affect of contracting out on FS fire management, and specified that agencies include, in any reports to the Appropriations Committees on competitive sourcing, information on the costs associated with sourcing studies and related activities.

Grazing Management (by Carol Hardy Vincent)

The BLM had expected to publish new grazing regulations in the *Federal Register* in mid-July, but on August 9, 2005, the agency announced its intent to prepare a supplement to the Final Environmental Impact Statement (FEIS). The agency expects to develop the supplement in the fall of 2005 for public review and comment. The FEIS, which was issued on June 17, 2005, analyzed the impact of proposed changes to grazing regulations as well as of two alternatives. (See [http://www.blm.gov/grazing/].) BLM asserts that regulatory changes are needed to comply with court decisions, increase flexibility of managers and permittees, improve administrative procedures and business practices, and promote conservation. Some of the changes in the FEIS would (1) allow title to range improvements to be shared by the BLM and permittees, (2) allow permittees to acquire water rights for grazing if consistent with state law, (3) change the definition of "grazing preference" to include an amount of forage, (4) eliminate conservation use grazing permits, (5) extend the time to remedy rangeland health problems, and (6) reduce occasions where BLM is required to consult with the public. BLM did not address some controversial issues, such as revising the grazing fee. BLM expects to return to the consideration of related grazing policy changes when the regulatory process is completed. On September 28, 2005, a Senate subcommittee held an oversight hearing on the regulatory changes and other grazing issues

Legislation has been introduced to compensate livestock operators on federal lands. H.R. 411 seeks to require federal land management agencies to compensate holders of grazing permits when certain actions reduce or eliminate their permitted grazing, and alternative forage is not available. The bill also would authorize grazing permit holders to sublease their allotments under specified conditions. Other legislation provides for buying out grazing permittees generally or in particular areas, with the allotments then permanently closed to grazing. H.R. 3166 provides for payment to federal grazing permittees who voluntarily relinquish their permits, at a rate of $175 per AUM. The bill also provides for payments to counties in which the relinquished allotments are located, and authorizes permittees to opt for nonuse or reduced use throughout a term. Other examples include H.R. 3701, regarding lands included in Ecosystem Protection Areas that would be created under the legislation, and H.R. 3603, for certain lands in Idaho.

Hardrock Mining (by Aaron M. Flynn)

Reform of the General Mining Act of 1872, the law governing hardrock mining on federal lands, may be considered in the 109th Congress. The Mining Act authorizes a prospector to locate and claim an area believed to contain a valuable mineral deposit, subject to the payment of certain fees. At such time, mineral development may proceed. Comprehensive legislation to reform the development of these mineral resources, H.R. 3968, has been introduced in the 109th Congress. Among other things, this bill would require a royalty payment based on hardrock mineral production, resolve current disputes regarding the number of acres available for mine-associated mill sites, prohibit patenting of federal lands in most circumstances, and establish new standards for determining which federal lands are available for development.

National Forest Planning (by Ross W. Gorte)

New FS planning regulations were promulgated by the Clinton Administration in November 2000, but their implementation was delayed. On January 5, 2005, the Bush Administration issued two new rules. The first (70 *Fed. Reg.* 1022) removed the Clinton regulations, and the second (70 *Fed. Reg.* 1023-1061) finalized new FS planning regulations. The Clinton regulations established ecological sustainability as the priority for managing national forests. The Bush regulations seek to simplify planning in response to concerns about the feasibility of the Clinton regulations. Plans are to articulate desired conditions and goals, and most planning details have been moved to agency "directives," some of which were published on March 23, 2005 (70 *Fed. Reg.* 14637). The new regulations replace ecological sustainability as the main priority with a balance of ecological, economic, and social sustainability. The regulations do not address species viability, roadless areas, or many other specific topics. Because plans will guide activity decisions but not make decisions, the regulations allow plans to be categorically excluded from NEPA analysis (see below) and public involvement requirements. House Agriculture Committee hearings were held on the new FS planning regulations on May 25, 2005 (H. Serial 109-9).

Forest Service NEPA Categorical Exclusions (by Ross W. Gorte)

The Forest Service has historically identified certain activities as not having significant environmental impacts, and thus exempted them from analysis and public participation under the National Environmental Policy Act of 1969 (NEPA; P.L. 91-190, 43 U.S.C. §§4321-4347), except in extraordinary circumstances. Various regulations have expanded the exempt activities in recent years, including for biomass fuel reduction projects (68 *Fed. Reg.* 33814, June 5, 2003), for "small" timber sales (68 *Fed. Reg.* 44598, July 29, 2003), and for forest plans (70 *Fed. Reg.* 1023, Jan. 5, 2005; see above). The agency has also modified its application of extraordinary circumstances (67 *Fed. Reg.* 54622, Aug. 23, 2002). Previously, the rule specified that the presence of extraordinary circumstances would automatically preclude an action being categorically excluded; the new rule gives the responsible official discretion to determine whether extraordinary circumstances warrant NEPA analysis and public involvement in otherwise exempt projects. Several of the 2003 regulations were challenged. On July 2, 2005, a U.S. District Court ruled that five regulations violated the Forest Service Decision Making and Appeals Reform Act (§322 of P.L. 102-381; 16 U.S.C. §1612 note) by excluding decisions from the public comment and appeals process and for other reasons (*Earth Island Institute v. Ruthenbeck*, 376 F.Supp. 2d 994 (E.D. Cal. 2005). The agency has responded to the ruling and subsequent orders by suspending more than 1,500 permits, projects, and contracts. On October 18, 2005, Senators Bingaman and Harkin sent a letter to President Bush asserting that the agency took an "unnecessary and inappropriate response" by suspending non-controversial activities, such as firewood cutting permits, and asking for a more rational response.

Roadless Areas of the National Forest System (by Pamela Baldwin)

The Clinton Administration issued several rules affecting the roadless areas of the National Forest System (NFS). The principal rule (66 *Fed. Reg.* 3244, Jan. 12, 2001) resulted in a nationwide approach to management that curtailed (but did not eliminate) most roads and timber cutting in roadless areas. National guidance was justified as avoiding the litigation and delays when decisions were made at each national forest. The rule was twice enjoined. The Bush Administration issued a new final rule to replace the Clinton rule and allow governors 18 months to petition the FS for a special rule for roadless areas in all or part of their state (70 *Fed. Reg.* 25654, May 13, 2005). Until such a new regulation in response to a petition is finalized, the FS is to manage roadless areas in accordance with interim directives (69 *Fed. Reg.* 42648, July 16, 2004) that place most decisions with the Regional Forester, and the Chief of the FS, until each forest plan is amended or revised to address roadless area management. This returns decisions on roadless area management to the individual forest plans, basically reversing the Clinton nationwide roadless rule. The new NFS planning regulations (see above) do not address roadless areas, apparently leaving decisions involving them to the project level within each forest, unless a special rule is adopted for a particular state. California, New Mexico, and Oregon have sued to challenge the new roadless area rule. H.R. 3563 has been introduced to direct that roadless areas be managed in accordance with the 2001 regulations.

National Monuments and the Antiquities Act (by Carol Hardy Vincent)

Presidential establishment of national monuments under the Antiquities Act of 1906 (16 U.S.C. §§431, et seq.) sometimes has been contentious. President Clinton's establishment or enlargement of 22 monuments set off renewed controversy regarding presidential authority to proclaim monuments. The 108th Congress focused on land uses within monuments (e.g., recreation, off-highway vehicles, and commercial uses); the inclusion of non-federal lands in monument boundaries; and whether the President should be required to seek congressional, state, or public input or environmental reviews. A bill was introduced to limit the President's authority to designate national monuments and establish a process for input into presidential monument designations, but no further action was taken. Similar legislation has not been introduced in the 109th Congress.

LEGISLATION

H.R. 6 (Barton); P.L. 109-58

The Energy Policy Act of 2005, among other provisions, amends the Geothermal Steam Act of 1970 and the Mineral Leasing Act, requires the Secretary of the Interior to evaluate the oil and gas leasing process, and shields various energy-related activities on federal lands from review under NEPA. The conference report was filed July 27, 2005, and agreed to by the House and Senate. Signed into law August 8, 2005.

H.R. 297 (Rahall); S. 576 (Byrd)

These bills amend the Wild Horses and Burros Act to restore the prohibition on the commercial sale and slaughter of wild horses and burros. H.R. 297, introduced January 25, 2005; referred to Committee on Resources. S. 576, introduced March 9, 2005; referred to Committee on Energy and Natural Resources.

H.R. 411 (Renzi)

The Cattleman's Bill of Rights Act directs compensation for ranchers when federal actions reduce their allowed amount of grazing. Introduced January 26, 2005; referred to Committee on Resources and Committee on Agriculture.

H.R. 2993 (Porter); S. 1273 (Reid)

These bills foster the sale and adoption of wild horses and burros while strengthening protections. H.R. 2993, introduced June 20, 2005; referred to Committee on Resources. S. 1273, introduced June 20, 2005; referred to Committee on Energy and Natural Resources.

H.R. 3166 (Grijalva)

The Multiple-Use Conflict Resolution Act compensates livestock operators who voluntarily relinquish grazing permits. Introduced June 30, 2005; referred to Committee on Resources, Committee on Agriculture, and Committee on Armed Services.

H.R. 3447 (Udall, M.)

The Highway Claims Resolution Act of 2005 establishes an administrative process and criteria to resolve R.S. 2477 claims. Introduced July 26, 2005; referred to Committee on Resources.

H.R. 3563 (Inslee)

The National Forest Roadless Area Conservation Act directs that inventoried roadless areas of the national forests be managed in accordance with the 2001 regulations. Introduced July 28, 2005; referred to Committee on Agriculture and Committee on Resources.

H.R. 3710 (Markey)

Under certain circumstances, the bill would direct the suspension of royalty relief programs for oil and natural gas production from federal lands and authorize resulting revenues to be used for specified hurricane relief and low income energy assistance programs. Introduced September 8, 2005; referred to Committee on Resources, Committee on Transportation and Infrastructure, Committee on Energy and Commerce, and Committee on Education and the Workforce.

H.R. 3893 (Barton)

The Gasoline for America's Security Act would, among other things, require the President to designate certain federal lands, which might include lands under the jurisdiction of BLM or FS, as suitable for refinery construction or expansion, at which time an expedited permitting process would be available for a refinery sited in the designated area. Introduced September 26, 2005; passed by the House October 7, 2005.

H.R. 3968 (Rahall)

The Federal Mineral Development and Land Protection Equity Act of 2005 would require a royalty payment based on hardrock mineral production, resolve current disputes regarding the number of acres available for mine associated mill sites, prohibit patenting of federal lands in most circumstances, and establish new standards for determining which federal lands are available for development. Introduced October 6, 2005; referred to House Subcommittee on Energy and Mineral Resources.

FOR ADDITIONAL READING

CRS Report RS21917, *Bureau of Land Management (BLM) Wilderness Review Issues*, by Ross W. Gorte and Pamela Baldwin.

CRS Issue Brief IB10143, *Energy Policy: Legislative Proposals in the 109th Congress*, by Robert L. Bamberger.

CRS Report RL32393, *Federal Land Management Agencies: Background on Land and Resources Management*, by Carol Hardy Vincent, coordinator.

CRS Report RS21402, *Federal Lands, "Disclaimers of Interest," and R.S. 2477*, by Pamela Baldwin.

CRS Report RL30755, *Forest Fire/Wildfire Protection*, by Ross W. Gorte.

CRS Report RL32244, *Grazing Regulations: Changes by the Bureau of Land Management*, by Carol Hardy Vincent.

CRS Report RL32142, *Highway Rights of Way on Public Lands: R.S. 2477 and Disclaimers of Interest*, by Pamela Baldwin.

CRS Report RL32893, *Interior, Environment, and Related Agencies: FY2006 Appropriations*, Carol Hardy Vincent and Susan Boren, co-coordinators.

CRS Report RL30647, *The National Forest System Roadless Areas Initiative*, by Pamela Baldwin.

CRS Report RS20902, *National Monument Issues*, by Carol Hardy Vincent.

CRS Report RL32315, *Oil and Gas Exploration and Development on Public Lands*, by Marc Humphries.

CRS Report RL32936, *Omnibus Energy Legislation, 109th Congress: Assessment of H.R. 6 as Passed by the House*, by Mark Holt and Carol Glover, co-coordinators.

CRS Report RS21423, *Wild Horse and Burro Issues*, by Carol Hardy Vincent.

CRS Report RS22025, *Wilderness Laws: Permitted and Prohibited Used*, by Ross W. Gorte.

CRS Report RL31447, *Wilderness: Overview and Statistics*, by Ross W. Gorte.

CRS Report RS21544, *Wildfire Protection Funding*, by Ross W. Gorte.

REFERENCES

[1]   Funding has risen substantially over the past decade or so. Average spending for FY1994-FY1999 was $1.07 billion annually.

In: Progress in Environmental Research
Editor: Irma C. Willis, pp. 175-202

ISBN 978-1-60021-618-3
© 2007 Nova Science Publishers, Inc.

Chapter 5

TIME COURSE STUDY ON THE MANIFESTATION OF LETHAL EFFECTS IN *ELLIPTIO COMPLANATA* MUSSELS EXPOSED TO AERATION LAGOONS: A BIOMARKER STUDY

F. Gagné, C. Blaise and C. André

Environment Canada, St-Lawrence Centre, 105 Mc Gill, Montréal,
Québec J5Z 2S9. Canada

ABSTRACT

The time course in the expression of sublethal effects that precede mortality events in mussels exposed to municipal wastewaters is not well understood. Our study thus sought to examine such a time course in biomarker responses of freshwater mussels exposed to two final aeration lagoons for the treatment of domestic wastewater. Mussels were caged in two aeration lagoons (AL1 and AL2) for different time periods (1, 15, 30, 40 and 60 days) and examined for biomarkers representative of biotransformation, metabolism of endogenous ligands, tissue damage and gametogenesis. Results showed that mortality events occurred after day 29 of exposure, gradually reaching 30% and 45% mortality for AL1 and AL2, respectively. These aeration lagoons were estrogenic, as evidenced by the rapid induction of vitellogenin-like proteins, and they were also good inducers of cytochrome P340 3A activity, considered as one of the major metabolizing enzymes of pharmaceutical products. Several biomarkers were expressed before the manifestation of mortality (vitellogenin-like proteins, aspartate transcarbamoylase, metallothioneins, monoamine oxidase, heme peroxidase, cytochrome P450A activity, glutathione S-transferase, DNA strand breaks and mitochondrial electron transport activity), while others were expressed during mortality events (xanthine oxydoreductase, cytochrome P4501A1 activity, lipid peroxidation, gonad lipids and cyclooxygenase). A factorial analysis revealed that temperature-dependent mitochondrial electron transport activity, hemoprotein oxidase, monoamine and aspartate transcarbamoylase had the highest factorial weights. Furthermore, canonical analysis of biomarkers revealed that reproduction (gametogenesis) was more significantly related with those linked to tissue damage and biotransformation, while energy status (i.e. increased energy expenditure and decreased lipid reserves) was not significantly related to gametogenesis or tissue damage in AL1, the least toxic and estrogenic lagoon. In AL2, the more toxic and estrogenic lagoon, energy status was significantly correlated with tissue damage and gametogenesis, in addition to biotransformation activity and metabolism of endogenous substrates. Domestic wastewaters treated in aeration lagoons display a complex pattern of sublethal responses in mussels prior to the manisfestation, in some instances, of mortality events.

Keywords: Domestic effluents, mussels, sublethal/lethal effects, biomarkers, kinetics

INTRODUCTION

Municipal effluents are recognized as important sources of pollution for aquatic ecosystems. Their degree of contamination is diverse and includes, for example, micro-organisms, heavy metals, polyaromatic hydrocarbons and pharmaceutical and personal care products (PPCPs) (Chambers et al., 1997; Kummerer, 2001; Andreozzi et al., 2003). Among other effects, this cocktail of contaminants makes such effluents potent endocrine disruptors to aquatic organisms such as bivalves and fish. Indeed, mussels placed for one year in the dispersion plume of a major urban effluent had elevated levels of the egg-yolk protein vitellogenin and displayed marked feminisation (Blaise et al., 2003; Gagné et al., 2001). These mussels also exhibited a number of neuro-endocrine alterations such as decreased monoamine oxidase, serotonin and its ATP-dependent synaptosome transport (Gagné et al., 2004). Dopamine levels in the gonad were decreased, but ATP-dependent synaptosome transport was increased, indicating enhanced turnover of this catecholamine. Moreover, mussels exposed to contaminants found in municipal effluents also show signs of genotoxicity, increased oxidative stress or inflammation, increased metallothioneins, and increased metabolisms of various xenobiotics (i.e. increased phase 1 and 2 biotransformation activity) (Ciccotelli et al., 1998; Gagné et al., 2004; Gowland et al., 2002; Peters et al., 1998).

Mussels chronically exposed to municipal wastewaters could eventually display signs of poor health leading to disease and mortality. However, the interplay between the expression of early biological effects and the manifestation of tissue damage and mortality is not well understood in feral invertebrates. To address this issue, various biomarkers of biotransformation activity of endogenous and exogenous compounds, energy status, gamete activity and tissue damage were examined in mussels exposed to municipal contamination. Biotransformation activity was determined by measuring total hemoprotein oxidase activity, cytochromes P4501A and P4053A and glutathione S-transferase activities. While total hemoprotein oxidase activtiy is a global measure of oxidative metabolism, cytochrome P4501A was used as an effect biomarker for polyaromatic hydrocarbons (Canova et al., 1998) and cytochrome P4503A for hydroxylated cyclic hydrocarbon exposure linked to many pharmaceutical products. Indeed, the latter cytochrome P450 was identifed as one of the major enzymes for the metabolism of many pharmaceutical drugs and possesses 6β-testosterone hydroxylase activity (Vignati et al., 2005). Our study sought to examine other metabolism parameters as well. Xanthine oxidoreductase is implicated in the salvage pathway of purines and the scavenging of oxygen radicals (Cancio and Cajaraville, 1999) such that its activity could be influenced by the occurrence of xanthine-related compounds like caffeine, drugs to treat gout (i.e., allopurinol) and theophylline in municipal effluents. Monoamine oxidase is the rate-limiting enzyme for the catabolism of monoamines such as dopamine and serotonin and its activity could be modulated by various contaminants such as the so-called MAO inhibitors used in the treatment of depression, for example. Change in heavy-metal metabolism was also examined by tracking changes in metallothionein (MT), a well-known biomarker implicated in the sequestration and detoxication of metals in mussels (Isani et al., 2000). The freshwater unionid *Elliptio complanata* is a sexually dimorphic ovoviviparous invertebrate that broods eggs rich in vitellogenins (Vtg). The synthesis of Vtg was shown to be controlled, at least partially, by β-estradiol in oysters (Li et al., 1998). Gamete activity was determined by following the gonado-somatic index (i.e. gonad-to-soft-tissue weight ratio), Vtg-like proteins and aspartate transcarbamoylase activity. The latter consists of the rate-limiting enzyme for pyrimidine synthesis to support DNA production during gamete development (Mathieu, 1987). The concept of cellular energy allocation was proposed as a highly relevant biomarker for predicting change at the population level (De Coen and Janssen, 2003). Exposure to contaminated environments leads to a metabolic cost in homeostasis, survival and reproduction for challenged organisms. Indeed, invertebrates surviving in contaminated habitats had increased energy expenditure as determined by increased mitochondrial electron transport activity and reduced lipid reserves (Smolders et al., 2004). Moreover, the susceptibility of organisms to temperature increments is exacerbated by pollution, as shown by increased temperature-dependent mitochondrial electron transport activity in clams collected at known polluted sites (Gagné et al., 2006a). In the assessment of tissue damage, the production of lipid peroxidation, DNA strand breaks and tissue inflammmation were examined. Protection against the toxic by-products of oxygen is assured by a series of enzymatic and nonenzymatic antioxidants that can be disrupted by various contaminants, leading to oxidation of endogenous macromolecules like lipids, proteins and nucleic acids (Vasseur and Leguille, 2004; Wills, 1987). The formation of DNA-adducts and increase in DNA repair activity are associated with the production of alkali-labile sites leading to DNA strand breaks (Olive, 1988). The onset of inflammation is associated with increased cyclooxygenase activity in tissue where necrosis and infection occur and is not related in spawning processes (Canesi et al., 2002; Sastry and Gupta, 1978).

Our study thus sought to shed light on the question of biomarker expression before and during the manifestation of mortality events in mussels exposed to the last aeration ponds that treat domestic wastewaters. An attempt was made to relate changes in early warning biomarkers to tissue damage and mortality observed in mussels.

METHODS

Mussel Collection and Exposure to Aeration Lagoons

Elliptio complanata mussels were collected by hand in the Richelieu River, Quebec, Canada, in May 2004; this corresponds to the period of late gametogenesis. They were placed in aerated tanks for two weeks at 15°C and fed *Selenastrum capricornutum* algal preparations (5–10 million cells/L). The mussels were then caged at N = 40 mussels per cage according to

Applied Biomonitoring's procedure (Salazar and Salazar, 2001). Three cages were then submerged at a depth of about two metres in two final aeration lagoons (AL1 and AL2) for the treatment of domestic wastewater. In a parallel study, mussels were caged and placed in an holding tank under a continuous flow of aerated, dechlorinated and UV-treated tap water at 15°C for the duration of the experiment as a negative control, to monitor for normal temporal changes in the biomarkers. Aeration lagoon AL2 is located in one town's wastewater treatment plant that is some 10 km from a second town where lagoon AL1 is located. The mussels were exposed to the wastewater for 62 days (July through end of August) in the summer of 2004 and a subset of eight mussels was collected on day 1, 15, 30, 40 and 62 for biomarker analyses. After the exposure period, the cages were recovered and the mussels were depurated in clean, dechlorinated water for 18 hr at 15°C. They were then dissected for gills, digestive gland and gonads and afterwards homogenized on ice in 10 mM Hepes-NaOH buffer, pH 7.4, containing 100 mM NaCl, 0.1 mM EDTA and 0.1 mM dithiothreitol with a Teflon pestle tissue grinder. A subsample of the homogenate was centrifuged at 3000 x g to obtain a subsample of the supernatant (S3 fraction) from the gonad and then at 15 000 x g for 20 min at 2°C to obtain the supernatant (S15 fraction). All were stored at -85°C until analysis. Total proteins were determined in all these fractions using the Coomassie blue binding principle (Bradford, 1976).

Gametogenetic Activity Evaluation

Gamete activity was examined with the following parameters: gonado-somatic index (GSI; weight of gonad/soft tissue weight), the relative levels of vitellogenin-like proteins (Vtg) and the rate-limiting enzyme aspartate transcarbamoylase (ATC) activity for pyrimidine synthesis. Sex was determined by microscopic examination of a gonad smear at 400X magnification. The presence of Vtg in gonad was determined by the organic alkali-labile phosphate (ALP) technique, developed by Gagné et al. (2001), using acetone fractionation instead of the t-butyl-methyl ether extraction procedure. Briefly, the gonad homogenate 15 000 x g supernatant was fractionated with 30% cold acetone and centrifuged at 10 000 x g for 5 min at 4°C. The pellet containing poorly soluble, high-molecular-weight proteins was dissolved in NaOH 1 M at 60°C for 30 min to release inorganic (alkali-labile) phosphates. Vtg-like protein levels were expressed as µg of alkali-labile phosphate (ALP)/mg of gonad protein. Gamete activity was determined by measuring the activity of ATC in the S15 fraction of gonad homogenate (Mathieu, 1987) using the spectrophotometric detection method for phosphate measurements in the presence of labile carbamoylphosphate (Herries, 1967). The data were expressed as µg phosphate/min/mg proteins using standard solutions of inorganic phosphate.

Metabolism of Monoamines, Xanthines and Heavy Metals

Metabolism of endogenous substrate and heavy metals was evaluated by tracking metallothionein, xanthine oxido-reductase and monoamine oxidase activity. Metallothionein (MT) levels were determined in the gills and digestive gland of mussels by the thiol spectrophotometric assay (Viarengo et al., 1997). The MT fractionation and assay procedure was performed on the S15 fraction. Standards of reduced glutathione (GSH) were used for calibration. The data are expressed as µmoles of GSH equivalents per mg protein. The metabolism of xanthines was measured by following xanthine oxidoreductase (XOR) activity in the gonad by the method of Zhu et al. (1994). The mixture contained 50 µL of 15 000 x g supernatant, 1 mM hypoxanthine, 0.1 mM aminotriazole (a catalase inhibitor), 2 µM dichlorofluorescein and 0.1 µg/mL horseradish peroxidase in 50 mM KH_2PO_4, pH 7.4, containing 10 µM molybdate. The reaction proceeded at 30°C for 0, 15, 30 and 60 min and the production of fluorescein was measured at 485 nm excitation and 520 nm emission. Enzyme activity was expressed as nmole of fluorescein formed/(min x mg proteins). Standard solutions of fluorescein were used for calibration. MAO activity was determined in the S3 fraction of the homogenate using tryptamine as the substrate. The supernatant was mixed with 100 µM tryptamine, 2 µM 2,7-dichlorofluorescin in 10 mM Hepes-NaOH, pH 7.4, containing 125 mM NaCl, 0.1 M aminotriazole and 0.1 µg/mL of horseradish peroxidase (Sigma-Aldrich Co., USA). The reaction proceeded at 30°C for 0, 15, 30 and 60 min and fluorescence was measured at 485 nm excitation and 520 nm emission. Enzyme activity was expressed as nmole of fluorescein formed/(min x mg proteins). Standard solutions of fluorescein were used for calibration.

Biotransformation Activity

Biotransformation activity in the digestive gland was determined by the following activities: hemoprotein oxidase, 7-ethoxyresorufin, resorufin benzyl ether, dibenzylfluorescein dealkalases and glutathione S-transferase activities in the S15 fraction of digestive gland homogenates. Hemoprotein oxidase, 7-ethoxyresorufin, and benzyloxyfluorescein dealkylase activities were determined as described elsewhere (Quinn et al., 2004b). Glutathione S-transferase (GST) was determined by the spectrophotometric method of Boryslawskyj et al. (1988) using 2, 4-dichloronitrobenzene as the substrate. Data are expressed as increase in absorbance at 412 nm/min/mg proteins.

Tissue Damage and Inflammation

Lipid peroxidation was determined in tissues according to the thiobarbituric acid reactants (TBARS) method of Wills (1987). TBARS were determined by fluorescence at 520 nm for excitation and 590 nm for emission using a fluorescence microplate reader (Dynatech, Fluorolite-1000). The data were expressed as µg of TBARS/mg of homogenate protein. Double and single DNA strand breaks were determined in tissue by the alkaline precipitation assay (Olive, 1988) with the fluorescence quantification of DNA strands method adapted for the presence of trace amounts of detergents (Bester et al., 1994). Salmon sperm DNA standards were added for DNA calibration. The results were expressed as µg of supernatant DNA/mg of proteins. Cyclooxygenase (COX) activity was measured by the oxidation of 2,7-dichlorofluorescin in the presence of arachidonate (Fujimoto et al., 2002). Briefly, 50 µL of gonad 15 000 x g supernatant was mixed with 200 µL of 50 µM arachidonate, 2 µM 2,7-dichlorofluorescin and 0.1 µg/mL of horseradish peroxidase in 50 mM Tris-HCl buffer, pH 8, containing 0.05% Tween 20. The reaction mixture was incubated for 0, 10, 20 and 30 min at 30°C, and fluorescence was measured at 485 for excitation and 520 for emission (Dynatech, Fluorolite 1000). The data were expressed as increase in relative fluorescence units/(min x mg proteins).

Cellular Energy Status

Mitochondria were prepared by sequential differential centrifugation. The S3 fraction of the homogenate described above was centrifuged at 9000 x g for 20 min at 2°C. The supernatant was then discarded and the pellet resuspended in 140 mM NaCl containing 10 mM Hepes-NaOH, pH 7.4, and 1 mM KCl. The mixture was centrifuged again as described above and the pellet resuspended in the same saline buffer. Total protein content in the mitochondrial fraction was then determined as described above.

Mitochondrial energy consumption was determined by measuring MET activity according to the p-iodonitrotetrazolium reduction dye method (Smolders et al., 2004; King and Packard, 1975). Briefly, mitochondria (100 µg/mL) were mixed with one volume of 0.1 M Tris-HCl, pH 8.5, containing 0.1 mM $MgSO_4$, 0.1% Triton X-100 and 5% polyvinylpyrrolidone for 1 min before adding 1 mM NADH and 0.2 mM NAPDH. The reaction was started by adding 1 mM of p-iodonitrotetrazolium. The reaction was allowed to proceed at 4°C and 20°C for 30 min and absorbance readings were taken at 520 nm at 10-min intervals. The data were expressed as change in absorbance/30 min/mg mitochondrial protein content. Temperature-dependent MET (MET_T) was determined as described elsewhere (Gagné et al., 2006c). The data were expressed as dye reduction rate (A520/min) at 20°C minus dye reduction rate (A520/min) at 4°C/total temperature change (20°C-4°C).

Data Analysis

For each type of analysis, a replicate number of N = 8-10 mussels was used for every time course assessment (days 1, 15, 30, 45 and 62) in the aeration lagoons and at the reference site (i.e., mussels held in laboratory holding tanks). Data were analysed using a factorial two-way analysis of variance (two-way ANOVA), with gender and site as the effect variables, after confirming for homogeneity of variance and normality in data distribution with Bartlett's test. If the data proved to be non parametric, they were log-transformed. Critical differences in initial time or controls (unexposed mussels) were appraised using the ranked-based Mann-Whitney U test. Correlation analysis was performed with the Pearson-moment correlation test. Significance was set at $p < 0.05$. The response pattern of the biomarkers was determined by the least-square method (Statistica 7 software). A canonical analysis was performed to identify trends between groups of biomarkers: biotransformation, metabolism of endogenous compounds, tissue damage, energy status and gametogenesis.

RESULTS

Aeration lagoons AL1 and AL2 treat the domestic wastewaters of two towns with respective populations of approximately 20 000 and 12 000 residents. These lagoons consist of four serially interconnected ponds undergoing constant aeration. Mussels were placed in the final aeration lagoons, the last treatment step before the effluent is released to the receiving river. Neither town has a large industrial zone, so contaminants are mostly domestic in nature. In the case of lagoon AL1, particles were flocculated by the occasional addition of ferric solutions at the first pond. The pH levels of the aeration lagoons were about 2 pH units (6.1 to 7.5) below that of the receiving water. The average temperature of the aeration lagoons was $23°C$, similar to that of the receiving water, but fluctuated from 8 to $27°C$ for short periods of time.

The final AL1 and AL2 aeration lagoons both incurred lethality on freshwater mussels. In AL1, 5% mortality was observed on day 29, gradually reaching 30% mortality on day 62. For AL2, 8% mortality was observed on day 30, reaching 45% mortality on day 62. Mussels placed in the holding tank laboratory water experienced no mortality events. No major changes in condition factor (mussel weight/shell length) or soft tissue ratio were observed in exposed mussels. Gametogenic activity was examined by tracking changes in gonado-somatic index, vitellogenin-like proteins and aspartate transcarbamoylase activity (Figure 1). In AL1, a factorial two-way analysis of variance revealed that GSI was significant for exposure time, with no apparent gender-based effect (Figure 1A). A significant interaction between exposure time and gender was observed. GSI dropped during the exposure period, with the index in males being more affected than in females for days 40 and 62. In AL2, a factorial two-way ANOVA revealed that duration of exposure, gender and the interaction of the two were all significant. The GSI rose on day 15 and returned to initial values afterwards. The GSI in males was higher than in females and the increase in GSI was stronger in males than in females. Levels of Vtg-like proteins were also significantly increased in both aeration lagoons (Figure 1B). A factorial two-way ANOVA revealed that exposure to the aeration lagoons was the main effect observed regardless of gender. The levels of Vtg-like proteins were significantly induced within the first few days of exposure, with the maximum response on day 30 (2.5-fold and 4-fold inductions for AL1 and AL2, respectively). The activity of ATC was also measured in mussels exposed to both aeration lagoons (Figure 1C). A two-way ANOVA for ATC activity in AL1 revealed that only exposure time was significant, with increased activity during the exposure period indicating increased gamete activity for both sexes. For the more toxic lagoon AL2, a two-way ANOVA revealed that ATC activity changed over time, being readily induced during the earlier days until day 15 and then gradually declining to reach initial values on day 62. A correlation analysis revealed that Vtg-like proteins were negatively correlated with GSI ($R = -0.43$; $p = 0.05$) in AL1. In AL2, pyrimidine synthesis enzyme (ATC) was positively correlated with Vtg-like proteins ($R = 0.5$; $p = 0.02$).

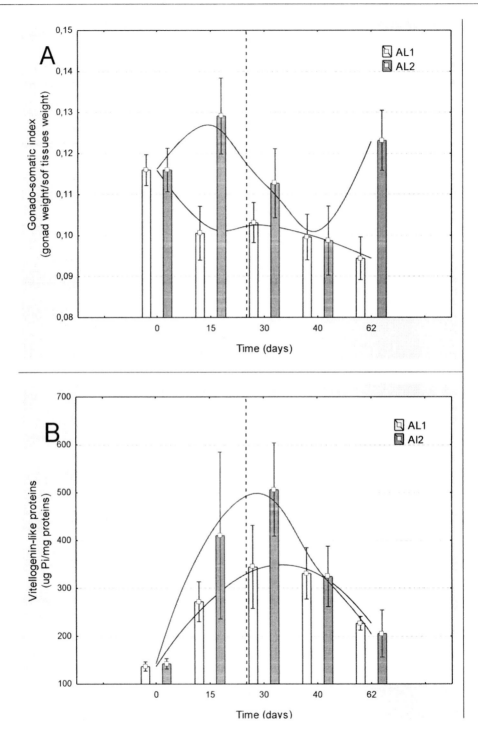

Figure 1. Kinetic response in gametogenic activity. Mussels were placed in two aeration lagoons for up to 62 days. They were analysed for GSI (A), vitellogenin-like proteins (B). The dotted line corresponds to the period when mortality events began.

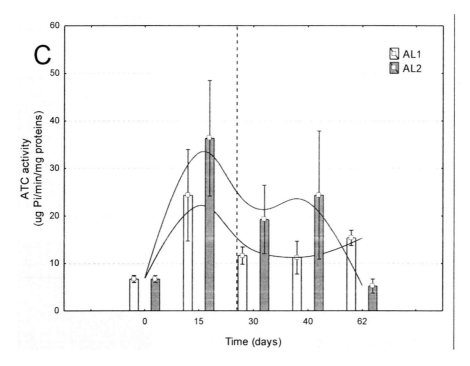

Figure 1. Kinetic response in gametogenic activity. Mussels were placed in two aeration lagoons for up to 62 days. They were analysed for aspartate transcarbamoylase activity (C). The dotted line corresponds to the period when mortality events began.

Levels of MT in digestive gland and gills, XOR activity and gonad MAO were determined in mussels exposed to aeration lagoons (Figure 2A–D). In AL1, a two-way ANOVA for MT in digestive gland revealed that exposure time and sex were significant. Males tended to have more MT than females. MT levels rose steadily over the exposure period in AL1. In AL2, MT in digestive gland varied with exposure time and gender, with significant interaction between these two variables. MT levels rose in mussels during the exposure period, with females responding more than males. MT levels in gills in AL1 remained constant with sex and exposure time, but increased significantly in males on day 30. In AL2, gill MT levels decreased after 30 days of exposure (Figue 2B). Xanthine oxidoreductase (XOR) activity was increased on day 62 in both males and females in AL1 and AL2 (Figure 2C). MAO activity in gonad tissues was increased on day 30 and returned to normal values afterwards in both aeration lagoons (Figure 2D). In AL1, correlation analysis revealed that gonad XOR was significantly correlated with MAO activity ($R = 0.46$; $p = 0.035$), Vtg-like proteins ($R = 0.32$; $p = 0.04$) and negatively so with GSI ($R = -0.52$; $p < 0.01$). MAO activity was negatively correlated with GSI ($R = -0.37$; $p = 0.04$). MT in digestive gland was also correlated with Vtg-like proteins ($R = 0.35$; $p = 0.02$) and with pyrimidine synthesis or ATC activity ($R = 0.53$; $p < 0.01$). No significant correlations were noted for these biomarkers in AL2. However, MAO activity was negatively correlated with GSI ($R = -0.6$; $p < 0.01$), and positively so with Vtg-like proteins ($R = 0.36$; $p = 0.03$) and pyrimidine synthesis ($R = 0.57$; $p < 0.01$).

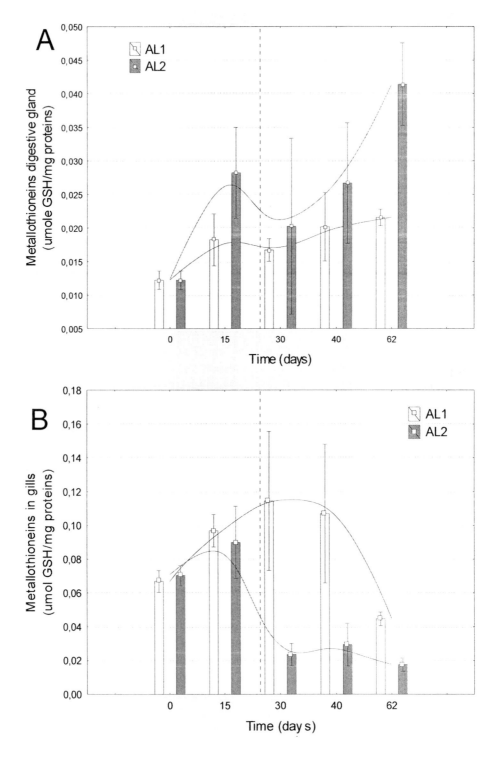

Figure 2. Kinetic response of various enzymes for amine, xanthine and heavy metal metabolism. Mussels were placed in two aeration lagoons for up to 62 days. They were analysed for MT in digestive gland (A), MT in gills (B). The dotted line corresponds to the period when mortality events began.

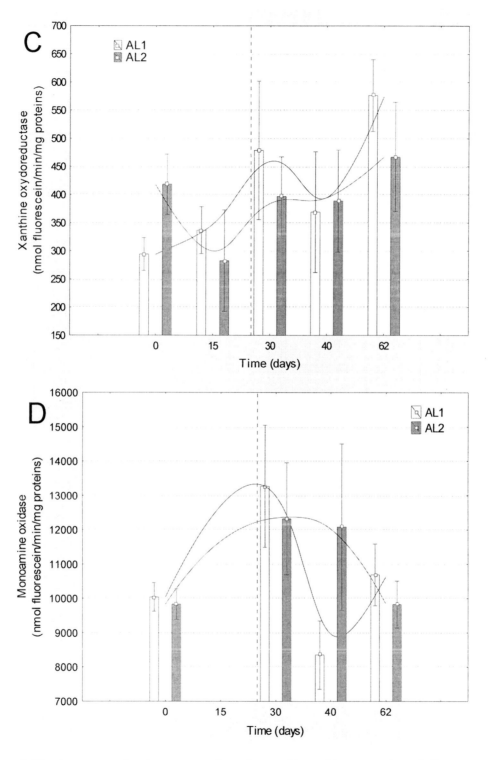

Figure 2. Kinetic response of various enzymes for amine, xanthine and heavy metal metabolism. Mussels were placed in two aeration lagoons for up to 62 days. They were analysed for MT gonad xanthine oxidase (C), and gonad MAO (D). The dotted line corresponds to the period when mortality events began.

Biotransformation activity was examined with the following endpoints: hemoprotein oxidase, EROD, cytochrome P4503A and GST activities (Figure 3A-D). Hemoprotein oxidase activity in AL1 changed over time, with no significant effects or interactions with gender (Figure 3A). Its activity rose within the first few days of exposure and returned to baseline values on day 62. Hemoprotein oxidase activity in AL2 changed significantly over time with the absence of effect and interaction with gender; its activity increased on day 40 and 62. The enzyme activity for polyaromatic hydrocarbon metabolism (cytochrome P4501A1 or EROD) in AL1 increased significantly on days 40 and 62, with no apparent effects or interaction with gender (Figure 3B). In AL2, CYP1A activity varied significantly with exposure time and gender. Males tended to have higher activity than females and the activity was significantly induced on days 30 and 40in both sexes. Cytochrome 3A activity, the pharmaceutical product and androgen steroid metabolizing enzyme complex in mussels exposed to the domestic aeration lagoons, was also examined (Figure 3C). Cytochrome P4503A activity in AL1 was readily increased on day 15 and remained elevated until the end of the exposure time. Neither effect nor interaction with gender was observed. A similar pattern was observed in AL2 for cytochrome P4503A-like activity, suggesting that exposure to final aeration lagoons readily induces this drug-metabolizing enzyme activity within the first few days of exposure *in situ*. The activity of the phase 2 conjugation enzyme GST was also tracked over time (Figure 3D). As with cytochrome P4503A activity, GST activity was readily induced in both aeration lagoons within the earliest days of exposure and remained elevated till the end of the exposure period (i.e. 62 days) for mussels placed in AL1. In the more toxic lagoon AL2, GST activity returned to initial values on day 45. Correlation analysis of these biomarkers for AL1 revealed that hemoprotein oxidase was significantly correlated with CYP3A activity ($R = 0.42$; $p = 0.04$), Vtg-like proteins ($R = 0.47$; $p < 0.01$), ATC ($R = 0.53$; $p < 0.01$) and GST activity ($R = 0.42$; $p = 0.007$). Cytochrome P450 3A activity was significantly correlated with GST activity ($R = 0.76$; $p < 0.01$) and Vtg-like proteins ($R = 0.59$; $p < 0.01$), and negatively correlated with GSI ($R = -0.51$; $p = 0.02$). EROD activity was negatively correlated with GST ($R = -0.43$; $p = 0.01$) and was not correlated with cytochrome P450-3A. In AL2, EROD activity was significantly correlated with CYP3A activity ($R = 0.51$; $p = 0.005$) and MAO activity ($R = 0.36$; $p = 0.02$). Heme protein oxidase was correlated with MT in digestive gland ($R = 0.37$; $p = 0.02$). CYP3A activity was also correlated with Vtg-like proteins ($R = 0.47$; $p = 0.01$) and MT in digestive gland ($R = 0.51$; $p < 0.01$). GST activity was correlated with MT in digestive gland ($R = 0.47$; $p < 0.01$).

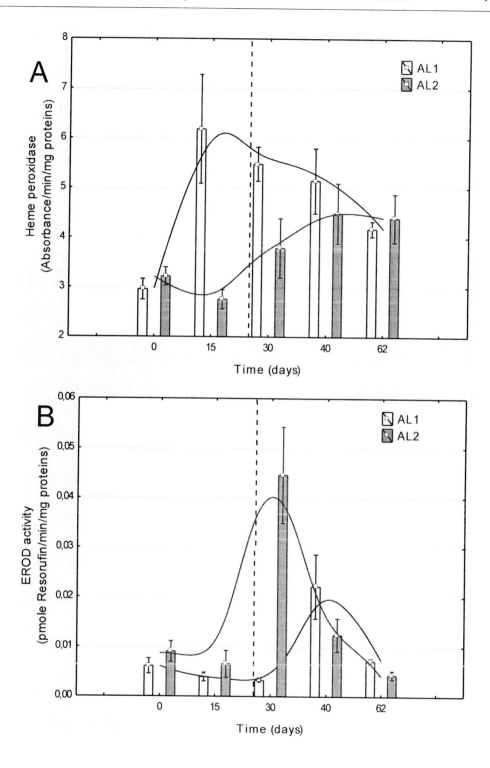

Figure 3. Biotransformation activity in mussels exposed to aeration lagoons.

Mussels were placed in two aeration lagoons for up to 62 days. They were analysed for total hemoprotein oxidase activity (A), EROD activity (B). The dotted line corresponds to the period when mortality events began.

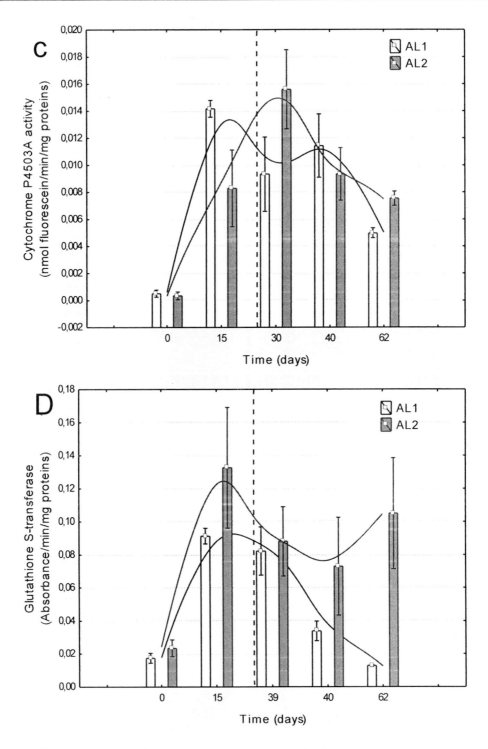

Figure 3. Biotransformation activity in mussels exposed to aeration lagoons.
Mussels were placed in two aeration lagoons for up to 62 days. They were analysed cytochrome
P4503A activity (C), and GST (D). The dotted line corresponds to the period when mortality events
began.

Tissue damage was also examined in mussels exposed to both aeration lagoons (Figure 4A–C). In AL1, LPO in the digestive gland was readily increased within the first few days of exposure (< 15 days) and remained so till the end of the exposure time (Figure 4A). No significant effects or interactions between gender and exposure time were observed. In AL2, LPO was also increased but dropped to initial values on day 62, with no significant effects or interactions with gender. DNA strand breaks were also examined in the same tissue (Figure 4B). In AL1, DNA strand breaks decreased over time with no apparent effect or interactions with gender. In AL2, DNA strand breaks rose on days 15 and 30 and returned to initial values afterwards. The marker enzyme for inflammation, COX activity, was significantly affected over time and displayed interaction with gender in AL1 (Figure 4 C). Its activity gradually rose over 45 days and females responded more than males on day 30. In AL2, COX activity gradually rose, peaking on day 62 (end of exposure time), with no apparent effect or interaction with gender. Correlation analysis for this biomarker set revealed that COX activity was somewhat significantly correlated with LPO ($R = 0.3$; $p = 0.04$). LPO in digestive gland was negatively correlated with GSI ($R = -0.32$; $p = 0.04$) and positively with CYP3A activity ($R = 0.4$; $p = 0.05$). DNA strand breaks were significantly correlated with GST activity ($R = 0.39$; $p = 0.035$) and negatively so with XOR ($R = -0.43$; $p = 0.018$). COX activity was negatively correlated with GSI ($R = -0.6$; $p < 0.01$), positively correlated with Vtg-like proteins ($R = 0.51$; $p < 0.01$), LPO ($R = 0.28$; $p = 0.05$), heme protein oxidase ($R = 0.52$; $p < 0.01$), MT in digestive gland ($R = 0.32$; $p = 0.025$), CYP1A ($R = 0.31$; $p = 0.03$) and CYP 3A activtities ($R = 0.39$; $p = 0.05$). In AL2, no significant correlations were obtained with biomarkers with the exception of COX activity, which was significantly correlated with heme protein oxidase ($R = 0.41$; $p = 0.01$), MT in digestive gland ($R = 0.57$; $p < 0.01$), CYP3A activity ($R = 0.46$; $p = 0.01$) and negatively so with MT in gills ($R = -0.56$; $p < 0.01$).

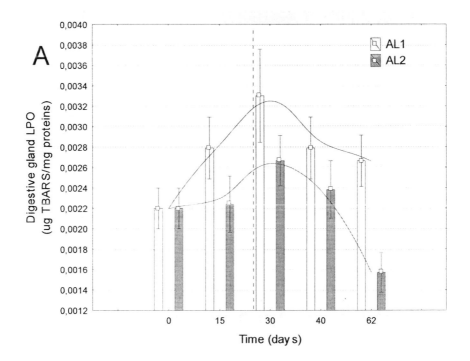

Figure 4. Time course of tissue damage in *Elliptio complanata*. Mussels were placed in two aeration lagoons for up to 62 days. They were analysed for LPO in digestive gland (A). The dotted line corresponds to the period when mortality events began.

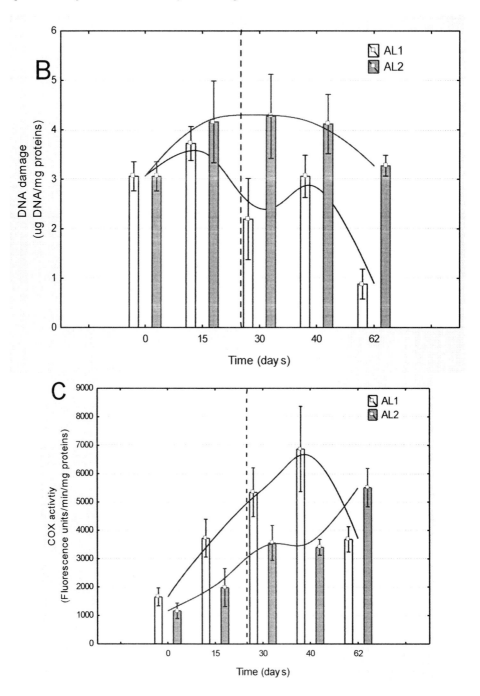

Figure 4. Time course of tissue damage in *Elliptio complanata*. Mussels were placed in two aeration lagoons for up to 62 days. They were analysed for LPO in digestive gland (A),digestive gland DNA strand breaks (B) and gonad COX activity (C). The dotted line corresponds to the period when mortality events began.

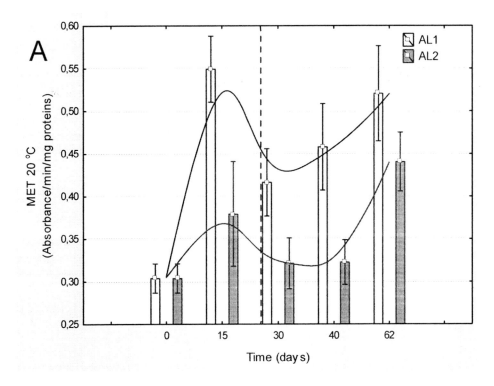

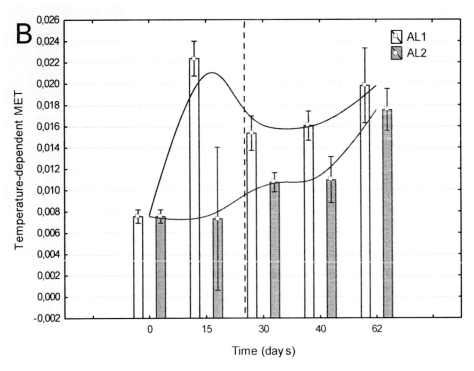

Figure 5. Change in energy expenditure and lipid reserves over time. *Elliptio complanata* mussels were placed in two aeration lagoons for up to 62 days. Their gonads were analysed for mitochondrial electron transport activity (A), temperature-dependent MET (B). The dotted line corresponds to the period when mortality events began.

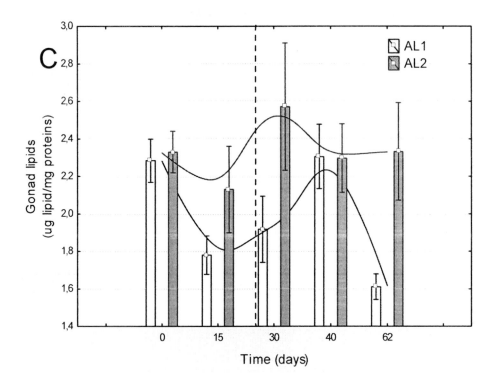

Figure 5. Change in energy expenditure and lipid reserves over time. *Elliptio complanata* mussels were placed in two aeration lagoons for up to 62 days. Their gonads were analysed for total gonad lipids (C). The dotted line corresponds to the period when mortality events began.

The mussels' energy statuses were examined by following mitochondrial electron transport (MET) activity (energy expenditure), temperature-dependent MET (MET_T) and total gonad lipid as a measure of energy reserve (Figure 5). MET activity was readily increased within the first days of exposure to both aeration lagoons and remained so till the end of the exposure period, indicating that mussels were expending more energy in these lagoons than those in the control tanks or at day 0 (Figure 5A). The response was strongest in AL1, where males expended more energy than females, but no significant interaction occurred between exposure time and gender. The same was true for AL2. Temperature dependence in energy expenditure was also examined in exposed mussels (Figure 5B). In AL1, MET_T also increased within the first few days of exposure and remained high till the end of the exposure period. Neither effect nor interaction was observed with gender on MET_T. In AL2, MET_T was significantly increased on days 45 and 62. Lipid reserves also changed significantly in gonad tissues in the less toxic lagoon (Figure 5C). Indeed, lipid levels dropped on days 15, 45 and 62, with no significant effect or interaction with gender in AL1. In AL2, lipid levels remained constant throughout the exposure period (i.e. no significant changes). Correlation analysis revealed that MET at 20°C in AL1 was significantly correlated with MET at 4°C (R = 0.57; p < 0.01), MET_T (R = 0.96; p < 0.01) and negatively with gonad lipid reserves (R = -0.33; p = 0.025). Gonad lipids were also negatively correlated with MET_T (R = -0.35; p = 0.018). MET at 20°C was also negatively correlated with GSI (R = -0.34; p = 0.04) and positively correlated with XOR (R = 0.41; p = 0.01). MET at cold temperatures was correlated with XOR (R = 0.5; p < 0.01) and MT in gills (R = 0.32; p = 0.04). Temperature-dependent MET was correlated with Vtg-like proteins (R = 0.34; p = 0.04). In AL2, MET at 20°C was significantly correlated with MET at 4°C (R = 0.45; p < 0.01) and MET_T (R = 0.66; p < 0.01). MET_T was negatively correlated with MET at 4°C (R = -0.37; p = 0.024). MET at 20°C was correlated with DNA strand breaks (R = 0.36; p = 0.03) and MET at 4°C was correlated with GST activity (R = 0.46; p < 0.01), gill MT (R = 0.68; p < 0.01), CYP 1A activity (R = -0.32; p = 0.03) and COX activity (R = -0.54; p < 0.01). MET_T was negatively correlated with MT in gills (R = -0.68; p < 0.01) and positively correlated with COX activity (R = 0.45; p = 0.01). Total gonad lipid levels were correlated only with MAO activity (R = 0.4; p = 0.01).

Biomarker data were analysed using factorial (Figure 6) and canonical (Figure 7) analyses in an attempt to examine the overall inter-relatedness of the various biomarkers and to identify which biomarker sets were particularly affected by the manifestation of mortality events. Mussels appeared to respond differently depending on the aeration lagoon, both of which produced different degrees of lethality. Indeed, factorial analysis for AL1 revealed that MET, MET_T, hemoprotein oxidase and ATC were the biomarkers with factorial weights > 65% explaining 40% of the total variance. Vtg-like proteins were closely related to MT in digestive glands, heme protein oxidase and ATC (gamete production), but negatively related to XOR in gonad and DNA damage. Energy production and temperature-dependent MET were negatively related to lipid reserves in gonad, suggesting a net loss of energy. EROD activity (a marker enzyme for polyaromatic hydrocarbons) was closely related to genotoxicity and GSI. In AL2, the biomarkers with factorial weights > 65% were MET at 4°C, MAO and Vtg-like proteins. MET at 20°C, MT in gills, XOR and MT in digestive gland were closely related to each other. ATC and Vtg-like proteins were significantly correlated. DNA damage was no longer associated with CYP1A activity, but rather with gamete activity. MET at 20°C, MT in gills, MT digestive and XOR in gonads were closely associated. GSI and Vtg were negatively related to each other, indicating that vitellogenin was initiated in spent gonad. A canonical analysis was performed in an attempt to identify which biochemical endpoints could predict physiological alterations in mussels (Figure 7). In AL1, the expression of biotransformation (defined as CYP1A, CYP3A, heme protein oxidase and GST activities) was highly correlated with biomarkers of gametogenesis (GSI, Vtg-like proteins and ATC). Activation of biotransformation activity was fairly correlated with biomarkers of tissue damage, which was in turn highly linked to gametogenesis. Energy status was influenced more by biomarkers of metabolism (xanthine, monoamines and metal) but also, to a lesser extent, by biotransformation activity. This pattern of response was somewhat different in the more lethally toxic lagoon AL2. Indeed, gametogenesis was more influenced by biomarkers of tissue damage, which are in turn influenced by metabolism and, to a lesser extent, by biotransformation. Energy status was now more influenced by either biotransformation or metabolism of various substrates. Moreover, energy status displayed more influence on tissue damage and gametogenesis in AL2 than in AL1. For both aeration lagoons, biomarkers of tissue damage (LPO, DNA strand breaks and COX) were most strongly associated with alterations in gametogenesis.

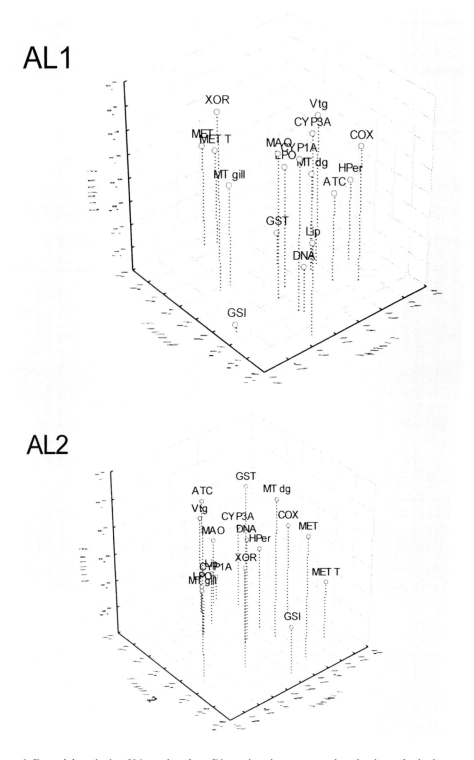

Figure 6. Factorial analysis of biomarker data. Biomarker data were analysed using principal component analysis; the three factors explained 45% of the total variance for both aeration lagoons.

AL1

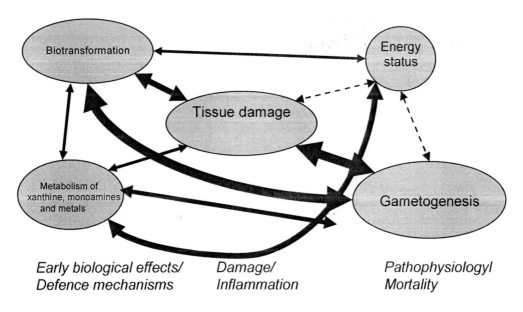

Early biological effects/   *Damage/*   *Pathophysiologyl*
Defence mechanisms   *Inflammation*   *Mortality*

AL2

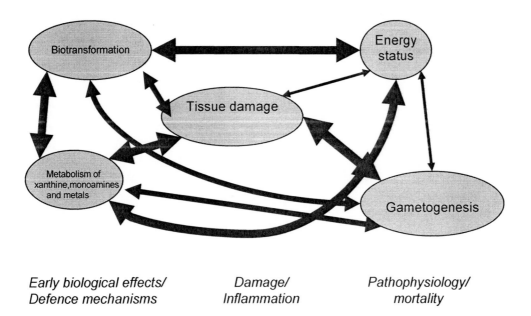

Early biological effects/   *Damage/*   *Pathophysiology/*
Defence mechanisms   *Inflammation*   *mortality*

Figure 7. Canonical analysis of various types of biomarkers.
Biomarkers were grouped as: early biological effect biomarkers (biotransformation and metabolism), biomarkers of tissue damage and the manifestation of pathophysiology (gametogenesis and energy status). The width of the arrow bars reflects the strength of the canonical correlation coefficient ranging from R = 0.77 to 0.3. The dashed bars indicate insignificant canonical correlations.

DISCUSSION

The biological wastewater treatment process employed consisted of four interconnected series of lagoons that gradually degrade wastewater by a combination of aeration (with hydraulic pumps), photodegradation (natural sunlight) and endogenous microfauna. The lagoons produce a series of sublethal effects leading to lethal effects, as evidenced by mortality events in caged mussels some 30 days after the onset of exposure. While the treatment in both aeration lagoons was similar, their respective sublethal response patterns differed in some cases, making it difficult to generalize on the hazards linked to all such aeration lagoons. In terms of lethality, lagoon AL2 proved to be more toxic than AL1 (45% vs. 30% mortality, respectively).

In this study, several biomarkers were expressed before the manifestation of mortality, while others were more strongly expressed after onset of mortality events (i.e. after day 30). Indeed, the following biomarkers clearly responded after day 30, when mortality events began: XOR, CYP1A, LPO, gonad lipid levels and COX activity. For one, early warning biomarkers reacting to sublethal insults were more strongly expressed before the manifestation of toxicity and usually displayed a U-shaped response pattern. For example, gill MT levels reached maximum values on days 15–30 and fell steadily, to below pre-exposure levels, when mortality events were observed. In another study, seabream exposed to a sublethal dose of malathion for up to 96 h revealed changes in antioxidant biomarkers (superoxide dismutase, catalase and glutathione peroxidase) long before the appearance of histopathological alterations, thus showing their usefulness as early warning systems (Rosety et al., 2005). In contrast, biomarkers expressed as mortality events unfolded failed to show a clear U-shaped response trend. This implies that sustained expression of these biomarkers underlines a manifestation of severe pathological change leading to mortality events. For example, increased LPO was observed after 30 days (during mortality events), but DNA strand breaks were reduced. The formation of DNA strand breaks was proportional to both DNA alterations (leading to alkali-labile sites) and the induction of DNA repair mechanisms (Olive, 1988). During mortality events, the significant decrease in DNA strand breaks could be the result of inhibition in DNA repair mechanisms or the formation of DNA cross-links with DNA (Olive et al., 1988). The response pattern of biomarkers was biphasic in some cases, which is consistent with progression of toxicity. During a field translocation experiment on the marine mussel *Mytilus galloprovincialis,* caged for one month in an industrialized (port) zone, a severe depletion was observed, after an initial rise, in antioxidant capacity (Frenzilli et al., 2004). Moreover, a depletion in antioxidant capacity concords with the appearance of cell damage, DNA integrity and lysosome membrane stability, for example (Regoli et al., 2004).

The expression of oxidative metabolism and conjugation was not always associated with clear evidence of toxicity. For example, in another study on caged eels and rainbow trout, induction of EROD activity was observed after 14 days in a polluted freshwater stream while GST was not significant (Fenet et al., 1998). However, the manifestation of tissue damage was observed in tilapia fish exposed for up to 90 days to treatment lagoons (one aerobic, one anaerobic) of a hog-processing plant (Lima et al., 2006). After day 15, LPO increased in fish exposed to both lagoons, but declined steadily on the 30th day to reach initial values on day 90. The fish exposed to these effluents showed increased DNA damage and reduced glutathione content. The LPO response pattern was similar to observations in the present study.

The induction of vitellogenins by municipal effluents is well documented in fish (Harries et al., 1996) and, more recently, in bivalves such as *Elliptio complanata* (Gagné et al., 2001) and zebra mussels (Quinn et al., 2004). However, kinetic studies of vitellogenin expression in the field are scarce for bivalves and fish. In the present study, the induction of Vtg-like proteins was significant within day 15, indicating the relatively high estrogen potency of aeration lagoons treating domestic wastewater. Levels of Vtg-like protein reached maximum values around day 30 and fell thereafter, but remained significantly elevated on day 62, indicating that endocrine disruption persisted throughout the manifestation of mortality events. The GSI did not follow the levels of Vtg-like proteins observed during normal gametogenesis without the influence of pollution or the injection of estradiol-17β (Li et al., 1998). However, Vtg levels were correlated with the drug metabolizing enzyme P450 3A (R = 0.57) and the inflammation enzyme COX (R = 0.51). CYP3A is recognized as a major drug metabolizing enzyme and of 6β-hydroxylase testosterone activity (Stresser et al., 2000). This enzyme complex was detected in *Mytilus galloprovincialis* mussels (Shaw et al., 2004). Hence, the increase in Vtg-like proteins suggests an association with the occurrence of estrogenic compounds (pharmaceuticals) and natural estrogens in these domestic effluents. This finding was supported by analysis of covariance, where Vtg-like proteins were more strongly associated with CYP3A activity than exposure time in both aeration lagoons. The estrogen 4-nonylphenol was shown to increase CYP3A activity in injected winter flounder (Baldwin et al., 2005), suggesting that the activation of the estrogen receptor pathway or steroidogenesis also increases CYP3A. However, E2 treatment in male sea bass caused a selective time- and dose-dependent inhibition of hepatic CYP1A-linked (EROD) and GST activities, but with no effect on the activity of CYP3A-linked 6ß-testosterone hydroxylase (Vaccaro et al., 2005). CYP3A activity was also considered as one of the major drug metabolizing enzyme systems (Crespi et al., 1997), where increase in activity is possibly associated with the occurrence of pharmaceutical products in effluents. However, an analysis of covariance revealed that CYP3A activity was more strongly associated with Vtg-like proteins than with exposure time, even though the latter remained significant. Thus, the increase in CYP3A activity appeared linked to both estrogenicity (by nonylphenol, but not E2) and perhaps by the presence of pharmaceutical compounds.

MT levels in female mussels were somewhat lower than in males exposed to these estrogenic aeration lagoons. Estradiol-treated fish injected with low doses of Cd or Zn did not respond by inducing Mt, suggesting suppression by estrogens (Olsson and Kling, 1995). In another study on plaice, hepatic MT levels were negatively related to the GSI, with the highest MT levels being found at the beginning of gonad development when vitellogenesis starts (Overnell et al., 1987). This finding was corroborated in the present study on mussels, with digestive gland MT being significantly correlated with Vtg-like proteins and pyrimidine synthesis ATC in the less toxic aeration lagoon AL1. Males appeared less affected than females by the estrogenic action of aeration lagoons. Indeed, MT in males was higher than in females in the less estrogenic and less toxic lagoon AL1, but MT in females was not significantly different than males in the more estrogenic and toxic aeration lagoon AL2. The activities of cytochrome P450-3A and GST were more responsive than that for P450-1A1 (EROD), indicating that more polar hydrocarbons (such as hydroxylated cyclic hydrocarbons) are more likely at play than the less-polar aromatic hydrocarbons in these lagoons. This is consistent with their role in the oxidative degradation of xenobiotics through aeration, microbial metabolism and photolysis. The occurrence of pharmaceutical and steroidal products in municipal wastewaters would likely contribute to the effects observed globally but the influence of other factors cannot be excluded.

In a recent study on rosy barb, males challenged by *E. coli* produced Vtg, with antibacterial and anti-hemagglutinating activities *in vitro* (Shi et al., 2006) but the implication of Vtg in immune defense in bivalves, however, needs confirmation. COX activity, a biomarker of tissue inflammation, however, was either significantly correlated with Vtg-like proteins (R = 0.51; p < 0.01) in AL1 or with CYP3A activity (which was correlated with Vtg at R = 0.48) in AL2. Hence, a possible interaction between tissue inflammation (infection) and vitellogenesis needs further investigation in bivalves. Furthermore, COX activity was also significantly correlated with MT levels in the digestive gland, suggesting that pro-inflammatory conditions contributed to the MT response in mussels. This was supported by the analysis of covariance of digestive gland MT with COX activity as the covariate, indicating that MT was more significantly associated with COX activity than with exposure time for both aeration lagoons. In another study, MT in phagocytes was readily induced by mycobacteria, but not with mercury in striped bass (Regala and Rice, 2004), which suggests that MT in addition to Vtg could be induced during a bacterial challenge in mussels exposed to aeration lagoon wastewaters.

Energy status was studied in relation to cellular energy expenditure (mitochondrial electron transport activity) and lipid reserves in the gonad. These physiological targets were considered ecologically relevant in invertebrates in that they have predictive value with respect to survival, reproduction and population maintenance (Smolders et al., 2004). More recently, temperature-dependent mitochondrial electron transport activity was proposed as a means to demonstrate a possible interaction between global warming and pollution. (Gagné et al., 2006a). MET at 20°C and MET_T were readily induced before and during the manifestation of mortality, which is in agreement with the reported predictive value for survival. Moreover, MET was negatively associated with gonad lipids and GSI, suggesting that increased energy expenditure was occurring at the expense of gonad lipids and size (weight). MET_T was negatively associated with gonad lipids, indicating that a temperature-dependent increase in MET could be deleterious to gonad energy reserves. MET_T was also positively associated with vitellogenin-like proteins production and LPO in the gonad, suggesting that temperature fluctuations could influence oxidative stress and vitellogenesis in bivalves exposed to urban pollution.

In the attempt to gain a systematic and pathophysiological view of the response patterns of mussels exposed to aeration lagoons, factorial (principal component) and canonical analyses were performed. In the less toxic and estrogenic aeration lagoon AL1, biomarkers of biotransformation and of tissue damage were more strongly correlated with gametogenesis than with energy status or metabolism of endogenous compounds. Energy status had only a weak but significant trend with biotransformation. In the more toxic and estrogenic aeration lagoon AL2, energy status was now strongly associated with biotransformation and metabolism of xanthines, metals and monoamines, but showed a significant trend with tissue damage and gametogenesis. Hence, the increased association between energy status (i.e. increase in MET with decreased lipids in the gonad) with tissue damage, biotransformation/metabolism and gametogenesis could reveal a more severe morbidity. In a field study with *Mya arenaria* clams, organisms at highly impacted sites displayed more severe damage at the gonad and the morphological levels (Gagné et al., 2006b) than in the digestive gland and gills. Clearly, as the quality of sites degrades, effects shift from tissues directly exposed to the environment (gills and digestive gland) to the gonad and at the individual levels (e.g. gonado-somatic index and condition factor). Lastly, energy status had more direct links with biotransformation and metabolism of xanthines, monoamines and metals in addition to the appearance of tissue damage and changes in gametogenesis in the more toxic lagoon.

ACKNOWLEDGMENTS

The authors are grateful for the technical assistance of Michel Arsenault (caging experiments) and Sophie Trépanier for the biomarker analyses. This project was funded by the St. Lawrence Centre of Environment Canada. The manuscript was edited by Patricia Potvin.

REFERENCES

Andreozzi, R., Raffaele, M., Nicklas P. (2003). Pharmaceuticals in STP effluents and their solar photodegradation in aquatic environment. *Chemosphere 50,* 1319-1330.

Baldwin, W.S., Roling, J.A., Peterson, S., Chapman, L.M. (2005). Effects of nonylphenol on hepatic testosterone metabolism and the expression of acute phase proteins in winter flounder (*Pleuronectes americanus*): Comparison to the effects of Saint John's wort. *Comp. Biochem. Physiol. 140C,* 87-96.

Bester, M.J., Potgieter, H.C., Vermaak, W.J.H., 1994. Cholate and pH reduce interference by SDS in the determination of DNA with Hoescht. *Anal. Biochem. 223,* 299-305.

Boryslawskyj, M., Garrood, A.C., Pearson, J.T. (1988). Elevation of glutathione-S-transferase activity as a stress response to organochlorine compounds in the freshwater mussel, *Sphaerium corneum. Mar. Environ. Res. 24,* 101-104.

Blaise, C., Gagné, F., Salazar, M., Salazar, S., Trottier, S., Hansen, P.-D. (2003). Experimentally induced feminisation of freshwater mussels after long-term exposure to a municipal effluent. *Fresenius Environ. Bull. 12,* 865-870.

Bradford, M.M. (1976). A rapid and sensitive method for the quantitation of microgram quantities of protein utilizing the principle of protein-dye binding. *Anal. Biochem. 72,* 248-254.

Cancio, I., Cajaraville, M.P. (1999). Seasonal variation of xanthine oxidoreductase activity in the digestive gland cells of the mussel Mytilus galloprovincialis: a biochemical, histochemical and immunochemical study. *Biol. Cell 91,* 605-615.

Canesi, L., Scarpato, A., Betti, M., Ciacci, C., Pruzzo, C., Gallo, G. (2002). Bacterial killing by *Mytilus* hemocyte monolayers as a model for investigating the signaling pathways involved in mussel immune defence. *Mar. Environ. Res. 54,* 547-551.

Canova, S., Degan, P., Peters, L.D., Livingstone, D.R., Voltan, R., Venier, P. (1998). Tissue dose, DNA adducts, and oxidative DNA damage and CYP1A-immunopositive proteins in mussels exposed to waterborne benzo(a)pyrene. *Mutation Res. 399,* 17-30.

Chambers, P.A., Allard, M., Walker, S.L., Marsalek, J., Lawrence, J., Servos, M., Busnarda, J., Munger, K.S., Jefferson, C., Kent, R.A., Wong, M.P., Adare, K. (1997). Impacts of municipal wastewater effluents on Canadian waters: A review. *Wat. Qual. Res. J. Canada 32*, 659-713.

Ciccotelli, M., Crippa, S., Colombo, A. (1998*).* Bioindicators for toxicity assessment of effluents from a wastewater treatment plant. *Chemosphere 37*, 2823-2832.

Crespi, C.L., Miller, V.P., Penman, B.W. (1997). Microtiter plate assays for inhibition of human, drug-metabolizing cytochromes P450. *Anal. Biochem. 248*, 188-190.

De Coen, W., Janssen, C.R. (2003). The missing biomarker link: Relationships between effects on the cellular energy allocation biomarker of toxicant-stressed *Daphnia magna* and corresponding population characteristics. *Environ. Toxicol. Chem. 22*, 1632-1641.

Fenet, H., Casellas, C., Bontoux, J. (1998). Laboratory and field-caging studies on hepatic enzymatic activities in European eel and rainbow trout. *Ecotoxicol. Environ. Saf. 40*, 137-143.

Frenzilli, G., Bocchetti, R., Pagliarecci, M., Nigro, M., Annarumma, F., Scarcelli, V., Fattorini, D., Regoli, F. (2004). Time-course evaluation of ROS-mediated toxicity in mussels, *Mytilus galloprovincialis*, during a field translocation experiment. *Mar. Environ. Res. 58*, 609-613.

Fujimoto. Y., Sakuma, S., Inoue, T., Uno, E., Fujita, T. (2002). The endocrine disruptor nonylphenol preferentially blocks cyclooxygenase-1. *Life Sciences 70*, 2209-2214.

Gagné, F., Blaise, C., Salazar, M., Hansen, P. (2001). Evaluation of estrogenic effects of municipal effluents to the freshwater mussel *Elliptio complanata*. *Comp. Biochem. Physiol. 128C*, 213-225.

Gagné, F., Blaise, C., Hellou, J. (2004). Endocrine disruption and health effects of caged mussels, *Elliptio complanata*, placed downstream from a primary-treated municipal effluent plume for one year. *Comp. Biochem. Physiol. 138C*, 33-44.

Gagné, F., Blaise, C., André, C., Pellerin, J. (2006a). Implication of site quality on mitochondrial electron transport activity and its interaction with temperature in feral *Mya arenaria* clams from the Saguenay Fjord. Environ. Res., In press.

Gagné, F., Blaise, C., Pellerin, J., Pelletier, E., Strand, J. (2006b). Health status of *Mya arenaria* bivalves collected from contaminated sites in Canada (Saguenay Fjord, Que., Canada) and Denmark (Odense Fjord) during their reproductive period. *Ecotoxicol. Environ. Saf. 64*, 348-361.

Gagné, F., Blaise, C., André, C., Salazar, M. (2006c). Effects of municipal effluents and pharmaceutical products on temperature-dependent mitochondrial electron transport activity in *Elliptio complanata* mussels. *Comp. Biochem. Physiol. 143*, 388-393.

Gowland. B.T.G., McIntosh, A.D., Davies, I.M., Moffat, C.F., Webster, L. (2002). Implications from a field study regarding the relationship between polycyclic aromatic hydrocarbons and glutathione S-transferase activity in mussels. *Marine Environ. Res. 54*, 231-235.

Harries, J.E., Sheahan, D.A., Jobling, S., Matthiessen, P., Neall, P., Routledge, E.J., Rycroft, R., Sumpter, J.P., Tylor, T. (1996). A survey of estrogenic activity in United Kingdom inland waters. *Environ. Toxicol. Chem. 15*, 1993-2002.

Herries, D.G. (1967). The simultaneous estimation of orthophosphate and carbamoylphosphate and application to the aspartate transcarbamylase reaction. *Biochem. Biophys. Acta 136*, 95-98.

Isani, G., Andreani, G., Kindt, M., Carpene, E. (2000). Metallothioneins (MTs) in marine molluscs. *Cell. Mol. Biol. 46*, 311-330.

King, F., Packard, T.T. (1975). Respiration and the activity of the respiratory electron transport system in marine zooplankton. *Limnol. Oceanog. 20*, 849-854.

Kummerer, K. (2001). Drugs in the environment: Emission of drugs, diagnostic aids and disinfectants into wastewater by hospitals in relation to other sources – Review. *Chemosphere 45*, 957-969.

Li, Q., Osada, M., Suzuki, T., Mori, K. (1998). Changes in vitellin during oogenesis and effect of estradiol on vitellogenesis in the Pacific oyster *Crassostrea gigas. Invert. Reprod. Dev. 33*, 87-93.

Lima, P.L., Benassi, J.C., Pedrosa, R.C., Dal Magro, J., Oliveira, T.B., Wilhelm, F.D. (2006). Time-course variations of DNA damage and biomarkers of oxidative stress in tilapia (*Oreochromis niloticus*) exposed to effluents from a swine industry. *Arch. Environ. Contam. Toxicol. 50*, 23-30.

Mathieu, M. (1987). Utilization of aspartate transcarbamylase activity in the study of neuroendocrinal control of gametogenesis in *Mytilus edulis. J. Exp. Biol. 241*, 247-252.

Olive, P.L., Chan, A.P.S., Cu, C.S. (1988). Comparison between the DNA precipitation and alkali unwinding assays for detecting DNA strand breaks and cross-links. *Cancer Res. 48*, 6444-6449.

Olive, P.L. (1988). DNA precipitation assay: A rapid and simple method for detecting DNA damage in mammalian cells. *Environ. Mol. Mutagen 11*, 487-495.

Olsson, P.-E., Kling, P. (1995). Regulation of hepatic metallothionein in estradiol-treated rainbow trout. *Marine Environ. Res. 39*, 127-129.

Overnell, J., McIntosh, R., Fletcher, T.C. (1987). The levels of liver metallothionein and zinc in plaice, *Pleuronectes platessa* L., during the breeding season, and the effect of oestradiol injection. *J. Fish Biol. 30*, 539-546.

Peters, L.D., Nasci, C., Livingstone, D.R. (1998). Variation in levels of cytochrome P4501A, 2B, 2E, 3A and 4A-immunopositive proteins in digestive gland of indigenous and transplanted mussel *Mytilus galloprovincialis* in Venice lagoons, Italy. *Mar. Environ. Res. 46*, 295-299.

Quinn, B., Gagné, F., Costello, M., McKenzie, C., Wilson, J., Mothersill, C. (2004a). The endocrine disrupting effect of municipal effluent on the zebra mussel (*Dreissena polymorpha*). *Aquat. Toxicol. 66*, 279-292.

Quinn, B., Gagné, F., Blaise, C. (2004b). Oxidative metabolism activity in *Hydra attenuata* exposed to carbamazepine. *Fresenius Environ. Bull. 13*, 783-788.

Regala, R.P., Rice, C.D. (2004). Mycobacteria, but not mercury, induces metallothionein (MT) protein in striped bass, *Morone saxitilis*, phagocytes, while both stimuli induce MT in channel catfish, *Ictalurus punctatus*, phagocytes. *Mar. Environ. Res. 58*, 719-723.

Regoli, F., Frenzilli, G., Bocchetti, R., Annarumma, F., Scarcelli, V., Fattorini, D., Nigro M. (2004). Time-course variations of oxyradical metabolism, DNA integrity and lysosomal stability in mussels, *Mytilus galloprovincialis*, during a field translocation experiment. *Aquat. Toxicol. 68*, 167-178.

Rosety, M., Rosety-Rodriguez, M., Ordonez, F.J., Rosety, I. (2005). Time course variations of antioxidant enzyme activities and histopathology of gilthead seabream gills exposed to malathion. *Histol. Histopathol. 20*, 1017-1020.

Salazar, M. H., Salazar, S. M. (2001). *Standard Guide for Conducting in situ Field Bioassays with Marine, Estuarine and Freshwater Bivalves*. American Society for Testing and Materials (ASTM), Annual Book of ASTM Standards.

Sastry KV, Gupta PK. (1978). Histopathological and enzymological studies on the effects of chronic lead nitrate intoxication in the digestive system of a freshwater teleost, *Channa punctatus. Environ. Res. 17*, 472-479.

Shaw, J.P., Peters, L.D., Chipman, J.K. (2004). CYP1A- and CYP3A-immunopositive protein levels in digestive gland of the mussel *Mytilus galloprovincialis* from the Mediterranean Sea. *Mar. Environ. Res. 58*, 649-653.

Shi, X., Zhang, S., Pang, Q. (2006). Vitellogenin is a novel player in defence reactions. *Fish Shell. Immunol. 20*, 769-772.

Smolders, R., Bervoets, L., De Coen, W., Blust, R. (2004). Cellular energy allocation in zebra mussels exposed along a pollution gradient: Linking cellular effects to higher levels of biological organization. *Environ. Poll. 129*, 99-112.

Vaccaro, E., Meucci, V., Intorre, L., Soldani, G., Di Belloc, D., Longo, V., Gervasia, P.G., Pretti, C. (2005). Effects of 17ß-estradiol, 4-nonylphenol and PCB 126 on the estrogenic activity and phase 1 and 2 biotransformation enzymes in male sea bass (*Dicentrarchus labrax*). *Aquat. Toxicol. 75*, 293-305.

Vasseur, P., Leguille, C. (2004). Defense systems of benthic invertebrates in response to environmental stressors. *Environ. Toxicol. 19*, 433-436.

Viarengo, A., Ponzanon, E., Dondero, F., Fabbri, R. (1997). A simple spectrophotometric method for metallothionein evaluation in marine organisms: An application to Mediterranean and Antarctic molluscs. *Mar. Environ. Res. 44*, 69-84.

Vignati, L., Turlizzi, E., Monaci, S., Grossi, P., Kanter, R.D., Monshouwer, M. (2005). An in vitro approach to detect metabolite toxicity due to CYP3A4-dependent bioactivation of xenobiotics. *Toxicology 216*, 154-167.

Wills, E.D. (1987). "Evaluation of Lipid Peroxidation in Lipids and Biological Membranes." In: Snell, K. and Mullock, B. (eds.), *Biochemical Toxicology: A Practical Approach*. IRL Press, Washington, USA, pp. 127–150.

Zhu, H., Banneberg, G.L., Moldeaus, P., Shertzer, H.G. (1994). Oxidation pathways for the intracellular probe 2',7'-dichlorofluorescin. *Arch. Toxicol. 68*, 582-587.

In: Progress in Environmental Research
Editor: Irma C. Willis, pp. 203-230
ISBN 978-1-60021-618-3
© 2007 Nova Science Publishers, Inc.

Chapter 6

RESPONSE OF THE AMOEBAE COMMUNITY TO SOIL CONDITIONS TRANSITING TO BADLANDS FORMATION

Salvador Rodríguez Zaragoza[*] *and Ernestina González Lozano*

Laboratorio de Microbiología, Unidad de Biología, Tecnología y Prototipos,
Facultad de Estudios Superiores Iztacala, UNAM.

ABSTRACT

The stress imposed on ecosystems has produced important biodiversity loss. In terrestrial ecosystems, loss of plant cover leads to soil degradation impairing the system capacity to recover after perturbations. Plant – microbe interactions are the key factor for productivity because of the beneficial role played by the microbial community living in the root zone. Soil degradation may negatively affect this microbial community and, indirectly affect the microbial predators such as the amoeboid protozoa. We aim to describe the changes in the amoebae community when soil is transiting to badlands formation and to correlate this variation with the soil physicochemical factors at the semiarid Valley of Tehuacán, México. Samples from the root zone of mesquite (*Prosopis laevigata*), the columnar cactus *Pachycereus hollianus*, and from the interspace soil were taken at 0 – 10 cm layer in both terraces. Samplings were done every four weeks from July to October 2002 to observe the variation change during the transition from dry to wet season. The physicochemical factors and the community of amoebae were significantly different between terraces and microenvironments. *Vahlkampfia*, *Platyamoeba*, *Mayorella*, *Hartmannella* and *Acanthamoeba* were the most frequent genera found in soil. However amoebae community was different between microenvironments. Species richness was smaller in soil forming badlands; even though plants played the role of fertility islands in this soil. Principal component analysis showed that pH, porosity and real density explained only 23.9 % of the amoebae community

---

[*] Corresponding Autor: Avenida de los Barrios # 1, Los Reyes Iztacala, Tlalnepantla, Estado de México, CP 54090, México. Tel: +55-5623-1223; Fax +55-5623-1225; E-mail: srodrige@campus.iztacala.unam.mx

variation. The response of amoebae to soil degradation was the reduction of species richness and change of the dominant species in the community. The physicochemical factors correlated poorly with the observed community variance.

INTRODUCTION

Biodiversity is one of the most important features related to ecosystem's capacity to provide goods and services (Usher et al., 2006). The variety and identity of species inhabiting a place will determine the capacity of the system to recover after perturbations, its resistance to changes and evolutionary fate (Holling, 1973; Potts et al., 2006). Belowground biodiversity is also very important to ecosystem functioning as plants depend strongly on soil food webs to obtain both, macro and micronutrients, and produce biomass. Plant cover protects soil from water and wind erosion and contributes to soil fertility. Once plant cover is reduced or removed, exposed soils become very sensitive to weather changes by receiving wind and rain erosive forces directly on their surfaces. Steady depletion of soil nutrients is one of the consequences of agriculture, as harvested products are exported for consumption away from the systems leading to soil impoverishment and degradation (Rodríguez Rodríguez et al., 2005). The former has been defined as the soil inability to sustain plant productivity.

Several human activities promote soil impoverishment, especially timber production, slash and burn agriculture, and plant recollection. Transformation of soils into badlands is the ultimate consequence of these activities. Soil degradation changes the landscape on a long term basis. However, in the short term it is unnoticed and becomes evident after the first signs of erosion are visible.

Drylands ecosystems are more vulnerable to perturbations than temperate ones because the time span of water availability is very short. It is only one or two months of rains in several shadow rain deserts and one or two events of rain during a whole year in the driest ones. Frequency and duration of rains determine the extent of biological activity in these ecosystems. These periods of water availability are known as "Windows of activity" (Noy-Mayr, 1973) because this is the effective time for biomass accumulation, or the opportunity for recovering after perturbation. This is also the main reason of drylands vulnerability to perturbations and explains their inability to withstand biomass losses imposed upon them by human activities (Zhi-Zhaoa et al., 2005).

Soil is a very dynamic system where biological activities promote nutrient cycling and sustain terrestrial ecosystems. Matter and energy are transformed through food webs that are controlled by inputs of detritus and dead plant matter (Paul and Clark, 1989). This process is responsible for plant nutrition and, through it, ensures the sustained productivity of ecosystems by building the soil's structural genesis, organic matter degradation and nutrient cycling (Griffiths et al., 2001).

Soil organic matter (SOM) is considered an important indicator of soil quality because it represents the nutrient status of the system. However, soil's organic matter as a whole changes very slowly, taking several decades to show some differences due to perturbations (Guo and Berry, 1998). On the contrary, microbial processes show relatively fast responses to perturbation than other soil properties (Jimenez et al., 2002) because their life cycles are completed in very short time (Rodríguez-Zaragoza, 1994). On the other hand, it has been possible to find out correlations of species occurrence with real density of soil, water content and organic matter in a temperate soil (Rodríguez-Zaragoza, 2000).

In general, the ecosystem stability is related to biodiversity (Philippi et al., 1998) because the niche widths of organism determine the speed of system's recovery and resistance to perturbations (Stout, 1980). Depending on the nature, strength and duration of the perturbing episodes, they may eliminate more sensitive species and benefit the more resistant ones (Odum et al., 1979). In this way, perturbations change the species assemblage of a community (Allison, 2004) and produce biodiversity reduction. Biodiversity loss also reduces soil capacity to produce biomass. Even though soil systems may be able to function with few surviving species, reduction in species richness makes systems more vulnerable to perturbations (Bamforth, 1995).

Fungi and bacteria carry out most of the nutrient mobilization and organic matter degradation in belowground food webs, while protozoa enhance organic matter mineralization and nutrient release by predation on these groups of microorganisms. Soil protozoa, increase the mineralization of organic matter by 60% and most of this mineralization is carried out by naked amoebae (Ekschmit and Griffiths, 1998; Griffiths et al., 2001). Microbial communities have the capacity to resist perturbations and still continuing performing their main functions in soil due to their high diversity, in other words, a high degree of functional redundancy (Griffiths et al., 2001).

Species diversity of bacteria and fungi has been estimated in thousands of species on earth, 99.9% of them are non cultivable (Mlot, 2004; Pedros-Alió, 2006), which still makes their study both technically difficult and expensive. On the contrary, soil protozoa are less diverse and can be managed for studying biodiversity in soils (Rodríguez-Zaragoza, 2000). However they are much more diverse in soils than other groups of Eukaryotes, excepting fungi. Naked amoeba represent from 50 to 90% of soil protozoa diversity, and are more abundant than ciliates in the soil matrix (Maizlish and Steinberger, 2004).

Amoebae can also play different functional role in soils as they feed on all groups of microorganisms, including multicellular ones (Rodríguez Zaragoza, 1994). Based on prey preferences, amoebae may be recognized as bactivorous, fungivorous, algivorous, carnivorous and omnivorous protozoa (Chakraborty and Old, 1985; Page, 1988; Rodríguez-Zaragoza, 1994; Rodríguez-Zaragoza et al., 2005 a). Carnivorous amoebae can even feed on nematodes and rotifers (Mast and Root, 1916).

Functional assemblages of species richness are based on the spatial-temporal variation of the soil physical and chemical parameters (Guo and Berry, 1998; Rodríguez-Zaragoza, 2000). Because of this, soil protozoa assemblage changes due to soil use modification as their reproduction rate, encysting capacity and mortality rate are affected during soil perturbation (Côuteaux and Darbyshire, 1998). Diversity of a functional group is important because it will perform much better and efficiently in a heterogeneous and changing environment than a species poor one (Ekschmitt and Griffiths, 1998; Griffiths et al., 2001).

Soil environment *per se* is very heterogeneous, and the diversity of microenvironments in these soils depends on the kind of plants growing up in the patches or the absence of vegetation. Plants strongly influence microbial community composition by means of their root exudates, which are mixtures of organic molecules released into soils. Root exudates allow plants to "choose" the microorganisms that may grow in their root canopy (Lynch et al., 2002). This selection is called the "rhizosphere effect" and is specific to each plant species (Appuhn & Joergensen, 2006). Therefore, comparison of species diversity of soil amoebae has to be done also by soil microenvironment. We chose the environments created by two of the most common plants at the terraces: mesquite (*Prosopis laevigata*) and a columnar cactus (*Pachycereus hollianus*). Both plants are present in preserved and badlands patches at the alluvial terraces allowing the comparison of the biodiversity of amoebae in similar microenvironments subject to different conditions. We expected the species richness of amoeba in badlands to be extremely low in comparison with the preserved patches, no matter the micro environmental differences (mesquite, columnar cactus or bare soil) as the ultimate effect of the extreme desiccation conditions.

MATERIAL AND METHODS

Tehuacán-Cuicatlán Valley is a shadow-rain desert located in central Mexico (240 km south of Mexico City) between 17° 39' to 18° 53' N latitudes and 96° 55' to 97 ° 44' W longitudes (Figure 1a). Zapotitlán is the Northwest corner of the Tehuacán valley with a surface of 400 km^2 at 1,480 meters above sea level. The annual mean temperature is 21 °C and annual mean precipitation ranges from 350 to 450 mm. Climate is classified as dry hot with summer rains (García, 1988). The geo-morphological units of alluvial terraces of Zapotitlán were originated in the Pleistocene; when sediment deposition was carried out by "El Salado" River. As consequence, soils are very deep with an incipient "A" horizon and a sequence of different "C" horizons. The textural classes are sandy loam with median content of organic matter and soils are identified as calcaric fluvisols and regosols (López et al., 2003).

Mesquital is the dominating plant association on the terraces. This association is made up mainly by legume shrubs, such as *Prosopis laevigata* (mesquite) and *Parkinsonia praecox* (palo verde), and the columnar cactus *Pachycereus hollianus* (Davila, 1997; Valiente-Banuet et el., 2001).

The alluvial terraces at the Basin of "El Salado" river form a mosaic of a fragmented landscape (Figure 1b), where patches of strongly eroded soils (badlands) coexist with patches of different degrees of preserved vegetation. Several patches of badlands (Figure 1c) have been formed by natural degradation processes, and are places almost devoid of vegetation (López et al., 2003)

Degradation and badlands formation at the terraces is taking place by the same process that formed them in the first instance. The river that once deposited the material in this part of the Valley is taking them to a different place. Erosion events are registered in the Valley dating back to 6 000 years before present (McAuliffe et al., 2001), as well as more recent events of erosion promoted by human over exploitation of the terraces 900 years ago (McAuliffe et al., 2001).

At present, terraces are used for seasonal agriculture, harvesting of edible fruits and insects, and wood recollection. All the former have made a patchy landscape where preserved vegetation is located nearby badlands.

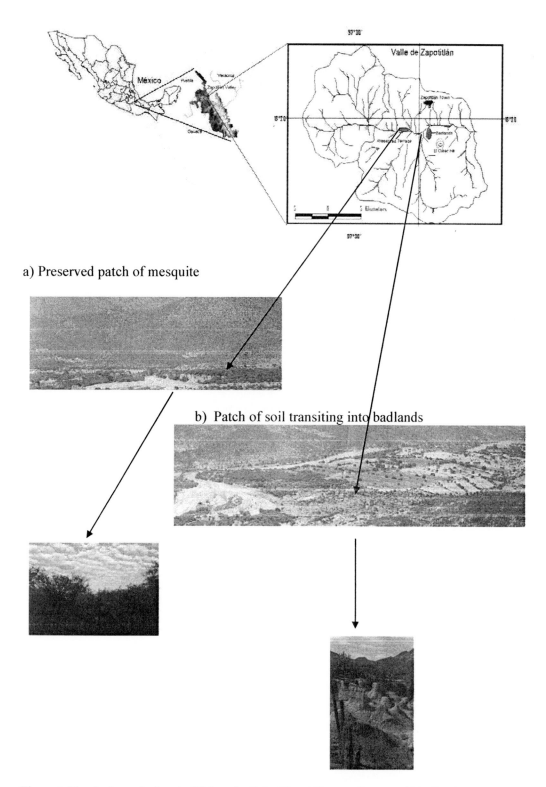

a) Preserved patch of mesquite

b) Patch of soil transiting into badlands

Figure 1. The shadow rain desert of Tehuacán-Cuicatlán: a) Preserved terrace. b) badlands.

The process of badlands formation can be recognized at different levels, from places where signs of eolic and hydric erosion start to be evident, to places where desiccation is high, soil has already disappeared, there are no vegetation and have no economic use at all (Morgan, 1995; Bouma and Imerson, 2000; Torri et al., 2000).

Sampling

Samples were taken at each terrace every four weeks, from the root canopy of 3 individual plants of *P laevigata* (Leguminosae), 3 individuals of *P hollianus* (Cactaceae) and from three sites of inter-space soil (as control) at 10 cm depth. We adjusted for sampling during months when water becomes available in soil. The moments of change in temperate ecosystems are very well known and seasonal changes are normally sampled in the middle of spring, summer, autumn and/ or winter. Below the tropic of Cancer, seasons change in a different way. Main changes in this shadow rain desert of Tehuacán are as follows: dry season, light showers (early June), dry season during July (known as canícula); stormy weather (August to early September), and no rains. This seasonal division was used instead of the classical ones outlined above. Thus, sampling was distributed in this window of time to allow detection of seasonal variation of water availability, lasting from Jun to October 2002. Soil was collected into plastic bags and taken to the lab, sieved in a 2mm diameter mesh, air dried and stored at 4 °C until needed for analysis. Samples for water content by gravimetry, (Anderson and Ingham, 1993) and biological analysis were accomplished before processing for storage.

Physical and Chemical Analysis of Soil

Soil moisture - a 5 g sub sample from each of four replicates from each site was weighed and dried at 105°C for 48 h for gravimetrical determination of water content.

Determination of pH was achieved by potentiometer (Waterproof pH tester 2) in a soil homogenate (rate soil/ water = 1:2.5).

Soil's apparent density was determined by volumetric method (López et al., 2003) and real density was determined by picnometer (López et al., 2003). Then porosity was calculated and expressed as percentage using the following formula: Porosity (%) = [1 − (Apparent density / real density)] X 100 (López et al., 2003).

Salinity was determined by the soluble salt tester brand Kelway model SST (Kiel instruments CO.) following the instructions of the manufacturer.

Soil organic carbon (OC) was determined by using the modified chromate and sulfuric acid oxidation methods of Walkley and Blak (Alef and Nannipiery, 1995).

Extractable phosphates (ortophosphates) were determined by the Olsen method as reported by Cajuste (1986)

Identification of Soil Amoebae

Identification of free-living amoebae was accomplished after cultivation in non-nutritive agar plates. Bacteria were not added to the medium, neither live nor dead, in order to avoid the overgrowth of bactivorous amoebae over those that feed on different sources such as yeast, fungi, algae, protozoa, and/or other organisms. The initial cultivation was performed by homogenizing 1 g soil sample in 10 ml soil extract (final dilution of 1:10). Homogenates were left untouched for 30 min for particle sedimentation. Then, supernatant was gently transferred onto bacteria-free non-nutritive agar plates (14 g non nutritive agar in 1000 ml of 1:5 dilution of soil extract). Soil extract was prepared by homogenization of 200 g of soil from sampling place in 1000 ml of distilled water and heated at 60 °C in a water bath for 2 h. Extract was autoclaved in standard conditions for 15 min after filtration in Wathman paper # 42. This is the stock solution; working solution was prepared by making a 1:5 dilution in sterilized distilled water. Then, 1 g of soil sample was homogenized in 10 ml of soil extract (working solution)

Amoebae were allowed to settle down on the agar plate for 2 h. Thereafter, the excess water was withdrawn to avoid ciliate and flagellate growth. Plates were incubated at 26°C and monitored every 24 hr to register growth of amoebae during 21 days. After this time no growth of new strains was registered. Amoebae were identified by morphology using phase contrast microscope, and based on the descriptions reported in the keys for Freshwater and Soil Amoebae (Page, 1976; 1988; Page and Siemensma, 1991; Patterson 1996; Lee et al., 2000). Shannon-Weaver Species diversity index was calculated with the PC ORD© software (McCune & Mefford, 1999).

Statistical Analysis

Physical and chemical values from the different microenvironments were analyzed by ANOVA; Tukey test was applied to distinguish differences between means. Sørensen' similitude index was used to estimate the similitude between microenvironments. These analyses were carried out with PC-ORD© Software Ver. 4 for Windows©. Principal component analysis was performed to look for influence of physical and chemical factors on the species occurrence.

RESULTS

Seasonal Variation of Physical and Chemical Factors

Soil organic matter (SOM) content was higher in the preserved terrace than in badlands. However, only *P laevigata* and *P hollianus* at preserved terrace were significantly higher (P< 0.0001) than other microenvironments, including the control soil at the preserved terrace. Pattern of SOM variation was similar among microenvironments no matter the terrace; the difference was the quantity of organic matter in each microenvironment. In general, soil under *P laevigata* had more organic matter, followed by *P hollianus* and control soil

(Figure 2a). Quantities of SOM varied in all microenvironments along this window of time, decreasing at the beginning during rains and returning to values comparable or higher than the initial ones at the end of the study.

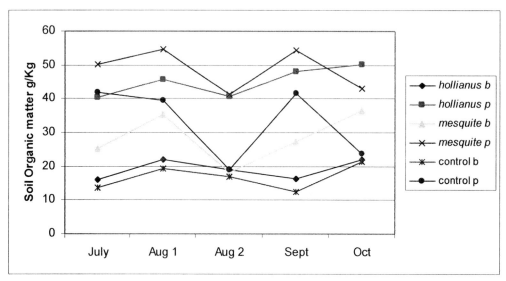

a) SOM content varied in all microenvironments during the study. Microenvironments are named in figure as hollianus (columnar cactus Pachycereus hollianus), mesquite (Prosopis laevigata) and control from preserved (p) terrace and badlands (b). The highest rainfall occurred during August and ended in early September. Time lag between samples is 4 weeks.)

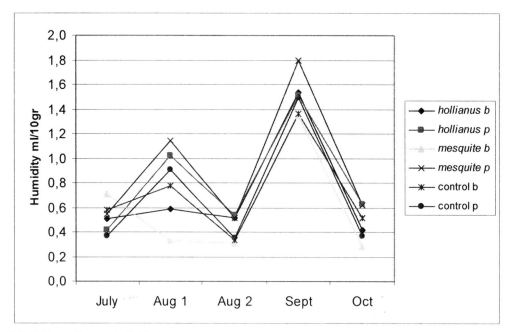

b) Soil Water Content (ml/ 10 g).

Figure 2. (Continued)

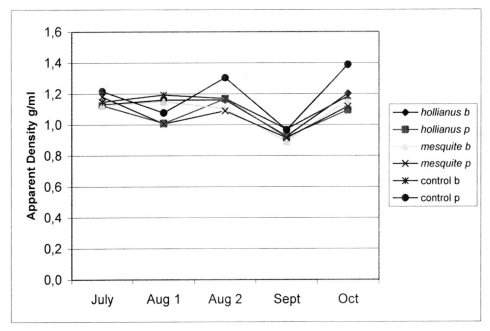

c) Apparent Density.

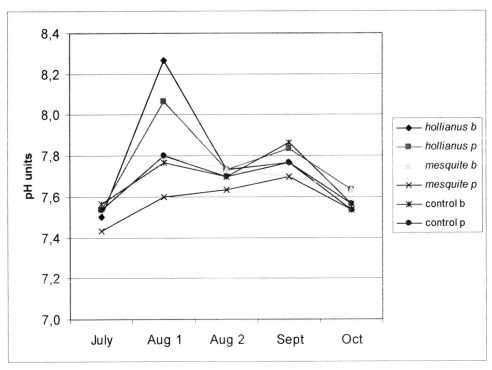

d) Seasonal variation of pH in soil from preserved terrace and badlands.

Figure 2. (Continued)

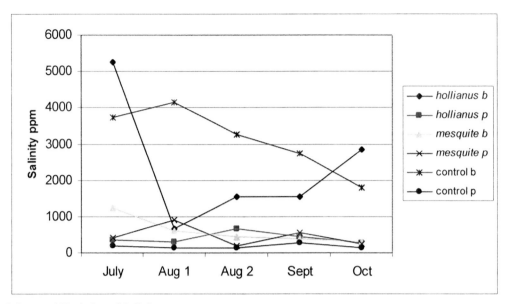

e) Seasonal Variation of Salinity.

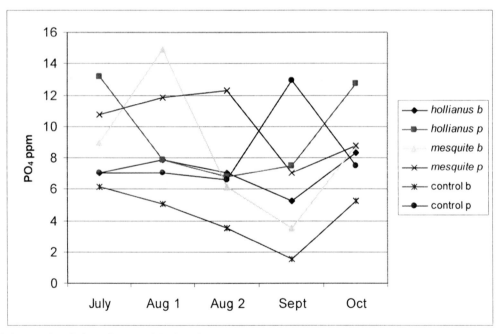

f) Seasonal variation of extractable phosphorous.

Figure 2. Seasonal Variation of Soil's Physical and Chemical Factors. Soil water content reflected the peaks of precipitation during the rainy season. Very similar patterns of water content were found between microenvironments. The highest water content was found in soil under *P laevigata* and *P hollianus* from the preserved terrace. Even though, they were not significantly different from each other. On the other hand, control soils from both terraces always showed the lowest values recorded along the study period (Figure 2b).

Seasonal variation of apparent density in all microenvironments followed similar patterns at both terraces (Figure 2c). Only the control soil at the preserved terrace was significantly higher (P< 0.001) than the other soils. Remarkably, apparent density diminishes in all microenvironments during late August and September, after the end of rains, and by October it increases again. This is in concordance with the negative correlation found between apparent density variation and soil moisture (r = -0.758).

Porosity at both terraces varied between 45 % (dry season) and 53% (wet season) and there were no significant differences between samples (P = 0.91) or microenvironments (P = 0.41) along the study period. The former implies that soil compaction will increase from wet to dry season and that the first rains may be more prone to runoff. However, late storms are more pervasive for water erosion and landslides.

Soil pH also varied along the time influenced by water availability. During the dry season the pH varied around 8.4 and it was significantly (P < 0.002) more neutral, around 7.4, during the rainy months. There were micro-environmental differences. In general, soil under *P hollianus* was more alkaline (P < 0.001), while *P laevigata* was more neutral (Figure 2d).

Soil salinity, measured as electrical conductivity, varied among microenvironments (P< 0.001), although it did not varied significantly along seasons (P = 0.77). *P hollianus* and control soil from badlands were significantly different from all others microenvironments. Control soil showed a peak of salinity at the middle of the rainy season, and then followed a steady decrease that was kept until the end of the sampling period (Figure 2e). On the other hand, salinity under *P hollianus* at badlands strongly decreased at the middle of the rainy season and it followed a steady increase up to the end of the sampling period (Figure 2e). It is worth to note that the highest value of electrical conductivity was found in soil under *P hollianus* at badlands, followed closely by the control soil at this same terrace. However, these two microenvironments showed different patterns along the change of seasons. Soil under cactus showed a strong diminution of salinity as soon as rains started, and then increased steadily, while the control soil diminished steadily along the rainy season. The pattern followed in soil under *P laevigata* at badlands was very similar to that one found in the preserved terrace. Although badlands showed slightly higher values than the preserved one, there were not significant differences between them (P = 0.4; Figure 2e).

Soluble phosphorous varied markedly along the sampling period (P< 0.005) and between microenvironments (P < 0.0003) where soils influenced by root canopies had significantly higher content of phosphorous. One of the problems for phosphorous availability in alkaline soils is the precipitation of this nutrient at high pH and its sequestration by forming complexes with calcium and sulfates. Such complexes immobilize this nutrient in the soil matrix making it unavailable to plants. The highest levels of soluble phosphorous were found in the preserved terrace and the lowest ones in the control soil from badlands (Figure 2f). The microenvironments of control soil and *P hollianus* showed a steady decrease of available phosphorous from dry to wet periods and an increase by October, contrary to the variations detected in *P laevigata* that were much higher, showing no discernible pattern. These changes were independent of the terrace and can be attributed to the plant effect.

Summarizing, the strongest differences between both terraces were found in the electrical conductivity, pH and content of organic matter. Regarding microenvironments, mesquite showed higher values of water availability, organic matter and phosphorous, while *P hollianus* and control soils showed very similar patterns for all the physical and chemical factors measured. Notwithstanding, *P hollianus* is better microenvironment to biological activity than control soil.

Biodiversity of Soil Amoebae Community

Soil amoebae inhabiting our study zone accounted for 85 species from 35 genera in these 6 microenvironments. Theoretically, there were between 104 to 108 species in these terraces as calculated from the species-area curve analysis. That means the present study listed about 79 to 81% of the species than can be found in all the microenvironments from these terraces.

From the observed species list (Table 1), 32 strains could not be assigned to a described species because the observed morphology did not match to the ones reported in the keys. They were left as "sp" following the genus to which they belong to. It is probable that most of them are ecotypes of species already described. However, several strains could be also new species.

The most common genus of soil amoebae in either terrace were *Vahlkampfia*, *Acanthamoeba* and *Platyamoeba*, meanwhile the rarest ones were *Hartmannella*, *Gymnophrys* and *Leptomyxa* (Figure 3).

Acanthamoeba genus was more frequent in preserved soils while *Vahlkampfia* was very common in badlands. Several genera of reticulose amoebae like *Biomyxa*, *Leptomyxa* and *Nuclearia* were exclusive of the preserved soil environments. The presence of genus *Mayorella* is remarkable because most of these species are non-cyst forming and were very common in badlands as well as in preserved soils.

There were 4 distinctive groups of amoebae revealed by TWINSPAN analysis: *Acanthamoeba* sp, *Platyamoeba stenopodia*, *Mayorella cultura*, *Mayorella penardia*, *Vahlkampfia* sp and *Vahlkampfia enterica* were present in all environments; *Gymnophrys* sp, *Vannella lata*, and *Vannella simplex* occurred only in soil under *P. hollianus*, no matter the terrace; *Acanthamoeba mauritaniensis*, *Mayorella bigema*, *Mayorella oclawaha*, *Tecochaos* sp, *Vannella cirrifera* and *Vannella mira* were exclusive of *P hollianus* at preserved terrace; and *Amoeba diminutiva*, *Cochliopodium bilimbosum*, *Mayorella cypresa*, *Pelomyxa* sp, and *Thecamoeba similis* were found only in preserved soil under mesquite. All other species were missing in one or more microenvironments.

In general, the microenvironments from the preserved terrace contained more species than the ones found in badlands (Table 2). When analyzed by microenvironment, root canopy of *P hollianus* from preserved terrace supported a comparable quantity of amoebae species as *P laevigata* did, while badlands showed more species under *P laevigata* than under *P hollianus* (Table 3).

Rains brought an increase of the species richness in all soil microenvironments as is evident in table 3. It can be observed that number of species increase during the stormy weather (early and late August) and decreased slightly after rains (September and October). The pattern of species richness in preserved soil was always richer than badlands during this seasonal change.

Shannon and Simpson's diversity indexes also shown mesquite as the best microenvironment for soil amoebae to live, soil under this plant had the highest diversity values even in badlands.

Table 1. Species presence in all microenvironments at *P. hollianus* (H), *P laevigata* (Mesquite; M) and control Soil (CS) at preserved terrace (P) and badlands (B)

Species	MB	MP	HB	HP	CSB	CSP	total
A castellanii	0	0	0	0	0	1	1
Acanthamoeba sp	1	0	0	0	0	0	1
Adelphamoeba galeacystis	0	0	1	0	0	0	1
Amoeba sp	0	0	1	0	1	0	1
Arcella sp	0	0	1	0	0	0	1
Cashia sp	0	0	0	1	0	0	1
Cochliopodium sp.	0	0	0	0	1	0	1
Echinamoeba sp	0	1	0	0	0	0	1
E exundans	0	0	0	0	0	1	1
Filamoeba nolandi	0	1	0	1	0	0	2
Gymnophris sp	0	0	0	1	0	0	1
Hartmannella sp	1	0	1	0	1	0	3
H cantebrigiensis	0	1	1	0	1	1	4
H vermiformis	1	1	1	0	1	0	4
Korotnavella hemistolonifera	0	1	0	1	0	0	2
Learamoeba sp	0	1	0	0	1	0	2
Mayorella sp	0	1	0	0	0	1	2
M cultura	0	1	0	0	0	1	2
M oscillosignum	0	1	0	0	0	0	1
Mastigamoeba	1	1	0	0	0	0	2
M. penardi	0	0	1	0	0	0	1
M. riparia	0	0	0	0	0	0	1
Nuclearia simplex	0	0	1	0	0	0	1
Paratetramitus jugosus	1	0	0	0	0	0	1
Platyamoeba placida	0	1	0	0	1	0	2
P pseudovannellida	0	1	0	0	0	1	2
P stenopodia	1	1	0	1	0	0	3
Rhizamoeba sp	0	0	1	1	0	0	2
Rosculus ithacus	1	1	0	0	1	0	3
Stachyamoeba sp	1	0	0	0	0	0	1
Techamoeba sp	0	0	0	0	1	0	1
Vahlkampfia sp	1	1	1	1	1	1	6
V aberdonica	0	1	0	0	0	0	1
V avara	1	1	0	0	0	0	2
V enterica	0	1	0	0	0	0	1
V hartmanni	1	0	0	0	0	0	1
V inornata	0	0	0	0	1	0	1
V ustiana	0	1	0	1	0	0	2
Vannella cirrifera	0	0	0	1	0	0	1
V lata	0	0	0	1	0	0	1
V platypodia	0	0	0	1	0	1	2
V simplex	0	0	0	1	0	0	1
Vexilifera sp	0	1	0	0	0	0	1

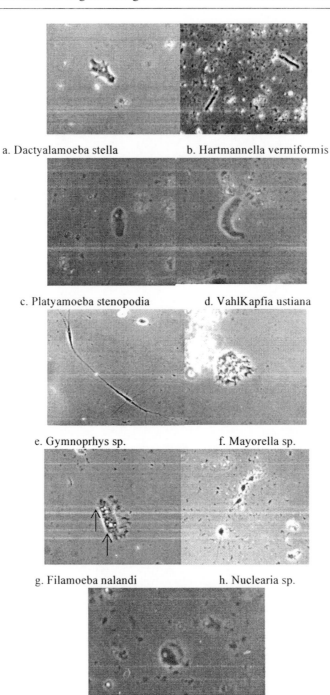

a. Dactyalamoeba stella   b. Hartmannella vermiformis

c. Platyamoeba stenopodia   d. VahlKapfia ustiana

e. Gymnoprhys sp.   f. Mayorella sp.

g. Filamoeba nalandi   h. Nuclearia sp.

i. Vannella platypodia

Figure 3. Species of soil amoebae commonly isolated from Zapotitlán. These pictures were taken with a Phase contrast photomicroscope Nikon AFX/DX at 40X magnification. Arrows in figure g show the filopodia that characterize and name this genus.

Table 2. Species richness and diversity indexes in the microenvironments of the terraces at Zapotitlán. Differences between microenvironments at badlands were higher than the ones found in the preserved terrace

	Pachycereus hollianus Badlands	*Prosopis laevigata* Badlands	Interespacio Badlands	*Pachycereus hollianus* Preserved	*Prosopis laevigata* Preserved	Interespacio Preserved
Species Richness	38	43	29	46	49	39
Shannon H index	3.341	3.451	3.11	3.657	3.623	3.455
Simpson D Index	0.9523	0.9589	0.9444	0.9367	0.967	0.962

The average diversity species richness is 40.7 for both terraces, the diversity index H is 3.42 and the Simpson index is 0.9622. This table shows that Cactus and control soil at badlands were always below the average.

Table 3. Variation of the species richness by sampling date and microenvironment. Microenvironments are separated by terrace (badlands or preserved) and ordered as cactus, mesquite and control soil

	Pachycereus hollianus Badlands	*Prosopis laevigata* Badlands	Interespacio Badlands	*Pachycereus hollianus* Preserved	*Prosopis laevigata* Preserved	Interespacio Preserved
July	10	8	4	13	16	12
August 1	12	20	14	19	26	18
August 2	21	18	10	15	23	14
September	9	13	9	14	19	9
October	12	13	4	15	13	9

It is worth to note that the highest number of species was reached during rainy season in August and microenvironments in the preserved terrace were always richer than badlands

The analysis of species ranks in badlands and preserved terrace shows almost identical graphs (Figure 4) quite similar between them; the identity of the species is what changes in each microenvironment, rather than the number of individuals, as has been shown above.

Sørensen analysis showed that mesquite in badlands and preserved soil had essentially the same assemblage of amoebae, sharing 100% of the information (Figure 5). Root canopy of mesquite shared 80% of the species information with control soil at preserved terrace, while *P hollianus* at badlands was still part of this cluster by sharing 60% of the information. Control soil at badlands was out of this cluster as it only shared 30% of the information and *P hollianus* from preserved terrace had a community assemblage of its own, that was completely out of likeliness with the other microenvironments (0% of shared information).

Principal Component Analysis (PCA) revealed that presence of amoebae species was related to real density, porosity and pH; although it was a poor correlation, because it only explained 23.9% of the observed variance (Figure 6).

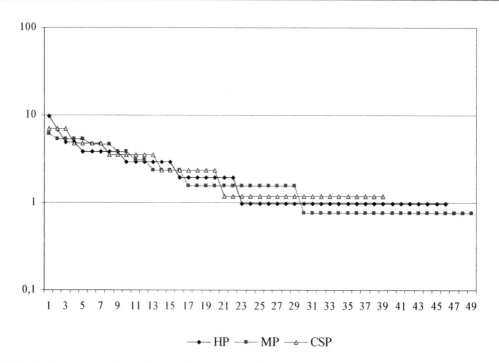

a) Relative importance of amoebae species at preserved terrace.

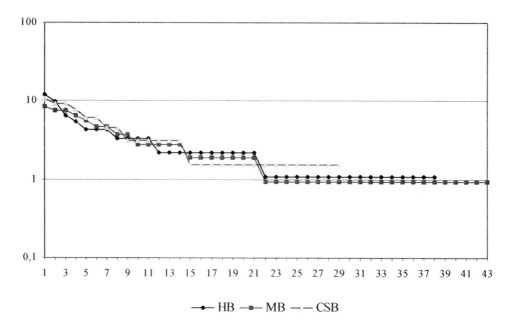

b) Relative importance of amoebae species at badlands.

Figure 4. Relative importance of amoebae curves: a) preserved and b) badlands. H = *P. hollianus* M = mesquite and CS = Control Soil; P = preserved terrace and B = Badlands

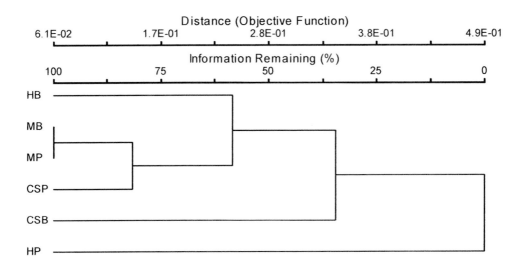

Figure 5. Sørensen analysis of amoebae communities in the microenvironments of *P. hollianus* (H), *P laevigata* (mesquite; M) and control Soil (CS) at preserved (P) and badlands (B).

Axis	Axis Value	% Varianza Value	% Acumulated varianza	Axis
1	7.518	8.844	8.844	5.026
2	6.648	7.821	16.666	4.026
3	6.165	7.252	23.918	3.526

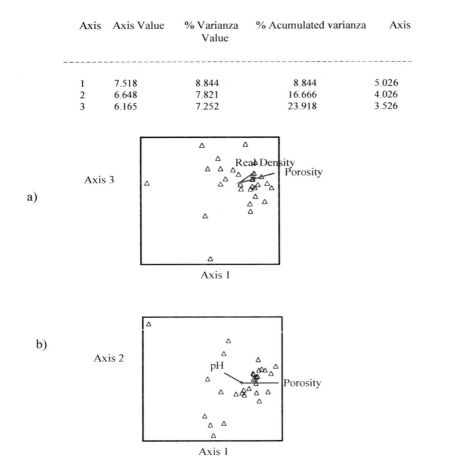

Figure 6. Principal Component Analysis between species presence and Soil Physical – Chemical factors. Lines represent the factor explaining the variance of species presence.

DISCUSSION

Several physical and chemical parameters were significantly different between microenvironment and between terraces. The apparent density, for example was significantly heavier in control soils than under cactus or mesquite. This could be a consequence of having a very limited amount of roots and organic matter, as these two factors change soil structure by favoring aggregates formation (Paul and Clark, 1989). Soil compaction has a negative impact on fungi and bacteria populations because of the reduction of surfaces for colonization and oxygen availability. Apparent density, however, resulted of minor importance for determination of species composition of the amoebae community. This factor alone explained less than 1% of the observed variance (Figure 6). On the other hand, apparent density was negatively correlated to water availability, being soils lightest during the wettest sampling dates. This can be explained based on the microorganisms' activities in soils. During the time where water becomes available, bacteria, fungi and plant roots exude molecules that play an important role as agglutination agents for soil aggregates (Bryant et al., 1982). Production of more aggregates increases the proportion of porous spaces in soils, allowing more water retention capacity. The variation of apparent density observed seasonally may be explained by changes in biological activity that depends on water availability, especially in arid soils.

Porosity had higher importance for species distribution as this factor explained 8.8% of the observed variance as indicated in PCA. Reduced porosity may be one reason for species impoverishment in compacted soil because of the limited spaces for transporting and feeding. The internal diameter of the bore pores determines also the size of the protozoa than can get into it (Darbyshire et al., 1989). This feature also limits the movement of bigger soil amoebae favoring the smallest species that can temporarily resist anaerobic conditions (Rodríguez Zaragoza, 1994).

Soil humidity was different between microenvironments; however, badlands were always dryer than the preserved soils. This pattern is the result of an important diminution of plant cover. Development of plant canopies protect soil from direct sunlight and, in absence of strong winds, keep the humidity in soil by providing lower temperatures because of shadowing, and help establishing a water saturation gradient from the air below the canopy to the soil surface.

Water limitation is one of the strongest forces determining biological activity in desert soil systems, in fact, windows of activity in such systems are entirely governed by water and nitrogen variability (Noy Mayr, 1973). This situation is similar at the scale of microorganisms. Farther than the most obvious impact of water films per se on the movement and survival of amoebae, water availability has an indirect effect on soil amoebae community. It influences plants' vigor and health, which is reflected in the quantity and quality of root exudates (Lynch et al., 2002) that select bacterial and fungi species (Starkey, 1958; Pal Bais et al., 2004) on which amoebae feed on (Bonkowski, et al., 2001).

The shape of canopy is also very important, and explains part of the water variation in soil below both plants. The main difference between mesquite and the cactus is that the former does not produce a canopy and its capacity to protect the underlying soil from desiccation is non existent. However, biological soil crusts may provide such protection to sunlight and desiccation (Belnap & Lange, 2001) instead of the plant's canopy. This factor was overlooked in the original plan of this study. However, because of its importance, presence and covering of biological soil crusts need to be considered in further studies to explain variation in water and nutrients availability in these soils. Variation of soil humidity explained less than 1 % of the species variation in soils as revealed by principal component analysis.

Notwithstanding, humidity *per se* is the trigger of biological activity in desert ecosystems and soil is not the exception. The above observation means that once water surpasses a threshold value, its quantity has no mayor effect on the species occurrence.

On the other hand, pH variation is one of the main effects humidity has in soils; it may change first as a sole consequence of water addition by dissolving many soil components. Thereafter, pH will become more acidic as consequence of roots and microbial respiration; however, such changes of pH are buffered by organic matter. This is an explanation why the preserved terrace showed less pH variation along the study, while the conjunction of low organic matter content, high desiccation due to lack of plant cover and low water penetrability explains the widest variations on pH and the highest electrical conductivity in badlands at the beginning of rains. This process is responsible for the differences detected between seasons and between microenvironments. The observed values of pH were more alkaline in September and October, just after rains, when humidity content is still high in soil. Salts infiltrate into the deeper soil layers because of being dissolved in incoming water. After rains, soil salts start their movement in reverse direction by capillarity, getting deposited on the soil surface when water evaporates and soil becomes dryer, during this process, pH turn even more alkaline because of the reduce metabolic activity of microorganisms (Gobat et al., 2004). Plants by themselves must have had important effects on pH changes in soils because of the different composition of their root exudates and different morphologies. Soil under *P hollianus* was alkaline, while soil than under *P laevigata* was nearer to neutrality along the change of season. Such observation can be attributable to the quantity of litter that mesquite shed around itself and to its root exudates, while *P hollianus* does not shed litter and its root system is completely different from that of mesquite (Flores-Martínez et al., 1998).

Plant effect is also evident on the variation of salinity (measured by electrical conductivity), as reported in Figure 2e, mesquite growing in badlands showed salt values slightly above those observed in the microenvironments of the preserved terrace, while both soil under *P hollianus* and control soil at badlands showed the highest salt content, reaching values that allows for establishment of only salt tolerant plants (Summer and Naidu, 1998). High salinity by itself can be a strong selective force determining protozoan communities in soil. However, its contribution to explain the observed species variation was negligible by PCA. This lack of effect can be attributable to the reported tolerance of free – living amoebae to a wide range of salinity (Hauer et al., 2001; Anderson et al., 2003).

Soil organic matter is very important for plant growth as it represents the storage of nutrients already used in biomass and, most of the time, are easier to extract from SOM than mobilization from their mineral forms, especially in the case of nitrogen or phosphorous. SOM represents also one way to withdraw CO_2 from atmosphere as the global tendency in soils is the long term accumulation of SOM (Davidson et al., 2000). This tendency was also observed in both terraces at Zapotitlán, even though there was an initial tendency of consumption of organic matter and, later on the season, there was an accumulation of SOM due to seasonal litter fall, especially after rains. The difference between organic matter content between badlands and the preserved terrace may be connected to the physical removal of the soil surface during the erosive process, which may be responsible of loosing the surface

SOM. Roots of mesquite provide a valuable source of organic matter in the layer immediately below surface soil at badlands. As can be seen in Figure 2a, SOM in soil under mesquite at badlands reaches percentages comparable to the ones found in microenvironments of the preserved terrace. Organic matter is one of the factors that need to be studied in further detail as it has tremendous impact in both the structure and biology of soils. On the other, hand "organic matter" groups a wide variety of molecules of very different nutritional values and degradability, from monosaccharide (carbohydrates) to humic acids and petroleum promoters. Consequently, this is a mixture of molecules with different solubility in water and recalcitrance that accumulates in soils (Lynch et al., 2002).

The entrance of organic matter into the soil matrix is accomplished mainly by litter decomposition and root exudates. Among the former, the main kind of molecules plants release onto the rhizosphere are sugars, amino acids, organic acids, lipids, low molecular weight proteins and vitamins, together with gases such as CO_2, methane and hydrogen (Frías-Hernández et al., 1999; Kent and Triplett, 2002). The quantity and quality of root exudates varies among plant species and even between the different physiological stages of the same plant (Apphun & Joergensen, 2006).

Porosity, pH and real density all together explained almost 25% of the variance of species occurrences observed in the study. The other 75% of variance of the amoebae community may be explained by biological factors such as the kind of root exudates, the identity of bacteria growing in the root canopy, presence of micro-invertebrate predators, and microbe - microbe interactions together with other physical and chemical conditions of soils.

Strong variations of extractable phosphorous observed between microenvironments and along the season may be a consequence of the microbial activity in soil. Under alkaline pH, phosphorous can be sequestered from the system by forming complexes with sulfates and or calcium that precipitate and make phosphorous unattainable for plants. Phosphate solubilization in these terraces is carried out mainly by bacteria from genus *Pseudomonas* and mycorrhiza from genus *Glomus* (Reyes-Quintanar et al., 2000). These are the conditions selecting the species of amoebae that thrives in these soils, where the physical and chemical differences play an important role in determining prey species presence.

One of the most important characteristics of communities is the existence of very few dominant species (those with the highest numbers of individuals or values of biomass), followed by many common ones and an elevated number of rare species (Swartz et al., 2000). Such a tendency was found in all the microenvironments at Zapotitlán (Figure 4). On the other hand, species rank curves were always concave in all microenvironments, which is characteristic of a stable community. This shape is characteristic of a community structure of soil amoebae in equilibrium with the environmental conditions. The stress introduced by perturbation would produce a more symmetrical shape (Hill and Hamer, 1998). Notwithstanding, the species assemblage of each terrace and microenvironment is different between each other, showing that velocity of the badland transformation process has been slow enough for community adaptation to the specific set of environmental variations.

One of the mechanisms a community has to equilibrate after perturbations is based on species' functional redundancy. Several species seems to have the same function in the system as bactivorous, fungivorous, omnivorous predators etc (Rodríguez Zaragoza, et al., 2005a). When environmental conditions eliminate one species, it favors other ones capable of carrying out the same function in the system. Then, no matter prey preferences of the substituting species, the functional equilibrium gets recovered. This mechanism help explain why rare species become dominants after perturbations and vice versa (Murray et al., 1999; Griffiths et al., 2001). The stability of amoebae community at badlands can be explained by the time required for this microbial eukaryotes to adapt after perturbation and because the natural process of soil transformation into badlands is very slow. Environmental changes take place at a velocity that allows the whole system to reach alternative steady states, giving enough time for species to fit into new conditions (Allison, 2004; Wolda, 1986).

Contrary to community resistance, species richness was different between terraces. Badlands show an important species loss in the control soil, meaning that root canopy of *P laevigata* and *P hollianus* are true refuges for amoebae to escape from the extreme conditions generated, besides being functioning as fertility islands in this terrace. In this sense, mesquite is better refuge for amoebae than *P hollianus* as the species richness of the legume is almost the same in both terraces. However, it has been found that roots of mesquite and columnar cactus (*Neobuxboumia tetetzo*) growing together are even better refuge for soil amoebae than growing in isolation (Rodríguez Zaragoza et al., 2005 b)

Species richness under *P hollianus* reaches close numbers at both terraces. However, the species composition of this cactus at the preserved terraces is very different from the other microenvironments as revealed by Sørensen analysis. This is probable due to the process of root growth of these cacti, because during the dry season roots are extremely thick, shallow and reduced to their minimum expression to withstand arid conditions (Flores-Martínez et al., 1998). During the wet season, roots get very active, growing and expand laterally to take as much water as they can, and promote the colonization of its root canopy, which can be done by the microbial "reserve" in bulk soil (Rodríguez Zaragoza and Garcia, 1997). Therefore, the pool of species that may colonize root canopies is restricted to the surviving species in bulk soil (Etienne and Olff, 2005). If soil has experimented species loss, as is the case in badlands, the cactus will have a very restricted set of possibilities because of the local extinction of many dominant, common and rare species. The surviving ones would be almost the same that can colonize mesquite and will produce more likeliness between the cactus and the former. Other process that may help alleviate local extinction of species is their transportation from nearby places, which may be by runoff, wind or animal faces (Rodríguez Zaragoza, 1994). We excluded long term airborne as a mechanism of microbial colonization because microbes have to survive the atmospheric killing factors (extreme drought and UV among many others) and reach new places in numbers high enough to establish a viable population, which is inversely related to the distance from which microbes are transported (Rodríguez Zaragoza et al., 1993; Rivera et al., 1994). However, plants may be able to select a specific array of species through its root exudates if soil preserves the pool of species (Appuhn & Joergensen, 2006). Then with wider possibilities of root canopy colonization, as was the case in the preserved terrace, the likeliness between mesquite and the cactus was non existent as could be observed in Sørensen analysis (Figure 5). It means that these two plants bear very different communities of amoebae when growing under good conditions.

Amoebae community in badland is dominated by bacterial feeding species, mostly from the *Vahlkampfia* genus, which implies a simplified trophic structure, heavily based on bacteria. Fungivorous and omnivorous amoebae are but very rare in this soil. This situation may be related to the scarcity of organic matter, as its content in badlands was really low despite being sampling the root canopy. In addition, it could have been lower diversity of organic molecules in soil, which is also related to the functioning of roots and fungal exudates (Starkey, 1958; Lynch et al., 2002). In badlands, mycorrhizas that infect legume and cactus roots tend to form more vesicular structures as an adaptation to survive soil dryness, while the same species living in the preserved soils tend to develop arbuscular structures (Reyes-Quintanar et al 2000). These morphological adaptations must be related to reduction of fungal exudates and other physiological changes in fungi (Beare et al., 1997). However, this variation of fungal exudates, if happening, might also help explain the different trophic structures of amoebae communities at badlands in comparison to preserved soils. During badland formation, litter accumulation is non existent, leaving the organic matter (OM) already integrated in soil and the dissolved OM as the main source of energy. The former may be imported by diffusion from the nearby sites during rain events. Thus, velocity of metabolizing high molecular weight OM is longer than the time water becomes available and the consequence is faster development of bacteria that are able to use low molecular weight OM over fungi using high molecular weight OM.

On the other side, soil at the preserved terrace water availability lasts longer, giving chance to structure a more complex microbial community in soil where fungi- feeding, algae-feeding, carnivorous, and the omnivorous amoebae could be observed. This community was dominated by the genus *Acanthamoeba* that are omnivorous amoebae. This is also a consequence of the variety of resources present in the preserved soil. All of this is in agreement with the broken stick model of McArthur that predicts the curve of species will acquire a geometric shape when developing in a much stressed habitat, where one or very few resources are the limiting factors. This means that there will be very few dominant species and many rare ones (Ludwig & Reynolds, 1988; Nummelin, 1998) as it was the case in badlands and the preserved terrace.

Food webs were also very different between microenvironments; community under *P hollianus*, in preserved soil, was composed of bacterivorous, fungivorous, algivorous, carnivorous and omnivorous amoebae. Only in this microenvironment *Gymnophrys cometa* and *Leptomyxa* sp were observed. These species are predators of multicellular organisms such as nematodes and rotifers. This has the implication that the microbial loop is longer under this cactus than under mesquite. On the other hand, the food web under mesquite was dominated by bacterivorous amoebae, where the absence of multicellular predators and few omnivorous amoebae are notorious. Fungivorous and algivorous amoebae like several species of *Nuclearia* were observed in soil under both plants but were absent in the control soil. The environment influenced by root canopy provided conditions for algae and fungi to develop that were missing in the control soil even during the wet season.

The importance of the microbial loop is that the longer the loop the more efficient cycling and transformation of matter and energy in the soil system (Coleman, 1994; Bonkowski, 2004). In this sense, nutrient cycling and energy use is less efficient in badlands when compared with preserved soil. As we are dealing with a community that has reach a steady state under badlands condition, we can not establish if this is the cause or the consequence of soil transformation into badlands. In general, badlands are loosing both structural and functional diversity of the soil amoebae community.

At the beginning of the study, we expected to find out at least the same biodiversity of amoebae at the alluvial terraces of Zapotitlán than the one found in the Negev desert (Rodríguez Zaragoza et al., 2005a), as both studies were carried out in soil desert environments and under legume plants (*Zygophylum dumosum* in the Negev, *Prosopis laevigata* in Zapotitlán). The main difference in these two deserts is the quantity of rain; the Negev receives only 30 mm a year, while Zapotitlán receives from 390 to 420 mm a year.

This fact alone allows thinking that resource restriction in Zapotitlán was loose enough to allow more species to live in the terraces. It was expected to find out at least the same quantity of amoeba than the ones below *Zygophylum dumosum* (95 species). However, we only observed 85 species. These results remark that limiting physical and chemical factors alone can not explain the diversity of microbial eukaryotes in desert soil.

Our evidence suggest the observed change of the amoebae community was promoted by the combined effect of water availability, soil's apparent density, pH and organic matter content (slightly less than 25% of the variance explained). There is still a huge variation source to be found in order to explain the differential species presence in soil. Even more, the exact determination of the main factors determining the community composition is still a matter of further experimentation, as the former numbers are only correlative evidence treated by multivariate analysis.

The observed change in the community structure is more dramatic if we consider that sampling sites are just hundreds of meters apart enhancing that such difference may be due to biological factors where the "rhizosphere effect" could be the leading one. It is important to consider this effect because amoebae, as all other microorganisms, are able to long term transportation by wind or animals and, potentially, they may be found everywhere or, at least, in all places fulfilling the overall ranges of vital factors needed to survive. These ranges represent their potential distribution or, in other words, the potential niche of the species. However, the environment selects what is going to be successful and what is going to fail as newcomer. This selection is the result of the combined action of physical and biological forces determining the survival and development of the set of species living in a specific habitat. This combined action restricts the actual distribution of a species, which is known as realized niche (Anderson, 1988). As per the evidence shown above, amoebae are not exceptions to this constrains. However, it has been recently proposed that biodiversity has a very different meaning for protozoa (Esteban et al., 2006). The report's main conclusion is that species dominant or rare in one environment are also dominant or rare globally (Esteban et al., 2006). If this would have been right for our system, we would have obtained the same community of amoebae under both plants, no matter the terrace. However, we found the opposite, several species that were common or dominant in the preserved terrace became rare under more stressful conditions. For example, *Nuclearia sp*, *Acanthamoeba mauritaniensis* and *Mayorella cypresa* were very common in preserved terrace and became rare or even disappeared from badlands. Up to this point we could only show that physical and chemical variation can explain only 23.9 % of the species occurrences in a desert soil. The amoebae community gets simplified as consequence of soil transformation into badlands, no matter the existence of fertility islands represented, in first instance, by mesquite and followed by *P hollianus*. The explanation of such changes are sustained in the theoretical framework of community ecology, supporting the conclusion that biodiversity of soil amoebae, as in the case of any other group of microbes has the same meaning than in higher organisms.

ACKNOWLEDGMENTS

This study was possible by the financial support of CONACYT-SEMANAT through the grant # 2002 – C01 – 0790.

REFERENCES

Alef, K. & Nannipieri P. (1995). Methods in applied soil microbiology and biochemistry. Academic Press, *Harcourt Brace & Company, publishers*, London. 576p.

Allison, G. (2004). The influence of species diversity and stress intensity on community resistance and resilience. *Ecological Monographs* 74: 117-134.

Anderson, R. O. (1988). Comparative Protozoology. Ecology, Physiology, Life History. *Springer Verlag*, New York.

Anderson, O. R., Nerad T. A. & Cole J. C. (2003). *Platyamoeba nucleolilateralis* n. sp. from Chesapeake bay region. *Journal of Eukaryotic Microbiology* 50: 57-60.

Anderson, J. & Ingham J. (1993). Tropical soil biology and fertility: Handbook of methods. 2° ed. Wallingford, U.K., *CAB International*.

Appuhn A, & Joergensen R. G. (2006). Microbial colonisation of roots as a function of plant species *Soil Biol. & Biochem.* 38: 1040–1051

Bamforth, S. S. (1995). Interpreting soil ciliate biodiversity. In: Collings, H. P., Robertson G. P. & Klug M. J. (eds.). The significance and regulation of soil biodiversity. *Klewer Academic Publishers*, Holanda. pp. 179-184

Beare M. H., Hu S., Coleman D C & Hendrix P F. (1997). Influences of mycelial fungi on soil aggregation and organic matter storage in conventional and no-tillage soils. *Applied Soil Ecology* 5: 211 – 219

Belnap, J. & Lange O. L. (2001). Biological soil crust: structure, function, and management. *Springer-Verlag*, Berlin.

Bischoff, P. J. (2002). An analysis of the abundance, diversity and patchiness of terrestrial Gymnamoebae in relation to soil depth and precipitation events following a drought in Southeastern U. S. *Acta Protozoologica* 41: 183-189.

Bonkowski M, Jentschke G, & Scheu S. (2001). Contrasting effects of microbial partners in the rhizosphere: interactions between Norway Spruce seedlings (*Picea abies* Karst.), mycorrhiza (*Paxillus involutus* (Batsch) Fr.) and naked amoebae (protozoa). *Applied Soil Ecology* 18: 193–204

Bonkowski, M. (2004). Protozoa and Plant growth: the microbial loop in soil revisited. *New Phytologist*, 162: 617 – 631

Bouma, N. A. & Imeson, A. C. (2000). Investigation of relationships between measured field indicators and erosion processes on badlands surfaces at Petrer, Spain. *Catena* 40: 147-171.

Bryant, R. J., Woods L. E., Coleman D. C., Fairbanks B. C., McClellans J. F. & Cole C. V. (1982). Interactions of bacterial and amoebal populations in soil microcosms whith fluctuating moisture content. *Applied and Environmental Microbiology* 43: 747-752.

Coleman D C (1994). The microbial loop concept as used in terrestrial soil ecology studies. *Microb. Ecol.*, 28: 245 – 250

Cajuste, L. J. (1986). El fósforo aprovechable en los suelos. Serie cuadernos de Edafología 6. Centro de Edafología. *Colegio de Posgraduados* Chapingo, México.

Chakraborty S. & Old K. M. (1985). Mycophagous amoebas from arable pasture and forest soil. In: Parker C. A., Rovira A. D, Moore K. J., Wong P. T. W., y Kollmorgen J. F. (eds.) Ecology and management of soil borne plant pathogens, *APS Press*. pp 14-16

Côuteaux M. & Darbyshire J. F. (1998). Functional diversity amongst soil protozoa. *Applied Soil Ecology* 10: 229-237.

Davidson E A, Trumbore S E, & Amundson R. (2000). Soil warming and organic carbon content. *Nature* 408: 789 – 790

Darbyshire, J F, Griffits B S, Davidson M S & McHardy W J. (1989). Ciliate distribution amongst soil aggregates. *Rev. Ècol. Biol. Sol.* 26: 47 – 56

Dávila, P. (1997). Tehuacan-Cuicatlán region, México. Pp 139-143 En: Davies, S. D. (ed.) Centres of Plant Diversity: Cambridge, *The World Wide Fund for Nature (WWF), The World Conservation Union (IUCN)*.

Ekschmit, K. & Griffiths B. S. (1998). Soil biodiversity and its implications for ecosystem functioning in a heterogeneous and variable environment. *Applied Soil Ecology* 10: 201-215.

Esteban G F, Clarke K J, Olmo J L , & Finlay B J. (2006). Soil protozoa—An intensive study of population dynamics and community structure in an upland grassland. *Applied Soil Ecology* 33: 137–151

Etienne, S., & Olff H. (2005). Confronting different models of community structure to species-abundance data: a Bayesian model comparison. *Ecology Letters* 8: 493-504.

Flores-Martinez A, Ezcurra E, & Sánchez-Colón S. (1998). Water availability and the competitive effect of a columnar cactus on its nurse plant. *Acta Oecologica* 19: 1-8

Frías-Hernández, J. T., A. L. Aguilar & V. Olalde (1999). Soil characteristics in semiarid highlands of central Mexico as affected by mesquite trees (*Prosopis laevigata*). *Arid Soil Research and Rehabilitation* 13: 305-312.

García E. (1988) Modificaciones al sistema de clasificación climática de Köppen 217 pp. *Instituto de Geografía*, UNAM, México.

Gobat, J.M., Aragno M. & Matthey W. (2004). The living soil. Fundamentals of soil science and soil biology. *Science Publishers* Inc.Enfield, New Hampshire, U. S., 602p.

Guo, Q. & Berry W. (1998). Species richness and biomass: dissection of the hump-shaped relationships. *Ecology* 79: 2555- 2559.

Griffiths, B. S., Bonkowski M., Roy J. & Ritz K. (2001). Functional stability, substrate utilization and biological indicators of soils following environmental impacts. *Applied Soil Ecology* 16: 49-61.

Hauer, G., A. Rogerson and O. R. Anderson (2001). *Platyamoeba pseudovannellida* n. sp., Naked amoeba whit wide salt tolerance isolated from the Saltom sea, California. *Journal of Eukaryotic Microbiology* 48: 663-669.

Hill, J. K. & Hamer K. C. (1998). Using species abundance models as indicators of habitat disturbance in tropical forest. *Journal of Applied Ecology* 35: 458-460.

Holling, C S. (1973). Resilience and stability of ecological systems. *Annual Reviews of Ecology and Systematics*. 4: 1- 23

Hughes Martiny, J B., Bohannan B J M, Brown J H, Colwell R K, Fuhrman J A, Green J L, Horner-Devine M C, Kane M, Adams Krumins J, Kuske C R, Morin P J, Naeem S, Øvreås L, Reysenbach A L, Smith V H & Staley J T. (2006). Microbial biogeography: putting microorganisms on the map. *Nature Reviews Microbiology* 4: 102 – 112

Jimenez, M., De la Horra A. M., Pruzzo L. & Palma R. M. (2002). Soil quality: a new index based on microbiological and biochemical parameters. *Biology and Fertility of Soils* 35: 302-306.

Kent, A. D. & Triplett E. W. (2002). Microbial communities and their interactions in soil and rhizosphere ecosystems. *Annual Reviews Microbiolog* 56: 211-236.

Lee, J., Hunter S. H. & Bovee E. C. (2000). An ilustrated guide to the protozoa. *Society of Protozoologists*. Lawrence, Kansas. 689p.

López, G. S., Muñoz D., Hernández M., Soler A., Castillo M., & Hernández I. (2003). Análisis integral de la toposecuencia y su influencia en la distribución de la vegetación y la degradación del suelo en la subcuenca de Zapotitlán Salinas, Puebla. *Boletín de la Sociedad Geológica Mexicana* tomo LVI, 1: 19 – 41.

Ludwig, J A & Reynolds J F. (1988). Statistical Ecology. A primer on methods and computing. *John Wiley & Sons*, New York , 337 p.

Lynch J M, Brimecombe M J, & De Leij F AAM. Rhizosphere. Encyclopedia of Life Sciences 2002 *Macmillan Publishers Ltd, Nature Publishing Group* / www.els.net

Mast S. O. & Root F. M. (1916). Observation on ameba feeding on rotifers, nematodes and ciliates, and their bearing on the surface-tension theory. *Journal of Experimental Zoology* 21: 33-49.

Mayzlish, E. & Steinberger. Y. (2004). Effects of chemical inhibitors on soil protozoan dynamics in a desert ecosystem. *Biol. Fert. Soils* 39, 415-421

McAuliffe J R , Sundt P C , Valiente-Banuet A, Casas A & Viveros J L. (2001) Pre-columbian soil erosion, persistent ecological changes, and collapse of a subsistence agricultural economy in the semi-arid Tehuacan Valley, Mexico's 'Cradle of Maize' *Journal of Arid Environments* 47: 47–75

McCune, B & Mefford M J. PC ORD. (1999). Multivariate Analysis of Ecological Data. Version 4. *MJM software Design*, Gleneden Beach, Oregon USA

Mlot, C. (2004). Microbial Diversity Unbound. *BioScience*, 54: 1064 - 1068

Morgan, R. P. (1995). Soil erosion and conservation. 2° edition. *Logran*, Essex. U.K. 198p.

Murray B. R., Rice B. L., Keith D. A., Myerscough P. J., Howell J., Floyd A. G., Mills K.& Westoby M. (1999). Species in the tail of rank-abundance curves. *Ecology* 80: 1806-1816

Noy-Meir I. (1973). Desert Ecosystems: Environment and Producers. *Annual Review of Ecology and Systematics*, 4: 25-51.

Nummelin, M. (1998). Log normal distribution of species abundance is not a universal indicator of rainforest disturbance. *Journal of Applied Ecology* 35: 454-457.

Odum, E. P., Finn T. J. & Franz E. (1979). Perturbations theory and the subsidy-stress gradient, *Bioscience* 29: 349-352.

Pal Bais H, Park S W, Weir T L, Callaway R M & Vivanco J M. (2004). How plants communicate using the underground information superhighway. *TRENDS in Plant Science* 9: 26 – 32

Page, F. C. (1976). An illustrated key to freshwater and soil Amoebae. *Freshwater biological Association*, Ambleside, Cumbria, U.K. 155p.

Page F. C. (1988). A new key to freshwater and soil gymnamoebae. *Freshwater biological Association*, Ambleside, Cumbria, U.K. 122 p.

Page F. C. & Siemensma F. J. (1991). Nackte rhizopoda and Heliozoea. *Gustab-Fisher Verlag*, Stuttgart. 297 p.

Patterson, D. J. (1996). Free-living freshwater protozoa. A color guide, *John Wiley and Sons*. Londres, U.K. 223p.

Paul E. A. & Clark F. E. (1989). Soil microbiology and biochemistry. *Academic Press* U. S., pp. 51-73.

Pedrós-Alió C. (2006). Marine microbial diversity: can it be determined? *TRENDS in Microbiology* 14: 257 – 263

Philippi, T. E., Dixon P. M. & Taylor B. E. (1998). Detecting trends in species composition. *Ecological Application* 8: 300-308.

Potts D L, Huxman T E, Enquist B J, Weltzin J F & Williams D G. (2006). Resilience and resistance of ecosystem functional response to a precipitation pulse in a semi-arid grassland *Journal of Ecology* 94: 23–30

Reyes Quintanar C.K., Ferrera-Cerrato R, Alarcón A C & Rodríguez Zaragoza S. (2000). Microbiología de la relación de nodricismo entre leguminosas arbóreas y *Neobuxbaumia tetetzo* en suelos no erosionados y erosionados de Zapotitlán de la Salinas, Puebla. In Alarcón, A. & Ferrera-Cerrato R. eds. Ecología, fisiología y biotecnología de la micorriza arbuscular. Colegio de Postgraduados, Montecillo, *Mundi Prensa*, México p.p. 56- 68.

Rivera, F.; Bonilla P., Ramírez E., Calderón A., Gallegos E., Rodríguez S., Ortiz R., Hernández D., & Rivera V. (1994). Seasonal distribution of air-borne pathogenic and free-living amoebae in Mexico City and its suburbs. *Water, Air & Soil Pollution*: 74: 65 - 87.

Rodríguez Rodríguez A, Mora J L, Arbelo C, & Bordon J. (2005). Plant succession and soil degradation in desertified areas (Fuerteventura, Canary Islands, Spain) *Catena*: 59: 117–131

Rodríguez-Zaragoza S. (1994). Ecology of free-living amoebae. *Critical Reviews in Microbiology* 20: 225-241.

Rodríguez Zaragoza S. (2000). Variación de la Comunidad de Amebas Desnudas en respuesta a la perturbación de un Suelo Forestal de Encino-Pino en Villa del Carbón, Estado de México. Ph. D. Thesis, *Escuela Nacional de Ciencias Biológicas*, Instituto Politécnico Nacional, México D.F. 121 p.

Rodríguez-Zaragoza, S.; Rivera F., Bonilla P., Ramírez E., Gallegos E., Calderón A., Hernández D. & Ortiz R. (1993). Amoebological study of the atmosphere of San Luís Potosí, SLP. México. *J. Exposure Analysis and Environ. Epidemiol.* 3 (suppl.1): 229 - 241.

Rodríguez-Zaragoza, S & Soledad G (1997). Abundance of Free living amoebae in the rizoplane of *Escontria chiotilla* (Cactaceae). *J. Euk. Microbiol.* 44: 122-126.

Rodríguez-Zaragoza, S., Mayzlish E. & Steinberger Y. (2005 a). Seasonal changes in free-living amoebae species in the root canopy of *Zygophyllum dumosum* in the Negev Desert, Israel. *Microbial Ecology* 49: 134-141.

Rodríguez-Zaragoza, S.,. Gaviria-González Ll, & Rivera V. (2005 b). Riqueza de especies de amebas desnudas en la rizósfera de *Neobuxbaumia tetetzo* y *Prosopis laevigata* que crecen en condiciones de degradación natural del suelo en el desierto del Valle de Tehuacán-Cuicatlán, México. *Revista de la Sociedad Mexicana de Historia Natural*, Vol 2 (1) 3ª época : 54 – 64

Schwartz, M. W., Brigham C. A., Hoeksema J. D., Lyons K. G., Mills M. H. & van Matengem P.J. (2000). Linking biodiversity to ecosystem function: implications for conservation ecology. *Oecologia*: 122: 297-305.

Starkey R L. (1958). Interrelations Between Microorganisms and Plant Roots in the Rhizosphere *Journal Series, New Jersey Agricultural Experiment Station, Rutgers, The State University, Department of Agricultural Microbiology*,New Brunswick, New Jersey. Vol. 22: 154 – 172

Stout D. J. (1980). The role of protozoa in nutrient cycling and energy flow. pp 1-59 En: M. Alexander (Ed) Advances in Microbial Ecology, Vol. 4, *Plenium Press*. New York.

Sumner, M. E. & Naidu R. (1998). Sodic Soils. Distribution, proprieties, management and environmental consequences. *Oxford University Press*, New York, U. S., 207p.

Torri, D., Calzolari C. & Rodolfo G. (2000). Badlands in changing environments: an introduction. *Catena* 40: 119-125.

Usher, M B, Sier, A R J, Hornung, M & Millard, P. (2006). Understanding biological diversity in soil: The UK's soil biodiversity research programme. *Appl. Soil Ecol.* 33: 101 – 113.

Valiente-Banuett, A.; Casas A., Alcántara A., Dávila P., Flores-Hernández N., Arizmendi M. C. Villaseñor J. L., & Ortega-Ramiréz J. (2001). La Vegetación del Valle de Tehuacan-Cuicatlan. *Boletin Sociedad Botánica de México*, 67: 24-74

Wolda, H. (1986). Seasonality and the community. In: Gee, J. H. R. and Giller, P. S. (Eds.) Organization of community, past and present. 27th Symposium of the British Ecological Society *Blackwell Scientific Publications*, Oxford, U. K. 576p.

Zhi Zhaoa W, Lang XiaoaH, Min Liub Z, & Li J. (2005). Soil degradation and restoration as affected by land use change in the semiarid Bashang area, northern China. *Catena*: 59 173–186.

In: Progress in Environmental Research
Editor: Irma C. Willis, pp. 231-243

ISBN 978-1-60021-618-3
© 2007 Nova Science Publishers, Inc.

Chapter 7

GRAZING REGULATIONS: CHANGES
BY THE BUREAU OF LAND MANAGEMENT[*]

Carol Hardy Vincent

SUMMARY

The Bureau of Land Management (BLM) is taking a two-pronged approach to grazing reform, by proposing changes to grazing regulations (43 C.F.R. Part 4100) and considering other changes to grazing policies. On June 17, 2005, BLM issued a final environmental impact statement (FEIS) that analyzes the potential impact of proposed changes in the regulations, a slightly different alternative, and the status quo. On August 9, 2005, BLM announced its intent to prepare a supplement to the FEIS. The delay is intended to allow the agency time to address public comment received after the closing date of March 2, 2004, primarily from the Fish and Wildlife Service. The agency anticipates developing the supplement in the fall of 2005, soliciting and reviewing public comment, and issuing it in final form in the spring of 2006. No deadline for the final rule has been announced.

BLM asserts that regulatory changes are needed to increase flexibility for grazing managers and permittees, to improve rangeland management and grazing permit administration, to promote conservation, and to comply with court decisions. The possibility of rules changes, and the particular changes proposed, have been lauded by some but criticized by others. The last major revision of grazing rules, which took effect in 1995 after a lengthy development process, was highly controversial. BLM is currently reexamining many of the changes made at that time.

The current proposal would make many changes. The BLM and a permittee could share title to structural range improvements, such as a fence. Permittees could acquire water rights for grazing, consistent with state law. The occasions on which BLM would be required to get input from the public on grazing decisions would be reduced. The administrative appeals process on grazing decisions would be modified and the extent to which grazing could continue in the face of an appeal or stay of a decision would be delineated. The definition of *grazing preference* would be broadened to include a quantitative meaning — forage on public land — measured in Animal Unit Months. Changes would be made to the timeframe and

---

[*] Excerpted from CRS Report RL32244, dated August 24, 2005.

procedures for changing grazing management after a determination that grazing is a significant factor in failing to achieve rangeland health standards. The current three-year limit on temporary nonuse of a permit would be removed, and permittees would be able to apply for nonuse of a permit for up to one year at a time. Conservation use grazing permits would be eliminated. BLM considered, but did not propose, certain changes due to adverse public reaction or other considerations.

BLM also is considering changes to its grazing policies, which the agency believes can be carried out under existing rules. Potential policy changes, to follow the rulemaking process, relate to: the establishment of reserve common allotments to serve as backup forage when permittees' regular allotments are unavailable; conservation partnerships between the BLM and permittee whereby permittees work to improve environmental health in return for certain benefits; voluntary allotment restructuring to allow multiple permittees to merge allotments; and landscape habitat improvement to promote species conservation and facilitate consultations under the Endangered Species Act.

HISTORY

The Bureau of Land Management (BLM) has proposed changes to grazing regulations (43 C.F.R. Part 4100) and is considering related policy changes. The last major revision of grazing regulations culminated in comprehensive changes effective August 21, 1995. The changes were the result of a several-year process of evaluating ideas and shaping alternatives, and occurred in the midst of a decades-long dispute over the ownership, management, and use of federal rangelands.

The 1995 changes were highly controversial, with criticism from many ranching interests that those new rules weakened grazing privileges and would reduce livestock grazing on federal lands, and from environmental organizations that the changes did not go far enough in protecting public lands. Supporters saw the changes as improving resource and range management and broadening participation in public land decisionmaking. Congress has considered many of the 1995 changes, as part of legislative proposals or committee oversight, and may examine the proposed regulatory and policy changes.

Among the changes made in 1995, many of which are being reexamined currently by BLM, are those that:

- separated grazing *preference* from *permitted use*, so that a permittee's[1] preference for receiving a grazing permit was not tied to a specific amount of grazing based on historic levels (described as *Animal Unit Months*, or *AUMs*);
- allowed permittees up to three years of nonuse of their permits;
- authorized the suspension or cancellation of a permit if a permittee is convicted of violating certain state or federal environmental laws;
- eliminated the express requirement that a permittee be engaged in the livestock business;
- replaced the term *affected interest* with *interested public*;
- allowed *conservation use* for the term of a grazing permit, thereby excluding livestock grazing from all or a portion of an allotment;
- required title of permanent structural improvements to be held in the name of the United States;

- required that water rights for livestock grazing be held in the name of the United States, to the extent allowed by state law;
- imposed a surcharge on a permittee who allows livestock not owned by the permittee or the permittee's children to graze on public land;
- eliminated Grazing Advisory Boards and replaced them with the broader interest Resource Advisory Councils; and
- adopted rangeland management standards called *Fundamentals of Rangeland Health*.

In issuing these changes, the Secretary of the Interior dropped the most contentious proposal — to increase the grazing fee — due to the rancor this issue generated.[2] However, dissatisfaction with the 1995 changes among ranching interests led to a lawsuit ultimately decided by the U.S. Supreme Court.[3] The regulations, challenged on their face, were upheld by the courts as not exceeding the authority of the Secretary, with one exception. The courts struck down the rule pertaining to conservation use for the term of a permit on the grounds that a grazing permit was for grazing and the Secretary could more appropriately accomplish conservation use through the land use planning process.

CURRENT EFFORTS TO CHANGE GRAZING RULES AND POLICIES

BLM is taking a two-pronged approach to this iteration of grazing reform on public lands, by proposing changes to grazing regulations and considering changes to grazing policies. Under this *Sustaining Working Landscapes* initiative, first announced in March 2003, BLM seeks to create *working landscapes* that are both economically productive and environmentally healthy. Changes to grazing regulations and policies could affect more than 18,000 grazing permits on 162 million acres of BLM land. The specific regulatory proposals and policy alternatives are discussed under separate headings below.

Conflict over livestock grazing on public lands has become common. Critics of the current reform effort assert that the 1995 regulations have not been in effect long enough to assess their effectiveness and that the policy issues are too vague to assess their potential effects. They also contend that BLM has not justified a need for regulatory and policy changes. One concern is that the changes would require more monitoring than is feasible, thus possibly preventing changes. Another is that BLM and the Forest Service (FS) are not developing joint rules, given that many BLM and FS lands are similar and adjoining and permittees often have permits for livestock grazing on both agencies' lands. There is also some disappointment among environmentalists that the reform effort does not encompass certain important issues such as altering grazing fees, controlling noxious weeds, retiring grazing permits, and establishing processes for identifying lands suitable for grazing.

Overview of Regulatory Process

BLM proposed changes to its grazing regulations on December 8, 2003 (68 *Fed. Reg.* 68451), and on January 2, 2004, issued a draft environmental impact statement (DEIS) analyzing the potential impact of the proposed changes. The DEIS also assessed the impacts of a slightly different alternative and of keeping the current grazing rules. Prior to proposing the changes, BLM reviewed more than 8,000 public comments on regulatory issues that were submitted in response to a March 3, 2003 advanced notice of proposed rulemaking.

BLM asserts that regulatory changes are needed to increase flexibility for grazing managers and permittees, to improve rangeland management and permit administration, to promote conservation, and to comply with court decisions. The possibility of regulatory changes has been supported by some livestock organizations and range professionals as helping both ranchers and the range. By contrast, others have criticized the proposed changes as removing important environmental protections and opportunities for public input.

In late January and early February of 2004, BLM held public meetings in the West and in Washington, DC, to gather public comments on the regulatory proposal and DEIS. The proposal and DEIS were open for public comment through March 2, 2004, during which time the agency received more than 18,000 comments. The BLM considered these comments, and on June 17, 2005, issued a final environmental impact statement (FEIS) on proposed changes and alternatives.[4]

The proposed revisions in the FEIS met with mixed reaction, like those in the earlier DEIS. A number of the key proposals are discussed under "Proposed Changes to Grazing Regulations" below. With regard to the environmental effects of the preferred alternative, the FEIS stated (p. ES-5) that "most of the proposed regulatory changes have little or no adverse effects on the human environment. Some short-term adverse effects may not be avoided because of increases in timeframes associated with several components of this proposed rulemaking." This statement has fueled concerns among environmentalists that the proposed changes could eliminate public land protections and lead to unsustainable grazing practices. The FEIS stated that to minimize the potential for adverse affects in the short-term, the BLM could "curtail grazing if resources on the public lands require immediate protection or if continued grazing use poses an imminent likelihood of significant resource damage." Further, the BLM asserts that the long-term outcome of the proposed changes would be better and more sustainable grazing decisions, and that the changes "would be beneficial to rangeland health."

A particular controversy surfaced recently over assertions by two members of the draft EIS team, a BLM hydrologist and a BLM biologist (both now retired), that their scientific conclusions were reversed by BLM because they did not support the new rules. Those conclusions apparently had asserted that the proposed new rules could harm water quality and wildlife, including endangered species. A BLM official is reported to have called the changes to the views of the two scientists a part of the standard editing and review process.[5] Further, a statement by the BLM contended that the EIS team found their work to be "seriously lacking in the quality expected from each contributor to the environmental impact analysis." The statement alleged that the conclusions of the two team members were "based on personal opinion and unsubstantiated assertions rather than sound environmental analysis. As a result, the work submitted by the two former BLM employees was rewritten."[6]

BLM initially intended to publish a final grazing rule in the *Federal Register* in mid-July, with an effective date in mid-August. However, on August 9, 2005, BLM announced its intent to prepare a supplement to the FEIS. BLM currently anticipates issuing a draft supplemental environmental impact statement (SEIS) in the fall of 2005, soliciting and reviewing public comment, and issuing a final SEIS in the spring of 2006. No deadline for a final grazing rule has been announced.

The delay is intended to allow the agency to address public comment received after the comment period ended on March 2, 2004, primarily the views of the Fish and Wildlife Service (FWS), according to BLM. In a 16-page draft comment submitted to BLM, the FWS asserted that the proposed changes would "fundamentally change the way BLM lands are managed temporally, spatially, and philosophically. These changes could have profound impacts on wildlife resources."[7] The FWS expressed overall concern that the proposed revisions would make grazing a priority over other land uses, which could be detrimental to fish and wildlife habitats and populations, for instance, management of sage-grouse habitat. The agency further contended that the proposed changes could "constrain biologists and range conservationists from recommending and implementing management changes based on their best professional judgment in response to conditions that may compromise the long-term health and sustainability of rangeland resources."[8]

While supporting some of the proposed changes, the FWS identified a number of areas of particular concern. They included potential effects of administrative inconsistencies between BLM and the Forest Service on their management of fish and wildlife resources across boundaries;[9] diminished requirements for public consultation on site-specific actions, which have the greatest potential for impacts to fish and wildlife; a phase-in of decreases (or increases) in livestock use that are greater than 10%, which may not be immediate enough to prevent irreversible harm to vegetation and wildlife; including an amount of forage in the definition of grazing preference, which may not account for other range attributes;[10] allowing shared title to range improvements, which could make it more difficult to reallocate land use, such as to provide quality habitat for wildlife; requiring monitoring of rangeland standards, which has not been achievable due to BLM funding and staffing limitations; and sharing of water rights, as water is the most important resource for fish and wildlife.

Proposed Changes to Grazing Regulations

BLM asserts that some changes would be substantive while others are clarifications, but it is not clear which potential changes BLM believes fall within each category. This adds to the uncertainty over which proposals are intended to, and likely to, make major changes in public lands grazing. There continues to be disagreement as to the extent of the environmental impact of the changes and whether that impact would be primarily beneficial or damaging in both the short- and long-terms. There also remains a difference of opinion as to the extent to which the regulatory effort should reinstate pre-1995 grazing provisions or substantially modify other current provisions.

Some of the key changes identified in the FEIS are discussed below. They involve ownership of range improvements and water rights, and opportunities for public input and appeals. Other discussed proposals pertain to terms and conditions of permits and rangeland health. These areas have been among the most controversial among affected interests.

Share Title to Range Improvements

BLM proposes reestablishing a pre-1995 rule allowing title to a structural range improvement, such as a fence, well, or pipeline, to be shared by the BLM and a permittee (or others) if it is constructed under a Cooperative Range Improvement Agreement. Title would be shared in proportion to each party's contribution to the cost of the improvement. Current regulations require documentation of a permittee's contributions to improvements and compensation if a permit is cancelled or passes to another. However, some advocate that ranchers should receive more direct compensation for improvements, would be encouraged to undertake and maintain improvements if they get title, and should be able to include improvements as assets to secure loans for grazing.

Opponents charge that shared title would create private rights on public land and could hinder action to correct grazing abuses. They contend that the government should hold title to improvements as they typically are important for other uses, such as recreation and wildlife habitat. Still others believe that improvements for grazing do not necessarily benefit other land uses, and thus permittees should not be rewarded with title.

Acquire Private Water Rights

The proposed regulations would allow permittees to acquire water rights, consistent with state law. Current rules require the federal government to follow state procedural and substantive law regarding livestock watering rights, but direct that title to the rights be held by the United States to the extent state law permits. Before 1995, practices as to water rights for livestock grazing varied and in some states could be acquired in the name of the permittee. Express language allowing private individuals to hold water rights is supported by some as providing an incentive for private water development on public land, and protecting permittees from being denied water. It is opposed by others who believe water rights should be in federal ownership to facilitate multiple uses and to preclude private claims for compensation for water rights, and because states typically do not allow grazing permittees on state lands to obtain water rights. Still others are concerned that public resources will be given away at no cost.

Reduce Requirements for Public Involvement

BLM proposes to reduce the occasions on which it is required to involve the public in its decisions. For instance, the agency would no longer be required to get input from the public regarding designation and adjustment of grazing allotment boundaries, the issuance or renewal of grazing permits, or modification of the terms and conditions of permits that are not meeting management objectives or the fundamentals of rangeland health. The agency also seeks to modify the definition of "interested public" so that only individuals, groups, and organizations who participate in the decisionmaking process on management of a specific allotment are maintained on the list of interested publics. Supporters maintain that the changes will prevent delays and facilitate timely decisions. Also, the agency views additional consultation as redundant, because the public already has opportunities to participate during the planning processes, reviews under the National Environmental Policy Act (NEPA),[11] and the development of reports used by BLM as a basis for increasing or decreasing grazing use and changing the terms of use (under 43 C.F.R. 4130.3-3(b)). The changes are criticized as restricting public input which could lead to ill-considered decisions. They are further opposed because decisions at the planning level are too general and broad to allow specific evaluation and comment. Still others contend that environmental reviews under NEPA are not required for some grazing decisions and where required are backlogged, and as a result public participation under NEPA often is delayed.

Modify the Administrative Appeals Process

The agency proposes to modify the administrative appeals process on grazing decisions and define the extent to which grazing should continue in the face of an appeal or stay of a decision. For instance, the proposed rule would provide that when a stay is granted on appeals to decisions involving renewing, modifying, suspending, or canceling a permit or on transferring preference, the affected permittee usually would continue grazing under the immediately preceding grazing authorization. Certain decisions would be required to be implemented immediately and not be eligible for a stay, including authorizations to graze temporary forage. The changes are sought to provide permittees with continuity of operations when a decision affecting their operations is appealed. They are opposed by some as limiting the ability of the public to participate in grazing decisions, reducing the flexibility of land managers to take certain actions based on what is best for resource conditions, and potentially continuing damaging grazing practices.

Broaden the Definition of Grazing Preference

Another proposal would broaden the definition of *grazing preference* to include a quantitative meaning —forage on public lands, measured in AUMs[12] — tied to a permittee's base property of land or water. The definition would continue to include a qualitative meaning —a superior or priority position to obtain a permit. The revised definition, which would be similar to pre-1995 rule language, is intended to link forage allocations to base property, give ranchers certainty as to the size of operations, and eliminate confusion as to the meaning of *preference*. Further, preference would include both *active use*, defined as use currently available for livestock grazing based on livestock carrying capacity and resource conditions, and *suspended use*, which is use that has been allocated for livestock grazing in the past but is currently unavailable. The new definition is opposed as infringing on the discretion of land managers to determine the extent of grazing that should be allowed.

Remedy Rangeland Health Problems

The proposal would require both assessments and monitoring of resource conditions to support agency determinations that grazing practices or levels of use are significant factors in failing to achieve rangeland health standards or conform with guidelines on an allotment. It would amend the timeframe and procedures for changing grazing management after a determination that grazing practices or levels are significant factors in failing to achieve standards or conform with guidelines. The change would allow a maximum of 24 months, rather than the current 12-month limit, for developing remedial changes in grazing practices. However, BLM could extend the deadline if responsibilities of another agency prevent completion within 24 months. Further, a change would phase in grazing increases or decreases of more than 10% over a five-year period, unless the changes must be made sooner under law (e.g., the Endangered Species Act (ESA)[13]) or the permittee agrees to a shorter period. BLM maintains that these changes will provide a sound basis for agency determinations and give the agency more time and flexibility in working with permittees who are not meeting the standards. They are opposed as potentially allowing damaging practices to continue and requiring excessive documentation even when damage is obvious. Opponents also claim that BLM lacks staff and funds to collect the necessary information formally.

Remove Limit on Permit Nonuse

The proposed rule would remove the current three-year limit on temporary nonuse of a permit by allowing permittees to apply for nonuse of all or part of a permit for up to one year at a time, for as many years as needed. The change is promoted as allowing for recovery of the land and providing flexibility to ranchers who may not be able to graze for reasons including financial hardship, drought, or overgrazing. Critics argue that the change does not address the underlying problem — permitting grazing that exceeds the capacity of allotments. Others are concerned that conservationists will obtain grazing permits and opt for extended nonuse. However, temporary nonuse is allowed only if authorized by BLM and for no longer than one year at a time.

Eliminate Conservation Use Grazing Permits

Regulations allowing BLM to issue long-term *conservation use* grazing permits would be eliminated to comply with court decisions that permits should be issued for grazing and conservation needs should be met through alternatives. Advocates of conservation use observe that the practice allows overgrazed land to be rested and that BLM should develop a legal alternative to the current language.

Other Proposed Changes

Other proposed changes include

- restricting BLM to taking action against a permittee convicted of breaking laws while engaged in grazing only if the violation occurred on the permittee's allotment;
- emphasizing that reviews under NEPA will consider the social, economic, and cultural impacts of proposed changes in grazing preference, in addition to the ecological impacts;
- increasing administrative fees for livestock crossing permits, billings, and preference transfers;
- providing that a biological assessment or evaluation by BLM under the ESA is not an agency decision for purposes of protests and appeals;
- specifying that BLM will cooperate with state, tribal, local, and county grazing boards in reviewing range improvements and allotment management plans on public lands;
- stating that the temporary changes that BLM can make within the terms and conditions of permits involve the number of livestock and period of use that would result in temporary nonuse and/or forage removal; and
- requiring BLM to document observations supporting a reduction in grazing intensity, and providing that reductions will be made through temporary suspensions of active use rather than through permanent reductions

Changes not Proposed

BLM considered but has not proposed many other changes to grazing regulations, according to the proposed rule and FEIS. For instance, the agency considered adopting rule language to support establishing and operating a new type of grazing unit, called a *reserve common allotment* (RCA), but did not do so because of negative public reaction to the idea.

However, the BLM continues to consider the issue of forage reserves as part of its consideration of policy changes. (See below under "Grazing Policy Changes Under Consideration" for a discussion of reserve common allotments.) The agency also considered allowing permit holders to temporarily lock gates on public lands, for instance to protect private property by preventing cattle from leaving grazing allotments and to minimize disturbances during lambing and calving seasons. The idea was opposed as preventing access by other land users, such as hunters and recreationists; giving a special privilege to permittees; and being currently prohibited by law.

BLM also did not propose altering the existing provisions under which a grazing fee surcharge is placed on permittees who allow livestock neither they nor their children own to graze on public land. The current surcharge provision was incorporated in 1995 to address concerns regarding the potential for a permittee to make a substantial profit when subleasing grazing privileges. BLM asserts that the current surcharge provision is equitable and that it does not want to address fee-related issues as part of the current reform effort.

Overview of Grazing Policy Process

On March 25, 2003, BLM first announced possible grazing policy changes as a complement to the regulatory changes being considered.[14] According to BLM, the focus is on policy changes that can be carried out under existing rules. The distinction between policies and regulations is not always clear, and when an agency must take action through formal rulemaking can be an issue.[15]

The agency seeks policy reforms to promote citizen stewardship of public lands, provide flexibility to managers of livestock grazing, and increase innovative partnerships. BLM has reviewed the advice and recommendations of its Resource Advisory Councils on policy ideas.[16] Final grazing policy changes will be addressed when the rulemaking process is "substantially completed," according to BLM.

GRAZING POLICY CHANGES UNDER CONSIDERATION

On March 25, 2003, BLM issued a press release announcing that policy changes under consideration include reserve common allotments (RCAs), conservation partnerships, voluntary allotment restructuring, conservation easement acquisition, and ESA mitigation.[17] BLM also examined the establishment of RCAs as a regulatory change, but did not propose rule language in this area. Some have asserted that other policy options under consideration might necessitate the adoption of new rules, which would require opportunities for public comment.

BLM solicited public feedback on the policy options under consideration through a series of public workshops. While some support for policy changes was expressed, many members of the public asserted that available information was inadequate to assess the policy changes, raised concerns about the outlined options, or viewed the initial schedule for considering policy and rules changes as too short. In response, BLM announced that it had extended the timeframe for developing policy changes, but did not issue a schedule for completing actions.

The agency also developed and published on its website more detailed information on RCAs, conservation partnerships, and voluntary allotment restructuring. It noted that conservation easements were no longer being pursued as a major policy tool, and that the concept of ESA mitigation had evolved to the broader notion of *landscape habitat improvement.*

Reserve Common Allotments (RCAs)

RCAs would serve as livestock forage for permittees while their normal allotments undergo rest or improvements, and might be used for unplanned needs, such as drought, fire, or flood. The BLM asserts that existing regulations allow the creation of RCAs but with impediments. RCAs are supported as encouraging improvements (such as a prescribed burn) and recovery from heavy grazing, and necessary in emergencies so that ranchers won't have to reduce herd size or sell out for lack of forage. Conservationists are concerned that this approach does not address what they view as the fundamental issue — overstocking or grazing unsuitable lands — and that RCAs will benefit ranchers who mismanaged their allotments. Livestock groups fear a reduction in grazing and loss of water rights through nonuse, coercion to participate, and use of RCAs as a subterfuge for conservation use. Key issues for both supporters and critics include how much land, and which lands, will become part of RCAs (e.g., vacant allotments, areas of nonuse); what will trigger their use; their term; how many permittees will be allowed to graze simultaneously; and how forage will be allocated.

Conservation Partnerships

The goal of conservation partnerships between permit holders and the BLM would be to improve environmental health. A permittee could enter into a performance-based contract with BLM to undertake projects to: restore streambanks, wetlands, and riparian areas; enhance water quantity and quality; improve wildlife or fisheries habitat; and support the recovery of threatened and endangered species, among other actions. In return, the permittee could receive management flexibility, increased livestock grazing, and stewardship grants to pay for investments in conservation practices. Advocates note that these arrangements would give permittees credit for improvements they have been making, encourage and reward good stewardship, and enhance the role of permittees in managing grazing allotments. Opponents contend that private property rights could be impaired, the amount of available funding is unclear, the extent of resource improvement is uncertain, permittees might receive benefits for little or no resource improvement, and partnerships may not be entirely voluntary. Differences of opinion exist as to a role for third parties, rewards for permittees, and dealing with intermingled private land.

Voluntary Allotment Restructuring

Voluntary allotment restructuring would allow two or more grazing permittees to merge allotments. One or more of the permittees would not graze temporarily, while the others grazed over the entire area, to achieve lighter grazing. Such restructuring is supported as improving range conditions while maintaining the economic viability of permittees. Concerns include that restructuring would reduce grazing and can already occur informally, operator to operator. Issues involve when restructuring would be used and whether and how to compensate ranchers who give up grazing privileges.

Conservation Easements

Conservation easements — land use restrictions — were being considered to preserve open space. Under this arrangement, BLM would place conservation easements on its land identified for disposal. Permittees would similarly restrict development on their private land in exchange for acquiring the BLM lands with the easements. These easements were advocated as benefitting the land, land managers, and permittees. However, BLM subsequently asserted that because they are limited in their ability to use conservation easements, such easements are not currently a major policy option. Easements have been opposed as reducing land values, limiting the management discretion of private landowners, not necessarily providing a public benefit, and encumbering land disposal.

Endangered Species Act Mitigation

BLM viewed the policy options listed above as providing opportunities to mitigate the effects of livestock grazing on species listed under the ESA. Mitigation banks also were contemplated to preserve or create habitat for listed species in exchange for mitigation credits. Such credits could be sold to other land users to offset the impacts of development on listed species. This idea raised concerns among livestock groups that grazing would be subordinated to conservation and private property rights could be weakened, and among environmentalists that permittees would be compensated for something the BLM already is obligated to protect. This concept is now being considered as *Landscape Habitat Improvement*, to promote species conservation and facilitate ESA consultations. Habitat management would be pursued on a landscape basis, perhaps involving lands under various ownerships, which presumes a larger geographic area than a grazing allotment. Grazing permittees could form partnerships to promote species conservation and maintain or improve habitat while continuing to graze public lands.

CONCLUSION

More than two years have passed since BLM notified the public of its consideration of changes to both grazing regulations and policies under its *Sustaining Working Landscapes* initiative. During this time, evaluations of possible regulatory and policy changes have been proceeding on separate tracks, and have met with mixed reaction.

Many of the key regulatory changes contained in the FEIS deal with provisions that took effect in 1995, during the last major revision of grazing rules. Among them are proposals to allow shared title to range improvements, allow private acquisition of water rights, reduce requirements for public involvement, modify the administrative appeals process, broaden the definition of grazing preference, change the timeframe and procedures for remedying rangeland health problems, remove the limit on permit nonuse, and eliminate conservation use grazing permits. The revisiting of issues dealt with a decade ago, together with other proposed changes, has been generally supported by livestock organizations and some range professionals who see benefits both to the range and those grazing on public land. By contrast, many environmental organizations and other range experts oppose the changes on the grounds that a need for change has not been demonstrated and the particular proposals could harm the environment.

Public comment on the proposed regulatory changes, together with the DEIS assessing their impact, was accepted through March 2, 2004. BLM evaluated the comments over many months, before publishing an FEIS on June 17, 2005. BLM postponed developing a final grazing rule to consider public comment allegedly received after the closing date, particularly from the Fish and Wildlife Service. The BLM is preparing a supplement to the FEIS, expected in final form in the spring of 2006. No date for the final grazing rule has been announced.

Public feedback on possible policy changes already has shaped the proposals under examination as well as extended the expected timeframe for considering changes. Key policy issues under consideration relate to RCAs, conservation partnerships, voluntary allotment restructuring, and landscape habitat improvement. Public reaction to policy changes could become more contentious once details of the changes are developed and announced to the public.

BLM has reviewed input from its Resource Advisory Councils on policy options. No timeframe for issuing policy changes has been announced. BLM expects to focus on final policy changes after the completion of the rulemaking process.

REFERENCES

[1]   The term permittee is used throughout to refer to both permittees and lessees, and permit refers to both permits and leases.
[2]   For more information on grazing fees, see CRS Report RS21232, *Grazing Fees: An Overview and Current Issues*, by Carol Hardy Vincent.
[3]   For more information on the legal challenge to the 1995 regulations on livestock grazing, see CRS Report RS20453, *Federal Grazing Regulations: Public Lands Council v. Babbitt*, by Pamela Baldwin.
[4]   The FEIS, together with a fact sheet and questions and answers on the new grazing regulations that are in development, are available at [http://www.blm.gov/grazing/], visited August 16, 2005.
[5]   Julie Cart, "Land Study on Grazing Denounced; Two Retired Specialists Say Interior Excised their Warnings on the Effects on Wildlife and Water," *Los Angeles Times*, June 18, 2005, sec. A, p. 1.
[6]   U.S. Department of the Interior, Bureau of Land Management, *Statement of the Bureau of Land Management re: Los Angeles Times Article of June 18, 2005*, unpublished draft received from BLM, Aug. 12, 2005.
[7]   U.S. Department of the Interior, Fish and Wildlife Service, *Comments on (1) Proposed Rule for Grazing Administration-Exclusive of Alaska (EC03/0049), and (2) Draft Environmental Impact Statement for the Proposed Revisions to Grazing Regulations for the Public Lands (EC04/0003)*, unpublished draft received from BLM, Aug. 12, 2005, p. 12.
[8]   Ibid., p. 1.
[9]   A goal of the 1995 regulatory reform was to increase consistency BLM and FS grazing administration, perhaps reducing administrative costs. The FWS expressed concern that

the current regulatory reform effort, in pertaining exclusively to BLM, could lead to inconsistencies between the BLM and the Forest Service in several areas.

[10] Under current BLM regulations, grazing preference is defined as having a superior or priority position against others for the purpose of receiving a grazing permit. The FWS expressed opposition to adding a quantity of forage to that definition without consideration of other features of range resources that are not quantifiable in terms of forage, such as species diversity and soil condition.

[11] P.L. 91-190; 42 U.S.C. §§4321-4347.

[12] An AUM is defined as the amount of forage necessary for the sustenance of one cow or its equivalent for a period of one month.

[13] P.L. 93-205; 16 U.S.C. §§1531-1540.

[14] The announcement took the form of a press release, now contained on the BLM website at [http://www.blm.gov/nhp/news/releases/pages/2003/pr030325_grazing.htm], visited on August 23, 2005.

[15] See 5 U.S.C. §551(4).

[16] BLM has two dozen Resource Advisory Councils (RACs) in western states to provide the agency advice on managing public lands. Each RAC consists of some 12-15 citizens representing diverse interests, including ranchers, environmental groups, tribes, academia, and state and local governments.

[17] For more information on policy options, see the BLM website at [http://www. blm.gov/nhp/efoia/wo/fy03/im2003-214ch1.htm] and [http://www.blm.gov/nhp/efoia/ wo/fy03/im2003-214.htm], visited on August 23, 2005.

In: Progress in Environmental Research
Editor: Irma C. Willis, pp. 245-261

ISBN 978-1-60021-618-3
© 2007 Nova Science Publishers, Inc.

Chapter 8

RURAL LIVELIHOODS: CONSERVATION, MANAGEMENT AND USE OF PLANT BIODIVERSITY: EXPERIENCES AND PERSPECTIVES OF THE WORLD AGROFORESTRY CENTER IN WEST AND CENTRAL AFRICA

P. Mbile[1,*], Z. Tchoundjeu, E. Asaah, A. Degrande, P. Anegbeh, C. Facheux and M. L. Mpeck, D. Foundjem-Tita and C. Mbosso

[1]Socio-ecologist, The World Agroforestry Centre, Africa humid tropics regional programme, 2067, Messa, Yaounde, Cameroon

ABSTRACT

Forest loss and fragmentation over the past decades in the west and central Africa region is having a direct effect on the habitats of valuable plants, driving species isolation, reductions in species populations and in some cases, increasing extinction rates of potentially useful plants. Furthermore, some tropical rainforest plants exhibit hampered seed germination or seedling establishment through hampered natural regeneration in disturbed ecosystems.

Nevertheless, these forests in west and central Africa remain important sites, habitats and sources of potentially useful plantdiversity. Many tropical tree species and their products have been documented regarding the roles they play as food, medicine and in terms of other services they provide to local peoples. The exploitation, uses and commercialisation of these tree products, constitute an important activity to people living around forests and beyond within the region. For some of these species, existing markets have expanded within and outside their wide ecological range as well as great potentials that exist for their further development at industrial level.

---

* p.mbile@cgiar.org. Tel/Fax: 237 223 75 60/237 223 74 34.

Since 1998, the World Agroforestry Centre, Africa humid Tropics, in partnership with several local and regional stakeholders in West and Central Africa have been implementing a Tree domestication programme aimed at diversifying smallholder livelihood options through the selection, multiplication, integration, management and marketing of indigenous trees/plants and their products, ensuring that they provide both livelihood and environmental services. As tree domestication itself depends on existing plant diversity, biodiversity at genetic, species and ecosystem levels have been important considerations in cultivar selection, farming systems diversification and contributing towards ecosystems resilience, respectively.

This tree domestication programme is being implemented in Cameroon, Nigeria, Gabon, Equatorial Guinea and more recently in the Democratic republic of Congo.

The programme started with the prioritisation of a range of indigenous fruit and medicinal tree species at local community levels. Emphasis then moved to capacity-building: training, follow-up and information dissemination focussing on a range of low-tech and adaptable propagation, marketing, selection, cultivation and management techniques for local level stakeholders, and training, backstopping and dissemination, for a range of regional government and non-governmental partners, in order to enhance ownership and adoption of the process.

So far, the programme has contributed firstly, in building of both natural assets of resource-poor farmers to increase their access to a diverse range of agroforestry trees and products, and human assets for perpetuating the knowledge and experience in the region, as well develop mechanisms for increasing and diversifying household revenue through better marketing of indigenous agroforestry tree products, protecting biodiversity on-farm and recognizing the value in maintaining both intra and inter-specific diversity on-farms. The programme has also developed multi-species, on-farm needs-based live gene banks as well as classical ones of regionally important high-value indigenous tree species in both Cameroon and Nigeria.

As the programme develops in the region, increasing emphasis is being placed on building strategic partnerships in order to achieve greater and more far-reaching impact by increasing the potential contribution that diverse agroforestry trees make to household revenue, and environmental management at farm and at landscape scale.

INTRODUCTION

In a global context according to CARPE (2001), annual deforestation rates are relatively low in central Africa (0.6% year[-1] between,1980-1990). However with forest losses averaging at between 200,000 and 220,000 hectares a year between 1990 and 2000 (Njib, 1999), in Cameroon, the potential for plant biodiversity loss remains high.

Moreover, although the highest rates of deforestation are not necessarily associated with the highest population densities, with a population growth rate of 2-3% per year (World Bank group, 2002) within Central Africa, the demand for agricultural land is increasing, as is the scale of forest transformation.

Forests are important sites of potentially useful plant biodiversity. Many tropical tree species and their products have been documented regarding the roles they play as food, medicine and in terms of other services they provide to low income people especially (Okafor, 1991, Falconer; 1990; Leakey and Newton, 1994). The exploitation, uses and commercialisation of diverse tree products, constitute an important activity of people living around forests and beyond (Ndoye *et al.*, 1998) within the region. For some of these species, existing markets have expanded within and outside their wide ecological range (Cunningham *et al.*, 1997; Tabuna; 1999). In addition, great potentials for their further development at industrial level exist (Leakey, 1999).

However, un-monitored forest loss and fragmentation in the region is having a direct effect on the habitats of valuable plants. Habitat loss is driving increased extinction rates of tropical plant species and is causing reductions in species populations (Wilcoe, *et al.*, 1986; Hudson, 1991; Forman and Gordon, 1989). Furthermore, Tropical Rainforests (TRF) plants suffer from hampered seed germination or seedling establishment through natural regeneration, due to environmental conditions imposed by forest conversion. They also face competition from helophytic ruderal herbs in these converted lands (Uhl, *et al.* 1988).

Therefore, while forest loss and fragmentation: conversion to farmland and degradation is causing plant biodiversity losses in one form or another they also indirectly hamper regeneration rates and help accelerate species rarity, isolation and extinction rates.

Faced with such global and regional challenges, an important contribution towards the conservation and management of indigenous plant biodiversity in the West and Central Africa sub-region need therefore include, a conscious, smallholder farmer livelihood-driven process, targeting both *ex-situ* and *in-situ* methods for plant biodiversity conservation and management, characterised by identification, collection, regeneration, value-adding and management approaches at farm and at landscape scale.

Since 1998, the World Agroforestry Centre in partnership with several local and regional stakeholders in West and Central Africa have been implementing a tree domestication[1] programme aimed at diversifying smallholder livelihood options through the conservation and management of indigenous trees/plants for both livelihood and environmental services.

This paper presents ICRAF's experiences and perspectives as they are on the ground and in the light of the conservation and management of plant biodiversity with current value to local people and how such knowledge and practices helps to develop a consciousness towards wider plant biodiversity management and the potential contributions the approaches can make towards optimizing the sustainable use and benefits from these resources for present and future generations.

INDIGENOUS TREES DOMESTICATION, AND PLANT BIODIVERSITY CONSERVATION AND MANAGEMENT: ICRAF'S EXPERIENCES

ICRAF's contribution towards the conservation and management of plant biodiversity in west and central Africa is driven and regulated by the livelihood needs, the strategies of resource-poor communities and by the generation of new knowledge on the importance of biodiversity.

The programme seeks to develop trees/plant products with a view to increasing their contribution to the livelihood of forest/tree-dependent peoples and their dependants, and to

---

[1] Bringing a wide range of tree species and subspecies into wider cultivation – numeric and geographical; and management by resource poor farmers. to address problems of food. medicines. nutrition. income and environmental management

the GDP while considering the conserving their productive base as a sin qua non for continuity of the practice.

To achieve this, the tree domestication programme seeks to identify both the existing natural potential with a view to a needs-based assessment and management of the resources, *in situ* and *ex-situ* as well as foster and organize the marketing of tree/plant products at the local, national and international levels as a mechanism through which local people can add greater value to biodiversity.

The thesis of ICRAF's tree domestication approach in the African humid tropics is captured in the World Agroforestry Centre's mission statement, which is to reduce poverty, improving food and nutritional security, and enhance environmental resilience in the tropics. It is this resilience, drawn from ICRAF's agroforestry tree domestication activities at farm and landscape scale which accords adaptive capacity to agricultural landscapes in the event of shocks and stresses resulting from intensification. The programme is thus characterized by research on livelihood aspects, opportunities and threats in dealing with inter and intra-specific diversity, options for species diversification and their adaptation at landscape scale. Thus, tree propagation, integration and management, germplasm dissemination pathways, trees and products processing and marketing, capacity building, partnerships, policy and networking are cornerstones of the process.

1. LIVELIHOOD ASPECTS: SOCIOECONOMIC ANALYSES AND PRIORITIZATION

At a broad socio-economic and farmer livelihood level, research work using participatory appraisal methods and approaches, focusing on issues of land tenure, participation, gender, and well being, reveal that, though land and tree tenure issues for instance, remain complex and variable, most households maintain secure access to land. Findings from fruit tree inventories in Cameroon and Nigeria, in particular, indicate strong relationships between tree numbers, diversity, density, and land use types. These however, were not related to wealth and educational levels of farmers.

Overall, although household decision-making rests with the largely male household heads, the participation of women and youths (currently variable), if increased, can significantly influence tree cultivation decisions. Women, youths and the elderly appear particularly attracted by the wider propagule-type options presented by vegetatively propagated trees. However, partly as a result of traditions, customs and low educational level of vulnerable groups, few checks and balances exist at the household socio-economic level for the moment, which can guarantee the longer-term economic autonomy of women and youths. Therefore, women, the old and other vulnerable groups remain prime custodians of intra-and inter-specific tree diversity. These they consider as a sort of insurance against uncertainties. Contrarily men and the less poor, tend to go for mono-cultures, and may consider maintaining higher tree diversities on · farm constraining to achieving higher productivity per hectare.

In setting priorities for agroforestry tree domestication, the choice of species for improvement is complex in both socioeconomic and biophysical terms. The clientele is very heterogeneous. It consists of many individual small-scale farmers with differing needs, which are difficult to generalize across a landscape. Farmers use many different species and little scientific knowledge is available about most of them. In order to determine the species for which improvement would have the highest impact, per unit of product, Franzel *et al.* (1996) developed a priority setting process, involving several steps. One of the steps encompasses assessment of clients' needs, which defines user groups and identifies their main problems and the agroforestry tree species or sub species that may best meet their needs and at what level of adaptation by the species to prevailing geo-physical conditions. In field surveys, farmers list the agroforestry species they grow and use, and rank them according to their preferences. Only the species that provide the most important products for solving the present and future problems of the clients are considered in the following stages. Researchers then refine the list further by ranking species on their research ability, expected rates of adoption, and non-financial factors that modify the objective of increasing financial value. Also, detailed data are collected from farmers and markets to estimate the value of products.

Synthesis of such information thus facilitates concrete prioritization of species per eco-zone or landscape. Tables 1; a, b & c below presents three examples of such ranking of 4 priority species in different parts of the humid tropics: the humid savannah, forest margin and high forest zones of Cameroon.

Table 1. Priority species in different eco-zones of Cameroon

a: Humid Savannah.

Species name	Priority ranking
Dacryodes edulis	1
Dacryodes edulis	2
Canarium spp.	3
Prunus africana	4

b: Tropical forest margin.

Species name	Priority ranking
Irvingia spp.	1
Dacryodes edulis	2
Ricinodendron heudelotii	3
Garcinia cola	4

c:Tropical high forest.

Species name	Priority ranking
Irvingia spp	1
Gnetum africanum	2
Baillonella toxisperma	3
Ricinodendron heudelotii	4

2. IMPROVING ACCESS AND KNOWLEDGE OF PLANT BIODIVERSITY AT LOCAL LEVEL

Increasing knowledge at local and regional levels on the conservation and management needs of plant biodiversity, as well as in biotechnology has been achieved by ICRAF through research and training, which enables the clients to consciously maintain and or build assets of plant's genetic resources useful for health, food and other culinary needs. ICRAF has developed technologies and approaches which facilitate the build-up of such natural assets especially where start-up costs have outweighed the discounted benefits to resource-poor farmers using classical tree improvement methods.

Asset-building by Increasing Plant Biodiversity through Technical Support

Plant/tree Multiplication and Appropriation of Knowledge via Capacity Building

Plant and tree biodiversity are very unevenly distributed within the humid tropics of Cameroon and west and central Africa in general. Nevertheless, the need to diversify revenue and subsistence mechanisms remains strong within poor rural communities throughout the region. Working on the domestication continuum[2] the World Agroforestry Centre is facilitating the adoption of a wide range of indigenous tree/plant species by resource-poor farmers through a process of capacity-building in propagation techniques and management, with the purpose of increasing local access to high quality plant genetic material. Following priorities-setting, targeted collections, selections, and the development of more efficient propagation methods at the village levels ensues.

Members of existing self-help groups are trained in techniques of vegetative and sexual propagation methods, seed collection, selection and germination methods; setting of cuttings, grafting, marcotting for vegetative propagation, and general nursery management techniques. The technical aspects of propagation have been extensively researched and reported (Tchoundjeu et al, 1999; 2002). Capacity building in propagation methods is not restricted to local populations at village level but includes students at MSC and PhD levels from African, European and American Institutions.

The range of species disseminated and researched, have also increased over the years in response to farmer needs, market opportunities, technological adaptability and subsistence requirements. Figure 1 below illustrates the trends in plant/tree species entering the domestication pipeline over the years, while Table 2 explains some of ICRAF's needs-based domestication approaches showing the relationships between plant/tree species and relevant approaches used.

Adoption of indigenous trees domestication approaches by farmers have led directly to the popularization of the techniques and approaches within the different countries in which it is practiced. Figure 2 illustrates the spread of nurseries practicing the various plant/tree propagation techniques over a period of 6 years (1998– 2003), within the humid forest zone of Cameroon.

---

[2] Considers plant/tree material from the wild to the completely transformed state or a cultivar.

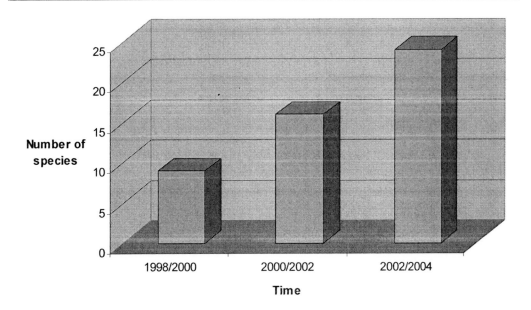

Figure 1. Trends in the species recruitment process for domestication.

It is perhaps early to evaluate the full potential social and economic impact of diverse indigenous trees produced and distributed from the village based nurseries within the degrading forest margins of Cameroon and supported and guided by ICRAF researcher.. Nevertheless, due to the networks within communities and within the extended family systems in Cameroon and Nigeria as depicted in Figure 3, it is feasible to appreciate the potential of participatory approaches on development and transformation impact of new technologies and or approaches such as agroforestry tree domestication.

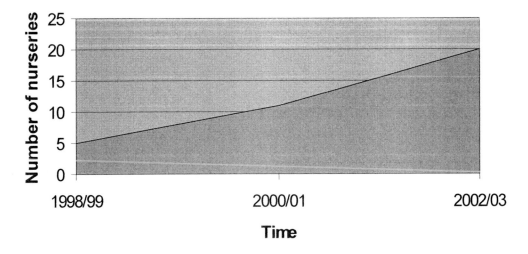

Figure 2. Spread of plant/tree nurseries in the humid forest eco-zone of Cameroon.

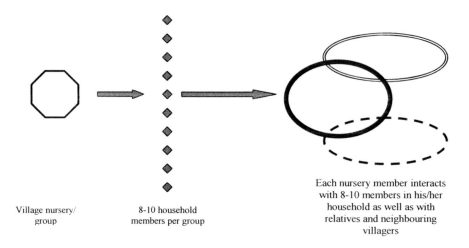

Village nursery/   8-10 household   Each nursery member interacts
group   members per group   with 8-10 members in his/her
                                                    household as well as with
                                                    relatives and neighbouring
                                                    villagers

Figure 3. Diffusion of knowledge from existing nursery groups through the family-community network..

Table 2. Needs-based domestication approaches for different plant/tree species

Plant/tree species characteristics.	Possible causes	Some domestication approaches
Hampered germination.	Converted lands, ecosystem change	Germination trials, targeted collections and vegetative propagation.
High variability in phenotypes	High intra species diversity	Range-wide collections, gene bank establishment, selection for marketable traits, vegetative propagation.
Rarity.	Restricted natural distribution, overexploitation.	Targeted collections, conservation through use, germination and propagation trials.
Good market potential.	High local use, part of local diets.	Conservation through cultivation, propagation trials and marketing research.
Difficult processing.	Unfamiliarity by local population, absence of technology.	Range wide surveys of indigenous processing methods. Development of appropriate processing techniques
High perishability.	Methods of consumption, nature of product.	Research on processing and transformation. Market intelligence studies.
Recalcitrant seeds.	High dormancy.	Germination trials, vegetative propagation.
Low regeneration rates.	Competition with predators and users ecosystem change.	Targeted collections, germination trials, and vegetative propagation.
Unsustainable harvesting methods.	High market potential. Lack of legal ownership.	Market-led production forecasting. Recommendations for policy reform, propagation trials and conservation through cultivation.
Difficult adaptability in new areas.	Low tolerance	Adaptability and propagation trials. Conservation through cultivation,
Endemic medicinal plants.	Limited natural distribution. Low tolerance	Targeted collections. Propagation trials, conservation through cultivation.

Asset Building of Plant Biodiversity through Transfer and Exchange with Direct External Inputs

Despite the tremendous efforts to increase farmer access to plant biodiversity at local levels through capacity-building mechanisms, the sheer immensity and range of useful plants that are potentially beneficial in increasing both direct income, indirect downstream benefits, the resilience in agricultural systems as well as upping the option value of on-farm plant biodiversity, have required that indirect methods as well as direct methods need be employed to increase farmers access to these resources.

These drastic measures have been necessary because, the ecosystems and livelihoods benefits of tree diversity notwithstanding resource poor farmers always discount the returns to investments in tree planting to the extent that the opportunity costs of committing scarce resources to tree planting today always outweigh the short-term benefits.

For instance, range-wide collection of plant genetic material to ensure ecological sound tree domestication processes are not only expensive but can be logistically complex (Simons et al, 1994). ICRAF has therefore developed a modest number of classical gene banks strategically located in both Nigeria and Cameroon as well as develop gene pool on farmers' fields to both conserve genetic materials and to enable resource-poor farmers have unregulated access to them directly.

Table 3. Classical gene banks of indigenous trees developed by ICRAF in west and central Africa

Location	Species	Type of collection	Number of families	Number of provenances
Mbalmayo, Cameroon	*Irvingia wombolu*	Targeted	30	2
Yaoundé, Cameroon	*Dacryodes edulis*	Targeted	20	4
Boyo, Cameroon	*Prunus africana*	Targeted	28	1
Onne, Nigeria	*Irvingia gabonensis*	Range-wide	385	93
Onne, Nigeria	*Irvingia wombolu*	Range-wide	69	13
Onne, Nigeria	*Irvingia robur*	Range-wide	18	3

Table 3 summarizes the classical gene banks developed by ICRAF in the region. Despite these efforts, ICRAF is putting increasing emphasis on 'on-farm' gene pools of plant genetic materials. The latter approach, though lacking the classical rigidity, constitutes the conservation of high-value and diverse plant genetic materials through use. In terms of management costs, these are generally lower than classical gene banks. Secondly, the impact on development in the short term is also quite high as farmers exercising ownership carryout management and also protect the trees and plants.

On-farm Gene Pools

Range-wide characterization studies of indigenous fruits by ICRAF (Anegbeh *et al*, 2003; Leakey *et al*., 2004) in Cameroon and Nigeria have identified species and sub species populations in agricultural landscapes where subsistence farmers have inadvertently 'domesticated' indigenous trees.. These studies of genetic/phenotypic variability are providing insights into the value of farmer-farmer exchange of plant genetic materials and the opportunities therein for indirect development and management of on-farm gene pools of a diverse range useful plant species.

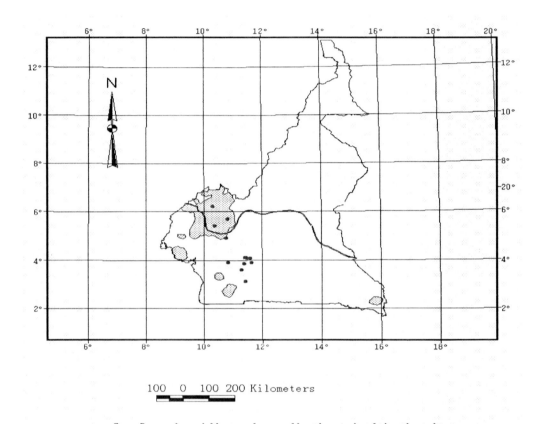

100   0   100 200 Kilometers

• Some Dacryodes edulis gerplasm collection and cultivation sites
∧∨ Northern limit of tropical humid forest (species' natural habitat)
∧∨ Cameroon frontiers
▦ Altitude >800 m (thoretically beyond tolerance range for species)

Source: Mbile *et al*, 2003.

Figure 4. Redistribution through use of an indigenous species: Case of *Dacryodes edulis* in Cameroon.

Farmer-to-farmer transfer of these domesticated, reproductive material of high-value fruit and medicinal trees, often from the natural forest habitats within agricultural landscapes (farms, agroforests and home gardens) affects the overall biodiversity of the species and has been described in Mbile et al (2003). These dynamics are being recognized as a cheap means of transferring and redistributing plant genetic materials around landscapes. Figure 4 illustrates the re-distribution of *Dacryodes edulis* in Camerooon (based on data from Germplasm collection sites and cultivation status surveys), despite the apparently limited temperature and precipitation range of its 'natural habitat'.

3. AGROFOREST DEVELOPMENT AND TREE PRODUCTS ASSESSMENTS

The World Agroforestry Centre Africa humid tropics, is putting a high premium on the in-situ and ex-situ management of plant diversity within what is generally termed agroforests. All trees produced in nurseries or acquired through purchase, exchange or indirectly from other partners, are integrated into existing and/or new systems. Such systems increasingly include fallows, secondary, degraded or community or communal forests. These increasingly complex agroforests serve both economic and ecological functions.

This potential sustainability of agroforests – both traditional and enhanced ones, is based on their flexibility both economically (e.g., flexible demand on labour) and ecologically (e.g., high diversity of species, different harvest periods for products) and the adaptation of products to a regional and international market while providing household needs. Agroforestry tree domestication technologies thus have the potential to make traditional as well as novel agroforest development more systematic, deliberate, productive and a major contributor to local and regional economic growth and environmental management in the African humid tropics. Among strategies employed by ICRAF to build knowledge and be able to improve the usefulness of these systems are on-farm biodiversity assessments and off-farm agroforestry products assessments.

'On Farm' Biodiversity Assessments

Socio-economic studies have stressed the importance of tree products to rural households (Ndoye, et al 1998). Despite this, few studies have focused on actual numbers and densities of trees in the different land use systems (cocoa, coffee, fallows, food crop fields, home gardens, etc) and there is little information on farmers' tree planting strategies in forest areas. ICRAF-AHT has examined the diversity, number and density of trees in relation to farm size, land use system, land tenure, proximity to the forest, market access and some household characteristics, such as well-being, age, sex and education of the head of household. These studies aim at understanding farmers' tree planting decision-making and practices (Degrande *et al.* 2006) and reveal that, land use systems, tenure and farm size seem to be the major factors, but off-farm availability of the resource, market access and household characteristics equally play a role in tree planting decision making. In general, the results show a high level of complexity and many factors are highly interrelated.

'Off-farm' Assessments

As part of ICRAF's integrated approach towards improving livelihoods and ensuring environmental management through the use and management of tree biodiversity as part of tree domestication technologies, the characterization of woodlands in agricultural landscapes and the assessments of trees and tree products in community and communal forests are being intensified. Both efforts are integrating indigenous knowledge systems into new technologies like geographic information systems (GIS) in order to establish a sound basis for the development of adaptable tree domestication approaches that improve the value of forests (ecologically) and also improve livelihoods (economically) by balancing use, knowledge and conservation of the intra and inter-specific diversity of multipurpose trees. ICRAF's current activities in the forest margins of Cameroon and in the Dja conservation site in eastern Cameroon, for instance, both seek to understand the implications of increasing the density and diversity of indigenous trees in woodlands, evaluate the value of forests in agricultural landscapes and develop local-knowledge-based methods for improving the biometric rigour of assessing agroforestry trees and tree products in community and communal forests (P Mbile et al, 2005). This research work is playing a major role in ICRAF's regional contribution in both the in-situ management of plant biodiversity for livelihood as well contribute towards sustainable management of community forests by local people at landscape or eco-regional scale.

4. INCREASING BENEFITS AND ADDING VALUE TO DIVERSE PLANT RESOURCES

The use of plant resources in west and central Africa goes beyond food and fruits. Plant medicines for instance have and continue to play an important role in local health care systems in the region. Furthermore, gains through local marketing of plant parts: foods, fruit, medicines for example, only succeeds in capturing a small proportion of the potential benefits that their marketing can bring to local communities. The World Agroforestry Center has thus embarked on extensive documentation of medicinal plants, their medicinal uses, propagation methods, systematic conservation through cultivation and the development of appropriate processing technologies and marketing strategies for both these medicinal plants and for other fruit and culinary products.

Medicinal Plants Development

Medical records from the Health Ministry in Cameroon from 1998 to date highlight malaria, typhoid, jaundice, sexually transmitted diseases (STDs), tuberculosis and stomach problems (such as diarrhea, dysentery, amoebae), as the most recurrent causes of morbidity (Facheux et al., 2003). Evidence from studies carried-out by ICRAF indicates that most of the key medicinal plants sold in the markets and or used by local practitioners on their patients, treat or provide the same remedy for the same diseases. ICRAF's interest in the conservation through cultivation and management of a modest list of medicinal plant species namely; *Anickia chlorantha, Baillonella toxisperma, Zanthoxylum gillettii, Pausinystalia yohimbe* and *Prunus africana* is therefore in response to their extensive local uses and good marketing potential at local and regional levels. ICRAF's work also reveals that the distribution network of medicinal plants as illustrated in Figure 5 below can be very complex at local, national and at international levels crossing borders and exchanging hands without any form of formal regulation. Although a National institute for medicinal plants was created in Cameroon in the mid 80s, regulation leaves much to be desired in the present economic climate.

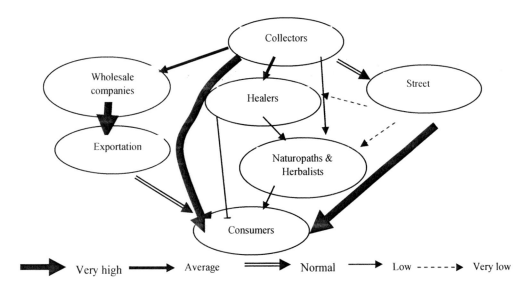

Figure 5. Medicinal plants general distribution channels (Source: Facheux *et al*, 2003).

Marketing

A large part of current data on trade in agroforestry tree/plants and their products for food and medicines are collected and synthesized at macro levels (FAO, 2003). While these provide indication of the economic potential of the sector at a higher level it does little in terms of the evaluation of productivity at local/regional levels, and of the contribution that the marketing of plants and their products make at the same levels. ICRAF has carried-out research to better understand the structure, conduct and performance of markets for agroforestry tree products in four countries of the African humid tropics (Cameroon, Nigeria, Gabon and Equatorial Guinea). Quantitative assessments for some high-value tree/plant products are under way at least for Cameroon to evaluate *in situ* productivity at local and regional levels. Though validation studies for the rapid reconnaissance market studies are being designed, preliminary results however, indicate that the markets are unstructured with little specialization, and enormous scope for rationalization. Results from the surveys indicate that farmers receive much less from the sales of forest products in rural areas as a result of poor market knowledge, perishability of products, seasonality of markets and products, and poor marketing skills.

Within the framework of a group marketing programme supported by the Belgian Directorate for International Development, ICRAF is in the region researching and developing mechanisms that help foster and organize the marketing of tree/plant products at the local and national levels as well as for export by groups of local producers. The main purpose is to enable producers exercise greater control over marketing activities, add value to the products through adaptable processing and transformation methods and increase their overall benefits from these activities. Current indications are that as methods for adding value to current products become more easily accessible, similar quantities of tree/plant and their products are likely to fetch better prices as opposed to past experiences, and producers can free-up time to deal with other tasks as well as reduce pressure on existing stocks under improved regulatory conditions. There is thus the likelihood for greater efficiency (less wastage) in resources management.

5. Some Institutional and Policy/Programme linkages

National/Regional Policy/ Programmes	Policy/Programme sector	Contribution of ICRAF's tree domestication (TD)	Specific on-going domestication approaches at ICRAF.
Cameroon National Forestry Action Plan (NFAP) (1995).	Programme of action: sub-sector 3 on non-timber forest products.	Increases the overall contribution of forest production to the GDP while conserving the productive potential.	- plant multiplication. - plant improvement - on-off farm assessments - value addition and marketing. - conservation through use.-
Cameroon National Environmental Management Plan (1996).	Develop economic landscapes around protected areas to ensure social and economic well being of local populations.	Contribute to sustainable management of community and communal forests around protected areas and in economic landscapes.	- in-situ assessments of AFTPs[3] in community forests (CF). - transformation and marketing of AFTPs - plant multiplication and selective enrichment - strengthening technical forestry aspects of SMP[4] of CF. - researching equity in AFTP use.
CORAF/WECARD agenda.	To Promote Biotechnology development, conservation/management of genetic resources and empowerment of NARs in post war countries.	Improvement of plant biodiversity (PGR) through selection, collection, propagation strategies. Conservation and management of PGR through use. Capacity-building for TD in RDC.	- range-wide plant germplasm collection activities. - selective and needs based cloning of plants/trees. - conservation/management through use on/off farm and in genebanks. - training and dissemination in RDC.
The CBD-ABS[5]	Equitable sharing of the benefits arising from the use of genetic resources	Increase access of local communities including vulnerable groups to plant biodiversity on and off farm.	- on-farm capacity building for multiplication and integration of plants. - off-farm assessments and enrichment of CF with plants. - researching equity aspects in AFTPs in CF. - improved marketing (processing/transformation) of plants and plant products) to add value and increase revenue

[3] Agroforestry Tree Products.
[4] Simple Management Plan.
[5] Access and Benefits Sharing mechanism of the Convention on Biological Diversity.

CONCLUSIONS

The participatory indigenous trees domestication activities of the World Agroforestry Centre have demonstrated considerable versatility, particularly in the productive management and conservation of biological diversity through the use of trees and tree systems as a potential strategy to provide important livelihood and environmental services. The main achievements during these formative years of technology development, adaptation, and local community involvement, demonstrate that participatory tree domestication will and can play an important role in the sustainable management and use of plant biodiversity. Nevertheless, a bold vision is required which should unite the continued development of the regenerative capacity of trees and trees systems, maintaining and managing biodiversity, and linking local people, tree products and markets as well as influencing the development of appropriate policy mechanisms which motivate people to conserve, enhancing markets and communication and dissemination of new scientific knowledge. As important, should also be the development and support for regional centres, networks and initiatives that facilitate exchange of both experiences as well as plant materials so that redistribution of materials and knowledge from areas of advantage to those of disadvantage can be a driving force behind the use of plant biodiversity to not only reduce poverty today but ensure nutritional and health security in particular, as well as environmental resilience at landscape scale, well into the future.

ACKNOWLEDGMENTS

The World Agroforestry Centre (ICRAF), Africa Humid Tropics tree domestication team wish to heartily thank the ICRAF Sahel Team for their contributions in editing this document. Special thanks also go to the International Fund for Agricultural Development (IFAD) and the UK Department for International Development providing financing during these formative years of the tree domestication programme. Thanks also go to the Belgian Government for is support to the marketing aspects of the project. Finally, thanks also go to the national research centres, civil society sectors and collaborating institutions in Cameroon, Nigeria, Equatorial Guinea, Gabon and Congo for their various support.

REFERENCES

Anegbeh, P. O., Usoro, C., Ukafor., Tchoundjeu, Z., Leakey, R. R. B. and Schreckenberg, K 2003. Domestication of Irvingia gabonensis: 3. Phenotypic variation of fruits and kernels in a Nigerian Village. *Agroforestry Systems* 8 (3) 213-218.

CARPE, 2001. *Phase 1 Results and Lessons Learnt* (1996-200). Taking action to Manage and Conserve Forest Resources in the Congo Basin. BSP. Washington DC.

Cunningham, M., Cunningham, A.B. and Schippmann, U. 1997. *Trade in Prunus africana and the implementation of CITES. Bundesamt* für Naturschutz, Bonn, Germany.

Degrande A., Schreckenberg K., Mbosso C., Anegbeh P., Okafor V., Kanmegne J. and Trivedi M., (under review by Forest Trees and Livelihoods). *Driving forces behind levels of fruit tree planting and retentions on farms in the humid forest zone of Cameroon and Nigeria*

Facheux C, Asaah E, Ngo-Mpeck M and Tchoundjeu Z. 2003. Studying markets to identify medicinal species for domestication: The case of *Enantia chlorantha* in Cameroon. *Herbalgram* (60) 38-46.

Falconer, J. 1990. The major significance of minor forest products: The local uses and value of forest in the West African humid forest zone. Community Forestry Note 6. FAO, Rome.

FAO, 2003. Non-Wood forest Products Assessments.

Forman, R.T.T. and Godron, M. 1986. *Landscape Ecology.* John Wiley and Sons. USA.

Franzel S., Jaenicke H. Janssen W., 1996. *Choosing the right tree: Setting priorities for multipurpose tree improvement.* ISNAR Research Report 8. International Service for National Agricultural Research. The Hague, The Netherlands. 87 p.

Hudson, W.E. (ed.). 1991: *Landscape linkages and biodiversity.* Island Press, Washington, D.C.

Leakey R.B., Tchoundjeu Z, Smith R.I, Munro R.C, Fondoun J.M, Kengue J, Anegbeh P, Atangana A, Waruhiu A. N, Asaah E, Usoro C, Ukafor V. 2004. Evidence that Subsistence farmers have domestication indigenous fruits *(Dacryodes edulis and Irvingia gabonensis)* in Cameroon and Nigeria *Agroforestry Systems,* 60(2): 101-111.

Mbile P, Tchoundjeu Z, Popoola L, Nchoutboube, J, G.Abanda. (in press) *Community-based stock assessment and monitoring system (CB-SAMS) for non-wood forest products in community forests in Cameroon.* Proceedings of the XXII IUFRO World Congress: Forest in the Balance: Linking Tradition and Technology. 8-13 August, 2005, Brisbane, Australia.

Leakey, R.R.B. 1999. Potential for novel food products from agroforestry trees. A review. *Food chemistry, 66,* 1-14.

Leakey, R.R.B. and Newton, A.C. 1994. Tropical trees: Potentials for domestication and the rebuilding of forest resources. Proceeding of ITE (Institute of Terrestrial Ecology) Symposium no 29 and ECTF (Edinburgh Centre for Tropical Forest) symposium no 1. Heriot-Watt University, Edingburgh.

Mbile P, Tchoundjeu Z, Degrande A, Asaah E and R Nkuinkeu. (2003.) Mapping the biodiversity of "Cinderella" trees in Cameroon. *Biodiversity* http://www.tc-biodiversity.org 4 (2) 17-21.

Ndoye, O., Eyebe, A. J. and Ruiz Perez, M. 1998. NTFP markets and potential forest resource degradation in Central Africa. *The role of research for a balance between welfare improvement and forest conservation.* In: proceeding of the conference on non-wood forest products for Central Africa. Edited by T. Sunderland and L. Clarke. Limbe Botanic Garden, Cameroon.

Njib Ntep.1999. Rapport national sur le secteur forestier. ONADEF. République du Cameroun.

Okafor, J.C.1991. Amélioration des essences forestières donnant des produits comestibles. *Unasylva,* 42(165),1-10.

Simons AJ, MacQueen DJ and Stewart JL. 1994. *Strategic concepts in domestication of non-industrial trees. In:* Leakey RRB and Newton AC, eds. *Tropical trees:* The potential for domestication and the rebuilding of forest resources. London, UK: HMSO. P91-102.

Tabuna H., 1999. Le marché des produits forestiers non-ligneux de l'Afrique centrale en France et en Belgique, papier occasionnel CIFOR, No 19, 31p.

Tchoundjeu Z., Avana M.L., Leakey, R.R.B., Simons A.J., Asaah E., Duguma B. and Bell J. M. 2002. - Vegetative propagation of *Prunus africana*: Effects of rooting medium, auxin concentrations and leaf area. *Agroforestry System,* 54 (3), 183 - 192.

Tchoundjeu, Z., Duguma, B. Tientcheu, M.L. and Ngo Mpeck, M.L. 1999. Domestication of indigenous agroforestry tree: ICRAF's Strategy in the humid tropics of west and Central Africa. Proceeding of the conference on Non-wood Forest Products for Central Africa. Edited by T. Sunderland and L. Clarke. Limbe.

The World Bank, World Development Indicators database, April 2002. www.devdata. worldbank.org accessed 04/04/03.

Uhl, L, R. Buschbacher and Serrao, E.A.S. 1988: Abandoned pastures in eastern Amazonia, I. Patterns of plant. Succession. *J. Ecol.* 76: pp. 663-681.

Wilcoe, D. S, McLellan, C. H. and Dobson, A. P. 1986: *Habitat fragmentation in the temperate zone.* In M. E. Soule (ed). Conservation Biology; The science of scarcity and diversity. Sinauer Assoc., Sunderland, Mass.:pp.273-286.

Wilcoe, D. S, C. H. McLellan and A. P. Dobson, 1986: *Habitat fragmentation in the temperate zone.* In M. E. Soule (ed). Conservation Biology; The science of scarcity and diversity. Sinauer Assoc., Sunderland, Mass.:pp.273-286.

In: Progress in Environmental Research
Editor: Irma C. Willis, pp. 263-278

ISBN 978-1-60021-618-3
© 2007 Nova Science Publishers, Inc.

Chapter 9

EPIDEMIOLOGY AND ENVIRONMENTAL POLLUTION: A LESSON FROM YOKKAICHI ASTHMA, JAPAN

*Katsumi Yoshida, Kunimasa Morio and Kazuhito Yokoyama**

Department of Public Health and Occupational Medicine, Mie University Graduate
School of Medicine, Tsu-shi, Mie 514-8507, Japan

ABSTRACT

Serious environmental problems occurred in the process of rapid growth of economy after World War II in Japan, including three water pollution (Kumamoto Minamata Disease, Niigata Minamata Disease, Itai-Itai Disease) and one air pollution (Yokkaichi Asthma), which had been called the Four Major Pollution in Japan.

Attention to Yokkaichi Asthma came from increasing patients of chronic obstructive pulmonary disease, such as bronchial asthma and chronic bronchitis, caused by very high concentration of SO_2 in the air. This problem occurred since around 1957, and was continued for about 20 years. By regulation of total emission of sulfur oxides from 1972, SO_2 concentration decreased greatly and the pollution problem was solved. In this article, many precious experiences in Yokkaichi Asthma are described.

1. INTRODUCTION

Japan lost about 80% of its industrial productivity in World War II by stoppage of import of industrial materials, such as oil, iron and bauxite, by attack of U.S. Navy submarines, and by large-scale bombing on about 200 cities by B29 bomber. The industrial production in 1945 of Japan decreased to the level of Taisho age (around 1910).

---

* Correspondences to: Kazuhito Yokoyama, MD, DMSc. Phone & Fax +81-59-231-5012; kazuhito@doc. medic.mie-u.ac.jp

However, the level of industrial production returned to that of prewar days in 1955. Till the end of the 20th century, annual growth rate of industrial production was exceeding 10~15%, and the period called "high economic growth" continued.

At this stage new industries were created to respond to technical innovations in U.S. or European nations. As a one leading field, large-scale petrochemical complex using Middle East crude oil was established in Yokkaichi, Mie Prefecture.

Since consideration given to environment was lacked in such great expansion of industrial production, serious environmental problems were caused, i.e. Yokkaichi Asthma (chronic obstructive pulmonary disease, COPD, by SO_x pollution), Minamata Disease (methyl mercury poisoning) of Kumamoto and Niigata, and Itai-Itai Disease (cadmium poisoning).

This article describes the discovery, way of aiding the patients, and solution of Yokkaichi Asthma.

2. THE DEVELOPMENT OF YOKKAICHI ASTHMA

Figure 1 shows the outline of oil complex built first in Yokkaichi (Shiohama district). From starting operations of the complex(1957), many patients had come to Shiohama Hospital, affiliated with Mie University Medical School, located in this district, complaining "a cough comes out," "sputum coming out," "a throat being painful," and "it being unable to sleep" by asthma attacks. The problems were also seen in Isozu, Mihama, and Akebono districts which were next to the complex. It was observed for about half a year including winter especially in Isozu district that was the lee of industrial complex.

The unusual frequent occurrence of asthma among the aged 50 years or above in Shiohama district raised questions, i.e., "Was it by chance?," "by unknown cause in this area?" It was thus necessary to conduct an extensive epidemiological investigation.

We investigated the Health Insurance Claim (called receipt) submitted to Yokkaichi City by each medical institution by the National Health Insurance Program [1]. Thus, with consent and assistance of Yokkaichi City, the incidence of illnesses of residents on respiratory diseases was surveyed, in 13 districts of the city.

Figure 2 shows the monthly rate of consultation of asthmatic disease among aged 50 years or above in Hobo (Mie) located far from the industrial complex and Shiohama districts. The rate of consultation showed changing in constant width at approximately around 0.5% level in Hobo. However, in Shiohama, the rate of consultation went up every month and increased 5~6 times at the end of study period.

Figure 3 shows the rate of consultation of asthma by sex and age. The ratio in the aged 50 years or above who lived near the industrial complex was particularly high, i.e. nearly 10 times of those in the control area.

Isozu district with the population of about 2,800, located in south of the industrial complex, showed the highest prevalence in Yokkaichi. Number of cases of asthma was expected to a few for this population size among the middle aged or elderly; only two cases were observed before operation of industrial complex. However, the number of patients reached 66 in 1963, as shown in Table 1, with unusual prevalence in the aged 50 years or above.

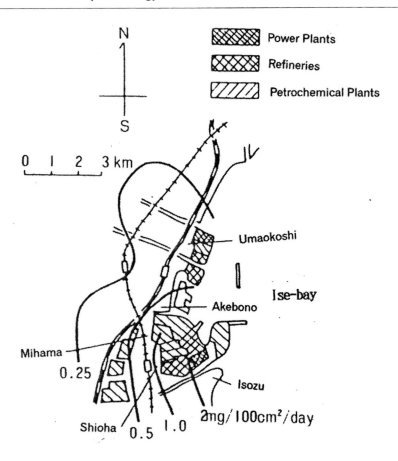

Figure 1. Shiohama oil industrial complex built first in Yokkaichi [1].

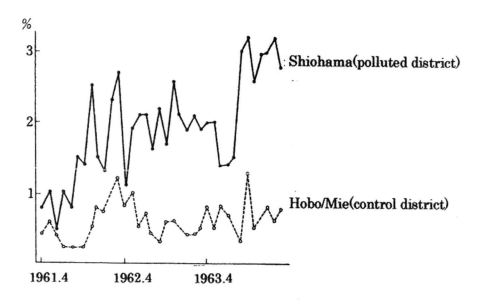

Figure 2. Monthly rate of consultation of asthma [1].

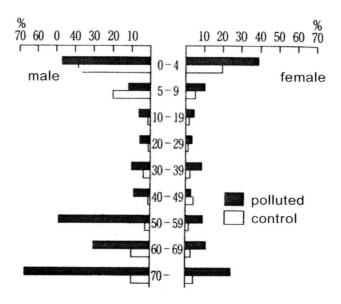

Figure 3. Annual rate of consultation of asthma by sex and age [2].

What was the reason for such high incidence of asthma? An investigation on occupation and life situation of patients revealed no particular causes. However, unusually high concentration of SO_2 was observed in Isozu district, as shown in Figure 4. Eleven observation points of air pollution were established in Yokkaichi; the amount of dust-fall, which was an index of air pollution in Britain, and SO_2 concentration (PbO_2 method) was measured. As shown in Figure 5, a strong "dose-response relationship" was seen between SO_2 concentration and incidence of asthma.

Table 1. Asthma patients in Isozu district (as of 1963) [3].

Age group	Population	Number of patients	Percentage
0-4 years old	261	1	0.38
5～9	276	3	1.09
10～19	509	1	0.20
20～29	612	4	0.65
30～39	467	8	1.71
40～49	260	8	3.08
50～59	210	17	8.10
60～69	148	10	6.76
70～79	47	13	27.66
80 years old or more	25	1	4.00
Total	**2,815**	**66**	**2.34**

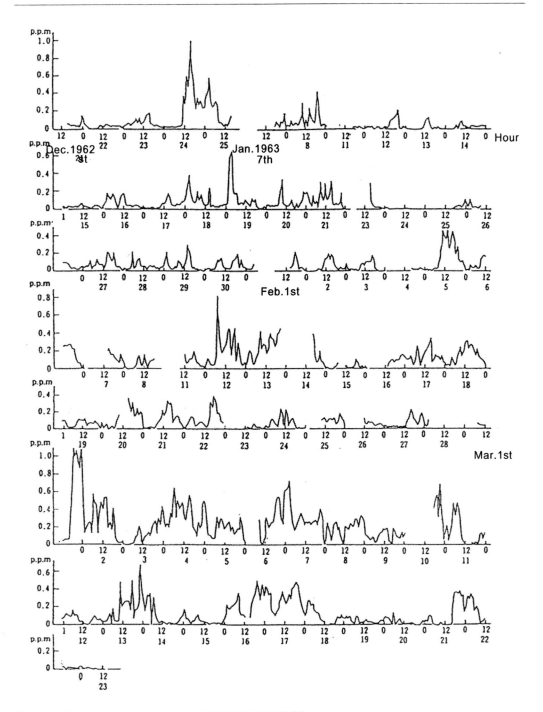

Figure 4. SO$_2$ concentrations in Isozu (1963.12-1964.3) [2].

The Middle East crude oil used in industrial complex had as high as about 4% of sulfur content. Therefore, SO$_2$ concentration around petroleum industrial complex unusually elevated. The total emission of SO$_2$ from the all factories were estimated at 120,000~200,000 t/year.

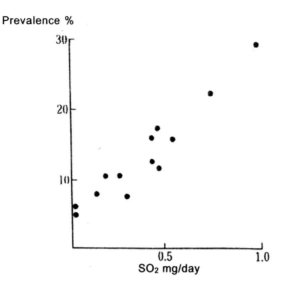

Figure 5. Yearly accumulated prevalences of asthma and sulfur oxides concentrations [1].
The prevalences over 50 years agers of enlistend to National Health Insurrance system in 13 districts
and sulfation-rates(PbO2-method,Mar. 63'-Apr.64')

3. CLINICAL FEATURES OF PATIENTS IN ISOZU DISTRICT

Table 2 shows the medical checkup results of asthma patients in Isozu district. As in this table, intracutaneous allergic reaction was negative for most of the cases. This rate was very low in contrast to the observation that intracutaneous allergic reaction to house dust was positive among 60% of asthmatic patients.

Most of the patients developed the disease less than four years after industrial complex operation, except for two cases with ten years of history and three unknowns. Nevertheless, in many patients, the forced expiratory volume one second percent was depressed gratefully. Patients with pulmonary P in electrocardiogram were seen, suggesting seriously impaired cardiopulmonary function.

Also, only seven cases had history of asthma in relations. On the other hand, the spouse had the same disease for four cases, which may be attributed to an environmental factor such as living in the same house, sharing pollution rather than heredity relations.

Table 3 shows the characteristics of asthma patients found in Isozu. As in Figure 6, asthmatic attacks increased especially during the period when SO_2 concentration was high. Moreover, when patients left Isozu district, they showed the remission.

The clean air rooms with charcoal air filter (98% of rates of SO_2 removal) were placed in Shiohama Hospital aiming at an improvement of clinical conditions of patients [4]. The result was shown in Figure 7.

The patients were divided into two groups. One was suffered from asthma before the operation of industrial complex (i.e., not related to air pollution). The other developed asthma after the operation started. As in Figure 7, effects of air filtration (SO_2 elimination) were remarkable in patients of the later group.

Table 2. Medical checkup results of asthma patients in Isozu district(January 1964) [2].

	Age Gender	Allergic reaction		Eosin in blood	Vital capacity	Forced expiratory volume one second percent (after inhaling alotec)	Views on electrocardiogram	Family history	Phonacoscopy	The number of years after the development of asthma	Peak flow
		Allergic reaction	Tatami matting								
1	52♂	-	-	0	60 (61)	100 (100)	Pulmonary P			4	240
2	73♂	-	-	8	96 (100)	45 (70)		Grandchild	Dry rale	3	187
3	57♀	-	-	3	70 (70)	60 (90)	Pulmonary P	Husband	Rough	3	100
4	58♂	-	-	4	66 (73)	55 (58)		Wife	Rough	3	180
5	72♂	-	-	2	66 (68)	50 (57)				10	186
6	81♀	-	-	6	58 (60)	66 (76)				4	120
7	70♂	-	-	1			Myocardial infarction		Rough	0.5	
8	59♂	-	-	4	95 (95)	67 (86)			Rough	1	400
9	63♂	-	-	0	68 (75)	88 (98)	Pulmonary P Coronary insufficiency		Rough	1	380
10	59♀	-	-	2	75 (82)	58 (58)			Dry rale	2	200
11	54♀	-	-	0	84	53		Husband	Unknown	1	260
12	32♂	-	-	6	91 (96)	60 (73)			Dry rale	1	
13	69♂	-	-	0	87 (100)	58 (64)			Rough	10	180
14	53♂	-	-	2	100 (105)	69 (81)			Rough	1	260
15	76♀	-	-	3	65 (65)	80 (82)		Younger sister	Rough	Unknown	100
16	49♂	+ + -	-	1	78 (78)	87 (94)	Pulmonary P	Wife	Rough	1	360
17	63♀	-	-	2	68 (68)	82 (87)		Grandchild		1	360
18	32♂	-	-	9	105 (105)	81 (83)			Dry rale	2	340
19	12♂	+	-	6				Grandfather	Dry rale	3	160
20	62♂	-	+	1	73 (73)	45 (75)		Grandfather	Dry rale	1	260
21	55♂	-	-	1	73 (80)	72 (89)	Pulmonary P		Rough	1	460
22	76♂	-	-	1	78 (80)	54 (60)				0.2	
23	63♂	+ +	-	7	98 (98)	53 (71)	Pulmonary P	Mother	Rough	0.2	340
24	8♂	-	-	10				Grandfather		5	120
25	61♂	-	-	1	74 (78)	62 (82)	Pulmonary P LAD LVH			3	335
26	37♂	-	-	8	95 (95)	66 (88)	Pulmonary P LAD		Dry rale	1	340
27	75♀	-	-	3				Unknown	Rough	Unknown	
28	54♂	-	±	2	92 (92)	88 (58)				1	270
29	28♀	-	+	5	80 (80)	60 (70)			Dry rale	3	320
30	46♂	-	-	6	78 (78)	66 (87)			Dry rale	3	385
31	68♂		-	6	94 (100)	62 (71)				1	480

Table 3. The characteristics of asthma patients in Isozu[2].

1. The incidence is relatively high among elderly and males.

2. The hereditary history is weak and intracutaneous reaction by house dust is almost negative.

3. There are many patients who smoke heavily (average: 23.1 cigarettes/day), but asthma also occurred with nonsmokers.

4. There is a great frequency of asthma attacks in periods when SO_2 is high.

5. When patients left the district, they began to have clear temporary remission.

6. The disease began to occur relatively suddenly in most cases.

7. Although patients react to a bronchodilator, the reaction is relatively low compared with classic asthma.

8. Leukocytosis is recognized, and patients often show abnormality in a liver function test (serum protein reaction).

9. Asthmatic sputum component is not recognized, and the degree of appearance degree of eosinophilic cell is low.

10. About one-thirds of patients showed pulmonary P in electrocardiogram (EKG), and had a tendency of moving to P wave in a relatively short period after the development of asthma.

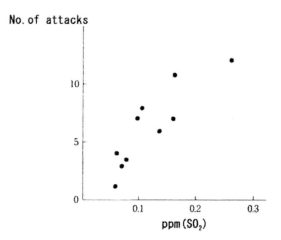

Figure 6. Number of asthmatic attacks and average weekly SO$_2$ concentrations in Isozu (13 cases, Jan.-Mar.1963) [1].

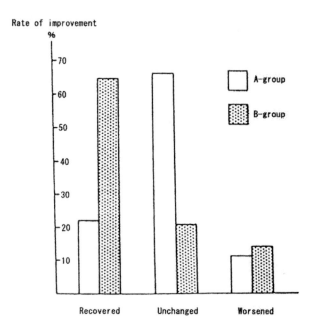

Figure 7. Pulmonary symptoms of the patients hospitalized in the clean air room[2].
A-group: Patients who suffered from asthma before the petroleum-complex started its operation
B-group: Patients who developed asthma after the operation started

4. INCREASE OF CHRONIC BRONCHITIS

Thus, bronchial asthma was increased by air pollution. The health problem was not only bronchial asthma but also chronic bronchitis characterized mainly by stubborn cough and sputum. In Japan, it is warm as compared with Britain and Northern Europe; chronic bronchitis was less seen. Thus, only few clinicians were concerned with this problem.

After finding the problem of air pollution in Yokkaichi, chronic bronchitis had been worthy of noticing. In Yokkaichi and Osaka, prevalence of chronic bronchitis was surveyed by the Ministry of Health and Welfare using British Medical Research Council's Committee on the Etiology of Chronic Bronchitis-questionnaires (BMRC).

Figure 8 shows the prevalence of chronic bronchitis by BMRC questionnaires. In Yokkaichi, differences in prevalence of chronic bronchitis between polluted and non-polluted areas were greater than those between polluted (Nishi-Yodogawa-ku) and non-polluted areas in Osaka.

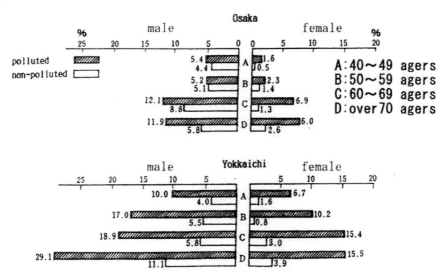

Figure 8. Prevalences of chronic bronchitis by BMRC-questionnaires[2].

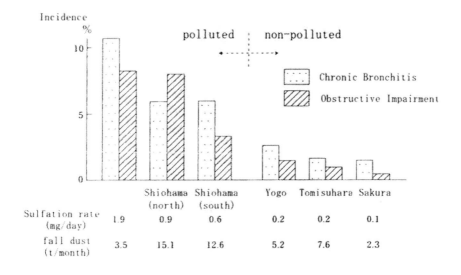

Figure 9. Health effects of air pollution estimated by BMRC-questionnaires and pulmonary function test[5].

Figure 9 shows the adverse health effects of air pollution evaluated by the BMRC questionnaires and pulmonary function test. They were increased with SO2 pollution.

5. INAUGURATION OF MEDICAL AID PROGRAMS

The Medical Aid Program for air pollution was established by Yokkaichi City for the first time in the world. Later, this program was expanded to the national one, resulting in "Pollution-Related Health Damage Special Measures Law."

Installation of the program was due to an increase of asthmatic and chronic bronchitis patients, who needed assistance for medical expenses. However, bronchial asthma and chronic bronchitis were the diseases which had existed since ancient times. It became a matter how their relations to air pollution were authorized.

It was a fact that there was an excessive out break of asthma in air pollution area in Yokkaichi. The distinction between these asthmas and classical ones by clinical conditions was impossible. Asthma was common disease in contrast to Minamata Disease, which was defined as methyl mercurially poisoning. Therefore, it was unable to make a specified standard for diagnosis.

Then, "Law Concerning Medical Service for the Atomic Bomb Exposed" was considered as a good guide to the program for Yokkaichi cases. Patients with cancer such as leukemia increased greatly in the atomic-bombed areas, Hiroshima and Nagasaki, in Japan. Leukemia and other cancers were seen before the exposure to atomic bomb; clinical conditions were not different between ordinal leukemia and atomic-bomb induced one. The situation was similar to the relation between air pollution and Yokkaichi asthma.

In "Law Concerning Medical Service for the Atomic Bomb Exposed," the areas where leukemia occurred excessively were specified; medical aid program was taken to the patients within these areas. Thus, medical expenses of patients who met the following three criterions were paid by the program for Yokkaichi cases.

1) With Specified Disease: Diseases (bronchial asthma, chronic bronchitis, pulmonary emphysema, and their complication), of which excessive occurrence in the polluted area had been confirmed epidemiologically.

2) In Specified Area: Areas where prevalences of the specified diseases had been increased.

3) During Specified Period: Three years of residence in the specified area (prevalence of the above diseases reached to the maximum, 10~15%, in populations 3 years after their migration to the specified areas).

Yokkaichi City decided that those who met the above 3 conditions should be recognized as the certified patients, and asked Ministry of Health and Welfare for the aid of medical expenses. However, it was not approved; Yokkaichi City had to carry out the program since June, 1965, by them.

For about one year later, public opinion against air pollution changed greatly. Ministry of Health and Welfare agreed to make payment as experimental and research expenses to the program of Yokkaichi City. Two years after, the "Pollution-Related Health Damage Special Measures Law" was enacted and enforced (1969).

6. YOKKAICHI POLLUTION SUIT

Medical aid programs for the patients were invoked as mentioned. However, remedy remained as a social problem for patients of bronchial asthma in the middle aged and elderly with impaired lung function. Improvement of clinical conditions could not be expected under the situation where air pollution was continued. Recurrent asthmatic attacks caused numbers of shocking incidents of suicide. As a result, a great number of people insisted that a powerful social movement was necessary for the elimination of air pollution.

Citizens' campaign raised a lawsuit against Yokkaichi pollution in September 1967 for the patient, who had been suffered from serious damage. They also wished to perform vastly and radical reform of the measure against air pollution. Nine patients of Isozu district agreed to become the plaintiff and the complaint was submitted to the Tsu District Court Yokkaichi branch.

The lawsuit was taken based on Article 719 (joint tort) and Article 709 (illegal act) of Civil Code. Here, two major issues were raised. One is the legal causal relationship between air pollution mainly due to SO_2 (possible cause) and the infringement of the right of health (having become sick) in 709 articles.

Yokkaichi City paid half of the medical bill of the patients certified according to the above-mentioned three criterions. The city carried out resident protection rather than investigation of the cause of disease. The causal relationship between the self-inflicted causing (development of air pollution) and the infringement (development of disease) in Article 709 of Civil Code needed arguments. The dispute in medical and law fields started on the ability of epidemiological causal relationship as legal causal relationship in Article 709 of Civil Code.

The second issue was that exhaust emission from the six companies (power plant which used petroleum with high sulfur content, petroleum-refining places, and petrochemical factories) were mixed, polluting Isozu and other districts. Therefore, it was impossible to relate each smokestack's exhaust emission to each patient. It was an issue whether this could be recognized as a joint tort determined by Article 719 of Civil Code.

These issues had hardly been discussed though they were fundamental about the responsibility for air pollution. Whether the opinion by the plaintiff accepted was important; this might affect measures against air pollution, especially environmental pollution control by the Government. Output of air pollutant was scarcely controlled by Air Pollution Control Law at that time.

The causal relationship which was argued in the Yokkaichi lawsuit attracted attentions. A great number of medical scientists and the Civil Code scholars concerned this matter and many editorials were published in mass communications and law magazines.

In such situation, Yoshida wrote a paper entitled "Epidemiological Causality and Legal Causality" in "Jurist", which was the most leading law magazine in Japan [6]. He insisted that the epidemiological causality satisfied causation of Article 709 of Civil Code enough. This gained popularity support of the Civil Code scholars. Judgment was given based on the epidemiological causal relationship; the plaintiff (patient side) won the case. The defendants (company side) obeyed this judgment and gave up the appeal.

7. IMPLEMENTATION OF TOTAL EMISSION CONTROL

The Yokkaichi pollution suit attracted much attention in Japan; nearness the trial finished, winning the case of a complainant was expected mostly. The conclusion of trial also led to change of measures against air pollution in Yokkaichi.

Mie Prefecture established the project team with Yoshida, director of Mie Pollution Control Institute (Public Health and Environment Research Division, Mie Prefecture Science and Technology Promotion Center, at present), as the person in charge to improve the air pollution.

Until then, only building high smokestacks had been the measure against air pollution. There were no essential measures for eliminating contaminations (e.g. desulfurization, and changing of fuel). High smokestacks decreased concentration of SO_2 near the industrial complex. However, they spread SO_2 in wide areas; pollution was increased in areas far from huge plants (industrial complex) because pollutants of high concentrations were combined.

To find the way out from this problem, the only possible solution was control of total SO_2 emission. It was necessary to cut down the total emission to reduce the environmental level lower than an acceptable limit which was the level by which no excessive COPD incidence was expected based on epidemiological survey (annual average of 0.017 ppm over the whole area in Yokkaichi City).

However, there was a problem of desulphurization, which was unsolved to practical use. The cost of desulphurization was higher than building high smokestacks. Moreover, there were about 400 smokestacks with varying height which had different amounts of emission in Yokkaichi district. Therefore, in order to cut down the emission from each factory to the proper output at fair burden, the influence of each smokestack had to be estimated as exactly as possible. It was necessary to repeat the air diffusion simulation to calculate emission reductions and to decide the permission of output. Whether the simulation was possible was a problem with technology of those days.

In order to carry out such simulation, there were two procedures. One was the repeating wind tunnel experiments, and the other was the air diffusion numerical calculation by computer. Because the former had a problem of facilities supply (Mitsubishi Shipbuilding of Nagasaki was the only institution which had a large-sized wind tunnel) and limitation to the experiment number of times, the latter was the only way in fact. Nonetheless, it was a problem whether calculated and actual values conform in the Yokkaichi district.

A trial calculation was done in Kyoto University Data Processing Center which was founded as a joint use facility among universities in Kansai area. As the result was seen useful, a decision was made to carry out the regulation of total emission using this method and to prepare the prefectural regulations.

In order to actually perform regulation, it was necessary to investigate the propriety of various types of regulation systems by repeated survey on the emission conditions of each smokestack and diffusion simulation based on the survey. Eight specialists were assigned to carry out it; regulation of total emission to each rise industrial source was started. The total emission control was started in Japan in 1972. This resulted in a great change to the air pollution of Yokkaichi. It was the first trial in the whole country.

Figure 10 shows year-on-year change in SO_x emission and fuel consumption and in SO_2 concentration. SO_2 concentrations at monitoring stations fell greatly with because of emission of sulfur oxides. Reduction of air pollution had reached the goal.

The goal of prefecture regulations was to reduce about 80% of the total emission. The imposition for compensation to each company was decided according to the emission of SO_2 by "Pollution-related Health Damage Compensation Law" enforced after the Yokkaichi judgment. The imposition was higher than the old "Pollution-Health Damage Compensation Law" which had been already abolished. Measures against air pollution, such as desulfurization of heavy oil and exhaust emissions, shift of fuel (to natural gas from heavy oil), and energy saving to reduce emission of pollutants by reduction in fuel consumption, were taken because of both the total emission control and the imposition system. A plant-and-equipment investment to measure against air pollution was increased. Figure 11 shows the sulfur oxides emission reductions and its means in Japan.

The project team submitted a proposal to the governor, asking for investigation on incidence of respiratory diseases in Yokkaichi, by the survey on the Health Insurance Claim (receipt) to clarify the effects of total emission control. The investigation was continued for about 15 years, and demonstrated that the new incidence of COPD, such as bronchial asthma and chronic bronchitis, fell with decrease of pollution. As shown in Figure 12, there were no significant differences in incidence of bronchial asthma and chronic bronchitis between the polluted and non-polluted areas since 1979, showing no excessive incidence in Yokkaichi [7].

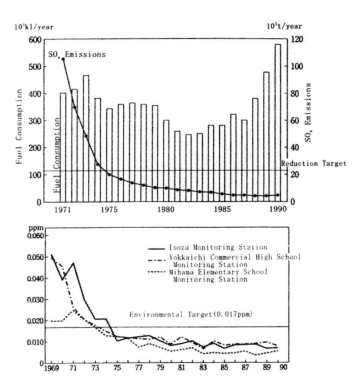

Figure 10. SO_x emission and fuel consumption (upper graph) and SO_2 concentrations (lower graph) [2].

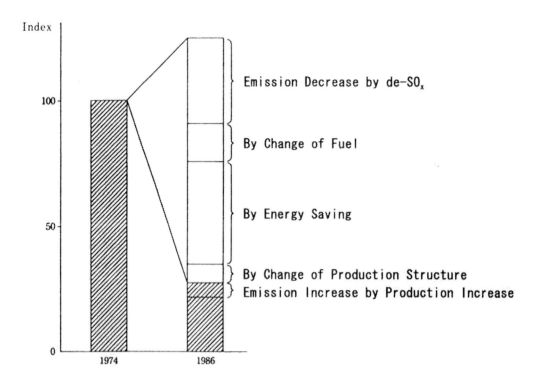

Figure 11. Sulfur oxides emission reductions and its means (in Japan) [2].

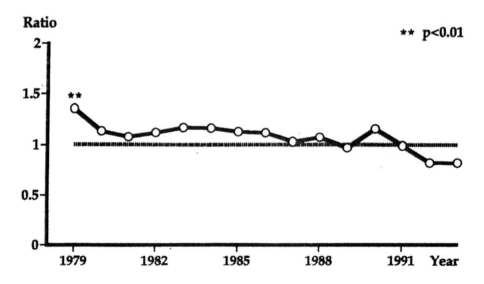

Figure 12. Ratio of annual incidences of chronic obstructive pulmonary diseases between the polluted and non-polluted districts[2].

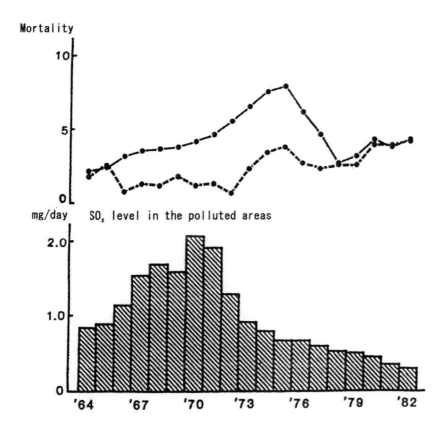

Figure 13. SO$_x$ and mortality from bronchial asthma and chronic bronchitis (3-year running average) [8].

Table 4. Changing of the mortality of bronchial asthma in youth(less than 20 years old) in Yokkaichi

Polluted Area	Mortality of Asthma
1963~1966	2.05
1967~1970	4.36
1971~1974	1.61
1975~1978	0.94
1979~1982	0
Non-polluted Area	**0.44**

Figure 13 shows the transition of SO$_x$ and mortality due to bronchial asthma and chronic bronchitis. By falling of pollution, with a delay of 3~4 years, the mortality rate of chronic bronchitis, which had increased every year, was also set to the almost same level as the non-polluted areas [8].

Also, as shown in Table 4, the mortality of bronchial asthma greatly decreased, which had been increased in youth with pollution.

Regulation of total emission started by Mie Prefecture was included into amendment of Air Pollution Control Law, two years after (1972), and carried out in about 40 heavy contaminated areas in Japan specified by the law. Similarly to Yokkaichi, concentration of sulfur oxides fell greatly in these areas; air pollution problems were solved completely in Japan.

CONCLUSION

This article was described the air pollution by starting operations of Japanese greatest petrochemical complex in Yokkaichi, health hazard of residents, and the measures against is problem. The air pollution caused bronchial asthma, chronic bronchitis, and pulmonary emphysema among residents especially the middle-aged and elderly. Intracutaneous allergic reactions were negative in these patients.

Also, when accommodated in the clean air room with charcoal air filter, the patients developed asthma after the operation of industrial complex showed improvement of clinical conditions more greatly than the patients suffered from asthma before the operation.

Mie Prefecture established the environmental quality standard to keep the rate of new development of COPD a normal level for each company. And regulation of total emission which reduced total emission of SO_2 of each company was enacted to reduce source of pollution. As a result, it was normalized that the rate of new development of COPD in polluted areas were decreased the same as natural increase rate in non-polluted areas.

REFERENCES

[1]   Yoshida, K., Oshima, H. & Imai, M.(1966). Air Pollution and Asthma in Yokkaichi. *Arch Environ Health*, 13,763-768.

[2]   Yoshida,K.(2002). *Yokkaichi Kougai,*Tokyo:Kashiwashobou.

[3]   Yoshida, K.(1984). Air Pollution and its Transition in Yokkaichi(review). *Report of the Environmental Science of Mie University,* 9,93-112.

[4]   Kashiwagi, H., Hattori, T., Soejima, K., Inoue, S. & Takahashi, S.(1983). The Studies on Effects of Clean Air Room on Asthmatic Attacks and Bloodgas in the Patients with Asthma caused by Air Pollution. *Report of the Environmental Science of Mie University,* 8,39-48.

[5]   Imai, M., Oshima, H., Takatsuka, Y. & Yoshida, K.(1967). On the Yokkaichi-asthma (Review). *Jap. J. Hyg.* 22(2),323-335.

[6]   Yoshida, K.(1969). Epidemiological Causality and Legal Causality. *Jurist*,440,104.

[7]   Imai, M., Kitabatake, M., Tomida, Y. & Yoshida, K.(1984). Improvement of Sulfur Oxides Air Pollution and its Effects on Health in Yokkaichi. *Mie Medical Journal,* 34(2),139-150.

[8]   Imai, M., Yoshida, K. & Kitabatake, M.(1986). Mortality from Asthma and Chronic Bronchitis Associated with Changes in Sulfur Oxides Air Pollution. *Arch Environ Health,* 41(1),29-35.

In: Progress in Environmental Research
Editor: Irma C. Willis, pp. 279-293

ISBN 978-1-60021-618-3
© 2007 Nova Science Publishers, Inc.

Chapter 10

BIOLOGICAL MONITORING: NATION-WIDE PRACTICE AND FUTURE DIRECTION

Toshihiro Kawamoto[*]

Department of Environmental Health, University of Occupational and Environmental
Health, Kitakyushu, Japan

Osamu Wada

The University of Tokyo, Japan

ABSTRACT

We aim here to review biological monitoring which has been carried out nation-wide in Japan during the last 15 years and its future directions. Biological monitoring is an assessment of overall exposure to chemicals that are present in our environment, including workplaces, through measurement of the appropriate determinants in biological specimens collected from humans. The measurement of lead concentration in blood (Pb-B) and *delta*-aminolevulinic (ALA) acid in urine of workers who handle lead, and urinary metabolites of workers who handle organic solvents (xylene, N,N-dimehyleformamide, styrene, tetrachloroethylene, 1,1,1-trichloroethane, trichloroethylene, toluene, and n-hexane) have been some of the indispensable items of occupational health examinations since 1989. During the last 15 years, the total number of Pb-B and ALA analyses decreased about $10 - 20\%$. On the other hand, annual analyses of urinary metabolites have been constant for the last 15 years (500,000 to 600,000 per year). Among analyses of urinary metabolites of organic solvents, the analyses of total trichlorocompounds and trichloroacetic acid have decreased, where as N-methylformamide and 2,5-hexanedione analyses have increased. The measurement results are classified into one of three categories, that is, distribution 1, 2 or 3. The percentages of distribution 2 (more than 1/3 - 1/2 of the biological exposure index (BEI) but less than BEI) and distribution 3 (over

---

[*] Correspondence to: Toshihiro Kawamoto, M.D., Ph. D., D.A.B.T. Department of Environmental Health, University of Occupational and Environmental Health; 1-1 Iseigaoka, Yahatanishi-ku, Kitakyushu 807-8555, Japan; Tel: +81-93-691-7243; Fax: +81-93-691-9341; E-mail: kawamott@med.uoeh-u.ac.jp

BEI) have deceased for the last 15 years. About 90% of institutes of occupational health examination had the samples analyzed in laboratories specializing in clinical chemistry outside of their institutes (outside orders) in 2004. The increase of outside orders resulted in an oligopoly by a few laboratories specializing in clinical chemistry. The development of new analytical methods has affected the measurement results. Especially, the change of ALA analysis from a colorimetric method to an HPLC method lowered measurement results. Also, Threshold limit value-time weighted average (TLVs-TWA) and occupational exposure levels (OELs) of some organic solvents have been altered during the last 15 years. In spite of such circumstances, the criteria for classifying distribution 1, 2 and 3, which are related to BEIs, have not been alterd because the criteria should be constant in order to compare the measurement results on the same scale over a long-term period. The determinants of occupational exposure should be specific to the occupational exposure itself. Therefore, chemical substances themselves and their metabolites are useful as determinants of biological monitoring in the occupational health field. It has been attempted to expand the number of chemical substances subjected to biological monitoring, and more than 15 determinants of exposure to organic solvents have recently been proposed. In addition, nonspecific early biological health effects which reflect combined exposure or long-term exposure are suitable as determinants of environmental exposure. Hemoglobin adducts, the Comet assay and urinary 8-hydroxyguanine have been reported to have some usefulness for monitoring the exposure to 7, 20 and 9 chemical substances, respectively, in humans.

ABBREVIATIONS

ACGIH	American Conference of Governmental Industrial Hygienists
ALA	*delta*-Aminolevulinic Acid
BEI	Biological Exposure Index
HA	Hippuric acid
HD	2,5-Hexanedione
MA	Mandelic Acid
MHA	Methylhippuric Acid
NMF	*N*-Methylformamide
OEL	Occupational Exposure Level
Pb-B	Lead in Blood
JSOH	Japan Society for Occupational Health
TLV	Threshold Limit Value
TLV-TWA	Threshold Limit Value-Time Weighted Average
TCA	Trichloroacetic Acid
TTC	Total Trichlorocompounds

INTRODUCTION

Biological monitoring is an assessment of overall exposure to chemicals that are present in our environment, including workplaces, through measurement of the appropriate determinants in biological specimens collected from workers or residents.

Partial amendments to the Japanese regulation of the prevention of lead poisoning and Japanese regulation of the prevention of organic solvent poisoninig were made in 1989. As a result the measurement of lead in blood (Pb-B) and *delta*-aminolevulinic acid (ALA) became indispensable items of the occupational health examination for workers who handle lead. Also, the measurement of urinary metaboliteds of workers who handle eight kinds of organic solvents (xylene, *N,N*-dimehyleformamide, styrene, tetrachloroethylene, 1,1,1-trichloroethane, trichloroethylene, toluene, and *n*-hexane) became mandatory. The relationship between chemicals and their determinants are shown in Table 1. Biological monitoring in Japan is now performed nation-wide.

In this review, we summerize the experience of biological monitoring carried during 15 years from 1990 to 2004 in Japan and its future direction

Table 1. Mandated biological monitoring in Japan

Chemical	Sample	Detarminant
Lead	blood	Lead (Pb-B)
Lead	urine	Delta aminolevulinic acid (ALA)
Xylene	urine	Methylhippuric acid (MHA)
N,N-Dimethylformamide	urine	*N*-methylformamide (NMF)
Styrene	urine	Mandelic acid (MA)
Tetrachloroethylene	urine	Trichloroacetic acid (TCA) or Total tichlorocompounds (TTC)
1,1,1-trichlororthane	urine	Trichloroacetic acid (TAC) or Total tichlorocompounds (TTC)
Trichloroethylene	urine	Trichloroacetic acid (TCA) or Total tichlorocompounds (TTC)
Toluene	urine	Hippuric acid (HA)
normal hexane	urine	2,5-Hexanedione (HD)

BIOLOGICAL MONITORING IN JAPAN

1. Number of Measurement Samples

The number of anayses of lead in blood and ALA in urine was around 140,000 each in 1991. The annual numbers of these anayses have decreased about 10 to 20% during the last 15 years (Figure 1). The annual changes of the number of both analyses have been correlated because Pb-B and urinary ALA are measured at the same time as indispensable items in the occupational health examination in workplaces. As workers undergo biological monitoring twice a year, the number of workers who handle lead is calculated to be 55,000 to 60,000 in 2004.

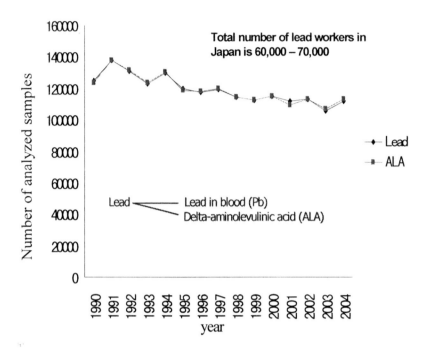

Figure 1. Changes with time of the number of analyzed samples (lead).

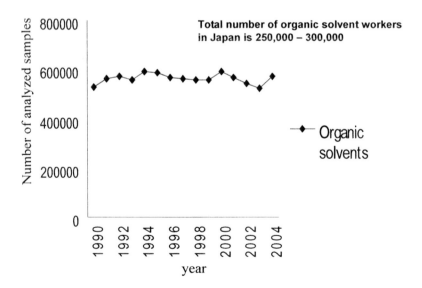

Figure 2. Sequential changes of the number of analyzed samples (organic solvents).

On the other hand, the total number of analyses of the urinary metabolites of eight organic solvents (hippuric acid (HA), methylhippuric acid (MHA), mandelic acid (MA), total trichlorocompounds (TTC), trichloroacetic acid (TCA), *N*-methylformamide (NMF), 2,5-hexanedione (HD)) has been constant (around 500,000 to 600,000) for the last 15 years (Figure 2). Since the biological monitoring of organic solvents has been carried out twice a year, the number of workers who handle the eight kinds of organic solvents was calculated to be 250,000 to 300,000 in the last 15 years. Among the analyses of urinary metabolites, the analyses of TTC and TCA have decreased, while MHA, NMF and HD have increased (Figure3). One of the biggest reasons why TTC and TCA measurements decreased is the phasing-out of the production and consumption of 1,1,1-trichloroetane according to the Montreal Protocol [1] and the restriction of trichloroethylene consumption by The Law Concerning the Protection of the Ozone Layer Through the Control of Specified Substances and Other Measure [2].

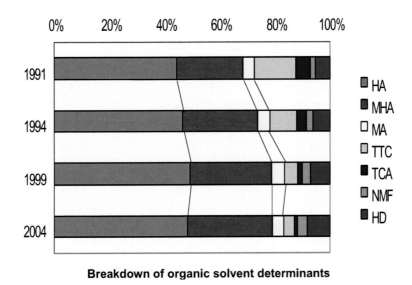

Breakdown of organic solvent determinants

Figure 3. Fraction of samples analyzed for each metabolite in 1991, 1994, 1999 and 2004.

2. Measurement Results

The levels of determinants are classified into one of three categories, that is, distribution 1, 2 or 3 according to values set by the Ministry of Health, Labour and Welfare. Fig. 4 shows the relationship between the concentrations of a chemical to which one is exposed and the level of its biological monitoring determinant. The Biological Exposure Index (BEI) represents the level of a determinant that is most likely to be observed in specimens collected from healthy workers who have been exposed to the chemical to the same extent as workers with inhalation exposure at the Threshold Limit Value (TLV). The criteria for classification into distribution 3 from as distinct from distribution 2 were set as the values corresponding to the BEIs recommended by the American Conference of Governmental Industrial Hygienists (ACGIH) in 1989 except that lead in blood was set at 40 ug/*d*l of blood. The values for classification into distribution 2 as distinct from the distribution 1 were set from 1/3 to 1/2 of the BEIs or the value used for classification of the distribution 3 versus 2 [3]. Determinants and their criteria for classification in the biological monitoring of lead and organic solvents are shown in Table 2.

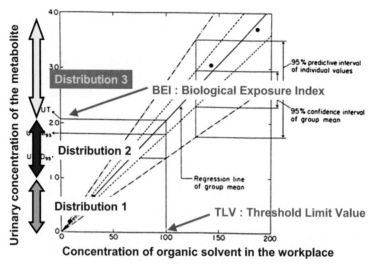

Lines fitted to the means, confidence limits of the means, and predictive limits of individual samples from the groups (UT, 2.05; UTLM₉₅, 1.80; UTLD₉₅, 1.38, g MHA/g creat)

Figure 4. Relationship between exposure-concentration of a chemical and the level of its biological monitoring determinant.

Table 2. Determinants and their criteria for classification in the biological monitoring of lead and organic solvents

Chemical	Determinant	Unit	Distribution		
			1	2	3
Lead	Pb-B	µg/dl	≦ 20	20 - 40	> 40
	Urinary ALA	mg/l	≦ 5	5 - 10	> 10
Xylene	Urinary MHA	g/l	≦ 0.5	0.5 - 1.5	> 1.5
N,N-Dimethylformamide	Urinary NMF	mg/l	≦ 10	10 - 40	> 40
Styrene	Urinary MA	g/l	≦ 0.3	0.3 - 1	> 1
Tetrachloroethylene	Urinary TCA	mg/l	≦ 3	3 - 10	> 10
	Urinary TTC	mg/l	≦ 3	3 - 10	> 10
1,1,1-Trichloroethane	Urinary TCA	mg/l	≦ 3	3 - 10	> 10
	Urinary TTC	mg/l	≦ 10	10 - 40	> 40
Trichloroethylene	Urinary TCA	mg/l	≦ 30	30 - 100	> 100
	Urinary TTC	mg/l	≦ 100	100 - 300	> 300
Toluene	Urinary HA	g/l	≦ 1	1 - 2.5	> 2.5
normal Hexane	Urinary HD	mg/l	≦ 2	2 - 5	> 5

Table 3. Changes in incidences of distribution 1, 2 and 3 in 1991, 1994, 1999 and 2004

Determinant	Year	Distribution1 (%)	Distribution 2 (%)	Distribution 3 (%)	Total
Lead	1991	126,775 (94.0)	6,353 (4.7)	1,724 (1.3)	134,852
	1994	111,812 (94.9)	4,842 (4.1)	1,134 (1.0)	117,788
	1999	108,638 (95.7)	4,069 (3.6)	842 (0.7)	113,549
	2004	108,736 (96.6)	3,186 (2.8)	667 (0.6)	112,589
ALA	1991	127,578 (96.9)	3,889 (3.0)	241 (0.2)	131,708
	1994	115,621 (96.7)	3,677 (3.1)	328 (0.3)	119,626
	1999	111,875 (98.4)	1,526 (1.3)	238 (0.2)	113,639
	2004	112,042 (99.1)	864 (0.8)	142 (0.1)	113,048
MHA	1991	145,687 (98.2)	2,283 (1.5)	350 (0.2)	148,320
	1994	173,393 (98.4)	2,292 (1.3)	438 (0.2)	176,123
	1999	186,293 (98.7)	2,174 (1.2)	254 (0.1)	188,721
	2004	203,754 (98.6)	2,639 (1.3)	171 (0.1)	206,564
NMF	1991	9,990 (96.0)	334 (3.2)	82 (0.8)	10,406
	1994	14,755 (96.6)	361 (2.4)	159 (1.0)	15,275
	1999	17,870 (96.1)	532 (2.9)	185 (1.0)	18,587
	2004	22,986 (96.5)	637 (2.7)	198 (0.8)	23,821
MA	1991	22,367 (87.3)	2,501 (9.8)	760 (3.0)	25,628
	1994	25,099 (90.5)	2,045 (7.4)	592 (2.1)	27,736
	1999	28,733 (93.3)	1,647 (5.3)	412 (1.3)	30,792
	2004	26,816 (94.0)	1,514 (5.3)	188 (0.7)	28,518
HA	1991	250,012 (90.3)	22,294 (8.1)	4,634 (1.7)	276,940
	1994	268,555 (89.5)	26,037 (8.7)	5,536 (1.8)	300,128
	1999	285,004 (90.6)	25,184 (8.0)	4,221 (1.3)	314,409
	2004	299,433 (91.8)	23,080 (7.1)	3,543 (1.1)	326,056
HD	1991	32,439 (98.2)	497 (1.5)	83 (0.3)	33,019
	1994	38,753 (98.7)	406 (1.0)	104 (0.3)	39,263
	1999	44,734 (99.3)	309 (0.7)	26 (0.1)	45,069
	2004	54,716 (99.6)	209 (0.4)	24 (0.0)	54,949

Trends of the measurement results of biological monitoring determinants with time are shown in Table 3. All determinants except hippuric acid (HA) showed a reduced percentage of distributions 2 and 3 from 1991 to 2004. The decrease of distributions 2 and 3 usually indicates that the number of workers with high exposure has decreased. The percentages of distributions 2 and 3 (over 1/3 - 1/2 of BEI) have deceased for the last 15 years. It can be concluded that occupational exposure to lead and the organic solvents tested has decreased after the commencement of nation-wide biological monitoring.

3. Increase of Outside Ordering of Measurements

Figure 5 shows two typical schemes of biological monitoring in Japan. Examiners from institutes for occupational health examination visit workplaces and collect specimens. Some institutes analyze the samples in their own laboratories. The other institutes have samples analyzed in laboratories specializing in clinical chemistry outside of their institutes (outside order). In 1999, more than 20% of the institutes for occupational health examination analyzed lead in blood in their own laboratories. However, this rate was reduced to only 10% 10 years later, when around 90% of the institutes for occupational health examination externalized the analyses (Figure 6).

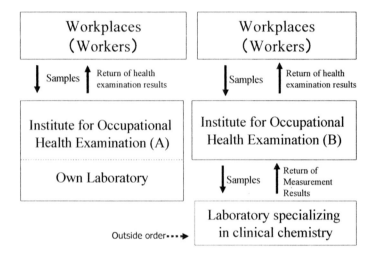

Figure 5. Two typical schemes of biological monitoring in Japan.

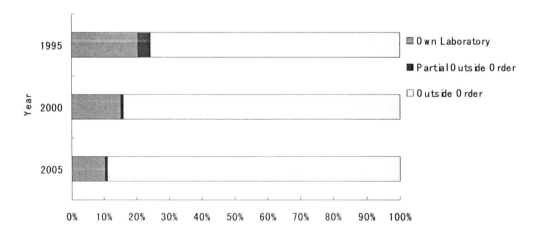

Figure 6. Increase of outside orders (Pb-B measurement).

However, the increase of outside ordering resulted in an oligopoly by a few laboratories specializing in clinical chemistry. Figure 7 shows the rates of outside orders for each determinant. The outside order rates for HD, NMF, TCA and TTC were higher than those of HA, MHA, MA, ALA and Pb-B.

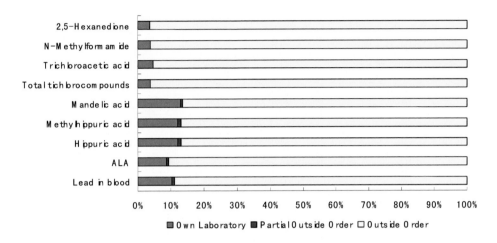

Figure 7. Rate of outside orders for each determinant in 2004.

4. Development of New Analytical Methods

During the past 15 years, new analytical methods have been developed. The biggest change was the alteration of ALA analysis from a colorimetric method to an HPLC method. Only one laboratory specializing in clinical chemistry among seven big laboratories [4] analyzed ALA by HPLC methods in 1990; however, all seven of these laboratories used HPLC methods to analyze ALA in 1999 (Figure 8). This alteration of the analytical method caused a big problem in comparing measurement results. In the colorimetric method which was developed by Tomokuni and Ogata [5], not only ALA but also aminoacetone causes colorimetric changes. However, the HPLC method measures only ALA. As a result, the measurement results obtained by the colorimetric method are always higher than those obtained by the HPLC method. Figure 9 shows the differences of measurement results obtained by the two methods. The BEI for ALA was set according to some field data obtained by the colorimetric method. Therefore, a new BEI for the HPLC method should be set rather than applying the old BEI to evaluate lead exposure. However, a new BEI has not been set for a long time. The number of urine samples with high ALA concentration has decreased drastically during the past 15 years, and the reason seems to be not only improvement of the working environment but also the alteration of the analytical method from the colorimetric one to HPLC.

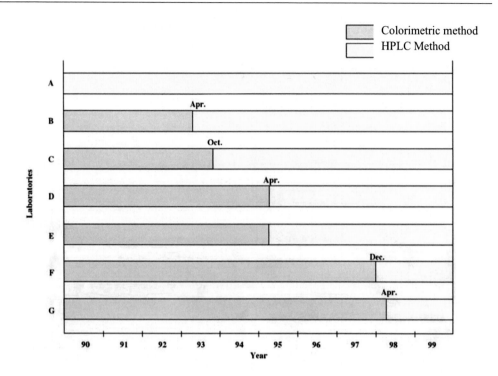

Figure 8. Transition of the analytical method for urinary ALA from a colorimetric method to an HPLC method.

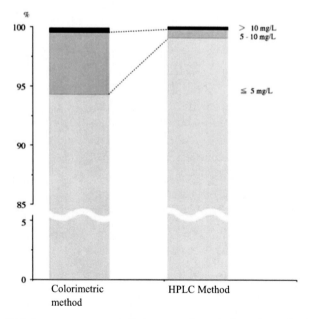

Figure 9. Comparison of ALA measurement results between the colorimetric method and HPLC method.

5. Alteration of TLVs-TWA (ACGIH) and OELs (JSOH)

After starting nation-wide biological monitoring, the TLVs-TWA and Occupational Exposure Levels (OELs) of some oraganic solvents, which are recommended by the ACGIH and the Japan Society for Occupational Health (JSOH), respectively, were altered (Table 4).

Table 4. Decrease of TLVs and OELs

Chemical substances	Standard (organization)	1990		2004	Year of change
Toluene	TLV-TWA (ACGIH)	50 ppm		(No change)	
	OEL (JSOH)	100 ppm	→	50 ppm (135 mg/m3)	('97)
Xylene	TLV-TWA (ACGIH)	100 ppm		(No change)	
	OEL (JSOH)	100 ppm	→	50 ppm (217 mg/m3)	('01)
Trichloroethylene	TLV-TWA (ACGIH)	50 ppm		(No change)	
	OEL (JSOH)	50 ppm	→	25 ppm (135 mg/m3)	('97)
Styrene	TLV-TWA (ACGIH)	50 ppm	→	20 ppm	
	OEL (JSOH)	50 ppm	→	20 ppm (85 mg/m3)	('99)

Although the criteria for classifying distributions 1, 2 and 3 are set according to the values of the BEIs, which usually decrease in conjunction with a decrease of TLVs-TWA, it was decided not to change the criteria for classifying distributions 1, 2 and 3. One reason for this is that the criteria for classifying distributions 1 and 2 are useful even if BEI is decreased. Another reason is that the criteria should be constant in order to compare the measurement results on the same scale over many years. Therefore, distributions 1, 2 and 3 now do not correspond to TLVs-TWA or OELs, but are used to classify groups for statistics.

FUTURE DIRECTION OF BIOLOGICAL MONITORING

There are three major directions for biological monitoring in the future. One is expansion of the chemicals subjected to biological monitoring. The next is countermeasures to overcome the problems in current biological monitoring operations. The last is to expand biological monitoring to environmental exposure.

1. Expansion of the Chemicals Subjecting to Biological Monitoring

The Japan Industrial Safety and Health Association (JISHA : Subordinate Institute of Ministry of Health, Labour and Welfare) reevaluated health examinations for organic solvents from 2002 to 2004 and submitted a report in 2005 [6]. In this report, biological determinants of 15 organic solvents were listed. This review attempted to survey the evidences in published reports which indicate the validity of the determinants.

Table 5. International comparison of biological monitoring operation
(●: Mandated item, ▲: Further examinationitem, ○: BEI is recommended)

PRT R No.	Chemical substances	CAS-No	Korea	Laws (Japan)	JSOH	ACGIH	DFG
15	Aniline	62-53-3				o	o
40	Ethylbenzene	100-41-4				o	
44	Ethyleneglycol monoethylether	110-80-5	choice				o
60	Cadmium and its compounds	-	●	▲		o	
63	Xylene	1330-20-7	●	●		o	o
69	Chromium(VI) compounds	-	choice			o	
93	Chlorobenzene	108-90-7	choice			o	o
99	Vanadium pentaoxide	1314-62-1	choice	▲		o	
100	Cobalt and compounds	-				o	
112	Carbon tetrachloride	56-23-5					o
120	3,3'-Dichloro-4,4'-diaminodiphenylmethane	101-14-4			o		
139	o-dichlorobenzene	95-50-1					o
140	p-dichlorobenzene	106-46-7					o
145	Dichloromenthane	75-09-2	choice			o	
172	N,N-Dimethylformamide	68-12-2	●	●		o	o
175	Mercury and its compounds	-	●	▲	o	o	o
177	Styrene	100-42-5	●	●		o	o
200	Tetrachloroethylene	127-18-4	●	●		o	
209	1,1,1-Trichloroethane	71-55-6	●	●			o
211	Trichloroethylene	79-01-6	●	●	o	o	
227	Toluene	108-88-3	●	●	o	o	o
230	Lead and its compounds	-	●	●	o	o	o
231	Nickel	7440-02-0	choice	▲			
232	Nickel compounds	-		▲			
235	Nitroglycol	628-96-6		▲			
236	Nitroglycerin	55-63-0					o
237	p-Nitorochlorobenzene	100-00-5	choice	▲			
240	Nitrobenzene	98-95-3		▲		o	o
241	Carbon disulfide	75-15-0				o	o
252	Arsenic and its compounds	-	choice	▲			
266	Phenol	108-95-2	Recommend	▲		o	o
283	Hydrogen fluoride	-	choice	▲		o	
294	Beryllium and its compounds	-		▲			
299	Benzene	71-43-2	choice			o	
303	Pentachlorophenol	87-86-5	choice	▲		o	
311	Manganese and its compounds	-	choice	▲			o

Table 5 show an international comparison of biological monitoring. The operation of biological monitoring varies according to the country. In order to set improve biological monitoring, international collaboration is urgently needed.

2. Countermeasures to the Problems of Current Biological Monitoring Operations

Nation-wide biological monitoring in Japan has many problems regarding the operation. The biggest problem is the time of sampling specimens from workers. Table 6 shows biological half-lives of some biological monitoring determinants. Heavy metals have long half-lives and no problems regarding sampling time, that is, we can collect specimens at any time. However, urinary determinants of organic solvents have very short half-lives. A short half-time implies that the appropriate sampling time is limited. For example, it is meaningless to analyze morning urine more than 12 hours after exposure is finished. Actually, some institutes of occupational health examination collect urine samples in the morning (before the work-shift) and thereby disrupt biological monitoring operations in Japan. In order to overcome such disadvantages of urinary metabolites, it is important to develop novel determinants which are independent of sampling time and reflect a relatively long-term exposure period.

Table 6. Biological half-life of each determinant

Deaterminants	Specimen	Half-time
Lead	blood	900 hrs
Cadmium	Urine	20 yrs
Hippuric acid	Urine	1.5 hrs
Metylhippuric acid	Urine	3 hrs
Mandelic acid	Urine	4 hrs
Trichloroacetic acid	Urine	75 hrs

3. Expansion of Biological Monitoring to Environmental Exposure

Environmental exposure is different form occupational exposure, as shown in Table 7. One of the important purposes of biological monitoring of occupational exposure is to clarify whether an occupational factor is involved in a certain health effect in workers. Therefore, biological determinants for occupational exposure need be specific to certain chemicals, i.e., parent chemicals, their metabolites and specific health effects. On the other hand, determinants for environmental exposure should generally reflect nonspecific, combined, and long-term exposure.

A working group was started in 2002 under the Ministry of Environment in order to develop new biological monitoring systems for environmental exposure. The group focused on three determinants, that is, hemoglobin adducts [7], the Comet assay and 8-hydroxyguanine. The results of literature reviews regarding three determinants are shown in Table 8. The validity of the determinants was reported as measure of exposure to some chemicals. However, significant elevations were recognized only in workers.

Table 7. Difference between occupational exposure and environmental exposure

	Occupational Exposure	Environmental Exposure
Number of exposed substances	Limited (1 – some)	Numerous
Targeting substances	Known	Unknown Secondary pollutants (ex. Oxidants)
Exposure time	Known	Unknown
Dose	High	Low

Table 8. Validity of hemoglobin adduct, the Comet assay and 8-hydroxyguanine in human studies (○: Significance has been reported in human studies)

PRTR No.	Chemical substances	CAS No.	Hb adducts	Come assay	8-OHdG
2	Acryamide	79-06-1	○		
7	Acrylonitrile	107-13-1	○	○	
25	Antimony and its compounds	-		○	
26	Asbestos	1332-21-4			○
42	Ethylene oxide	75-21-8	○		
56	1,2-Epoxypropane (Propione oxide)	75-56-9	○		
60	Cadmium and its compounds	-		○	
68	Chromium and chromium(III) compounds	-		○	
69	Chromium(VI) compounds	-			○
77	Chloroethylene (Vinyl chloride)	75-01-4		○	
82	2-Chloro-2',6'-diethy-N-(methoxymethyl)acetanilide (alachlor)	15972-60-8		○	
100	Cobalt and its compounds	-		○	
131	2,4-Dichlorophenoxyacetic acid (2,4-D; 2,4-PA)	94-75-7		○	
175	Mercury and its compounds	-			○
177	Styrene	100-42-5	○	○	○
179	Dioxin and derivatives	-		○	
199	Tetrachloroisophthalonitrile (chlorothalonil; TPN)	1897-45-6		○	
227	Toluene	108-88-3		○	
230	Lead and its compounds	-		○	○
231	Nickel	7440-02-0		○	○
241	Carbon disulfide	75-15-0		○	
252	Arsenic and its compounds	-		○	○
268	1,3-Butadiene	106-99-0	○	○	
299	Benzene	71-43-2		○	○
306	Polychlorinated biphenyl (PCB□	1336-36-3		○	
311	Manganese and its compounds	-			○
329	1-Naphthyl N-methylcarbamate (carbaryl; NAC)	63-25-2		○	
340	4,4-Methylene dianiline	101-77-9	○		

CONCLUSION

It has been more than 15 years since nation-wide biological monitoring was started in Japan. This monitoring in Japan was the first attempt in the world to perform legally required biological monitoring of all workers who handle lead or eight kinds of organic solvents in workplaces. During the last 15 years, the number of samples analyzed for lead decreased from 130,000/year to 110,000/year. Although the number of samples analyzed for organic solvents was constantly 500,000/year to 600,000/year, the share of each determinant was changed. The number of samples with high concentrations of the determinants decreased as time passed. This result indicates that the exposure levels of Japanese workers were reduced after starting biological monitoring. Despite the adoption of a new analytical method for ALA and changes of TLVs-TWA or OELs, the criteria for classifying the distributions have not been changed in order to be able to survey sequential changes using the same criteria. Two efforts have already been made forward the future directions of biological monitoring in Japan. In the occupational health field, conventional biological monitoring to measure chemical substances themselves and their metabolites will be expanded to include additional organic solvents. In addition, nonspecific early biological health effects which reflect combined exposure or long-term exposure should be suitable as determinants of environmental exposure.

REFERENCES

[1]   Albert, LA, El-Sebae, AH, Fairhurst, S; Gilbert, B; Jensen, AA; Kawamoto, T; Nielsen IR; Wahlstrom, B; Walentowics, R; Wood, G. Environmental Health Criteria 136. 1,1,1-*Trichloroethylene*. Geneva: World Health Organization; 1992.

[2]   Ministry of Environment, Government of Japan. The Law Concerning the Protection of the Ozone Layer Through the Control of Specified Substances and Other Measures (Law No. 53 of May 1988). 2006. PDF file: *http://www.env.go.jp/en/laws/global/ozone2.pdf*

[3]   Ogata, M; Numano, T; Hosokawa, M; Michitsuji, H. Large scale biological monitoring in Japan. *Kawasaki Journal of Medical Welfare*. 1996:2(1), 1-15.

[4]   Kawamoto, T; Kodama, Y; Kohno, K. Interlaboratory quality control and status of n-hexane biological monitoring in Japan. *Arch. Environ. Contam. Toxicol.* 1995:28,529-536.

[5]   Tomokuni, K; Ogata, M. Simple method for determination of urinary *delta*-aminolevulinic acid as an index of lead exposure. *Clin. Chem.* 1972:18,1534-1536.

[6]   Sakurai, H; *Retrace of specific health examination for organic solvents*. The 33rd Annual Meeting of Oargnic Solvebnt Poisoning, Nagoya, 2006.

[7]   Ogawa, M; Oyama, T; Isse, T; Yamaguchi, T; Murakami, T; Endo, Y; Kawamoto, T. Hemoglobin adducts as a marker of exposure to chemical substances, especially PRTR class I designated chemical substances. *J. Occup. Health.* 2006:48,314-328.

INDEX

B

C

G

N

Q

R

S

W

X

Y

Z